Partnerships in Birds

Oxford Ornithology Series
Edited by C. M. Perrins

1 *Bird Population Studies: Relevance to Conservation and Management* (1991) Edited by C. M. Perrins, J. -D. Lebreton, and G. J. M. Hirons
2 *Bird–Parasite Interactions: Ecology, Evolution, and Behaviour* (1991) Edited by J. E. Loye and M. Zuk
3 *Bird Migration: A General Survey* (1993) Peter Berthold
4 *The Snow Geese of La Pérouse Bay: Natural Selection in the Wild* (1995) Fred Cooke, Robert F. Rockwell, and David B. Lank
5 *The Zebra Finch: A Synthesis of Field and Laboratory Studies* (1996) Richard A. Zann
6 *Partnerships in Birds: The Study of Monogamy* (1996) Edited by Jeffrey M. Black
7 *The Oystercatcher: From Individuals to Populations* (1996) Edited by John D. Goss-Custard

Partnerships in Birds

THE STUDY OF MONOGAMY

Edited by
JEFFREY M. BLACK
The Wildfowl & Wetlands Trust
Slimbridge, Gloucestershire

Drawings by
MARK HULME

Oxford New York Melbourne
OXFORD UNIVERSITY PRESS
1996

Oxford Univesity Press, Walton Street, Oxford OX2 6DP
Oxford New York
Athens Auckland Bangkok Bombay
Calcutta Cape Town Dar es Salaam Delhi
Florence Hong Kong Istanbul Karachi
Kuala Lumpur Madras Madrid Melbourne
Mexico City Nairobi Paris Singapore
Taipei Tokyo Toronto
and associated companies
Berlin Ibadan

Oxford is a trade mark of Oxford University Press

Published in the United States
by Oxford University Press Inc., New York

A catalogue record for this book is available from the British Library

Library of Congress Cataloging in Publication Data
Partnerships in birds: the study of monogamy/edited by Jeffrey M. Black; drawings by Mark Hulme.
(Oxford ornithology series; 6)
Includes index.
1. Birds—Behavior. 2. Sexual selection in animals. I. Black, Jeffrey M. II. Series.
QL698.3.P38 1996 598.256—dc20 95–41167
ISBN 0 19 854860 5 (Hbk)
ISBN 0 19 854861 3 (Pbk)

Typeset by Footnote Graphics, Warminster, Wiltshire
Printed in Great Britain by
Biddles Ltd, Guildford & Kings Lynn

Preface

The idea for this book arose from an attempt to comprehend the evolution and maintenance of long-term monogamy in geese and swans, species in which mates maintain contact all day, each season, often for life. Somehow these pair members stay together in large foraging flocks and during gruelling migrations. The stimulus behind the book was sheer intrigue about the possible link between the partners' team work and reproductive performance and why this lifestyle varies across species. For example, seabird researchers have long claimed that their birds reproduce best when with a previous rather than a new mate. These pairs remain loyal each year without staying together on a daily basis. In fact, many seabirds get together just for the breeding season, reuniting on arrival at a common site after many months on the wing over the sea. The passerine lifestyles span both extremes of mate faithfulness; some reunite for a few weeks during the breeding season, and others maintain contact for the whole year. The possibilities for comparison are endless.

Fresh thinking about monogamy in birds, presented here, includes consideration of the conflicts of interest between female and male partners, whether divorce is a matter of choice or stochasticity, and how extra-pair behaviour may alter our perception of mating systems and sexual selection. Hopefully, the reader will agree that a renewed framework for continuing the study of monogamy has been established.

In addition to two or more critical 'internal' reviews by the contributors, each chapter was appraised by at least one 'external' referee. Therefore, many of the ideas and approaches that are presented were encouraged by a group of interested colleagues. Their detailed comments on draft chapters helped to improve the manuscripts and broaden our perspectives: Evan Cooch, John Coulson, Innes Cuthill, Nick Davies, Peter Dunn, Nick Fox, Mike Harris, Ben Hatchwell, Kate Lessells, Jürg Lamprecht, Jan Lifjeld, John McNamara, Carl R. Mitchell, Markku Orell, Simon Pickering, Jouke Prop, and José Veiga. For encouraging me in and through the project I thank Nick Davies, Patty Gowaty, John Marzluff, and Doug Mock. A special acknowledgement is due to Barbara and Michael Taborsky, Steve Nesbitt, and Thomas Tacha whose insights about Brown Kiwis and Sandhill Cranes also helped shape the ideas that are presented. I am particularly grateful to Evan Cooch, Sharmila Choudhury, Bruno Ens, and Richard Pettifor for help and discussions. Finally, I thank the WWT research staff, Judith May, and Angela Turner for assistance with producing the book, Mark Hulme for the lovely drawings, and my partner, Gilly, for steadfast encouragement.

Slimbridge April 1995 J. M. B.

Contents

Contributors

FRANK ADRIAENSEN, Department of Biology, University of Antwerp, UIA, B-2610 Wilrijk, Belgium.

RUSSELL P. BALDA, Department of Biological Sciences, Northern Arizona University, Flagstaff, Arizona, 86011, USA

T. R. BIRKHEAD, Department of Animal and Plant Sciences, PO Box 601, University of Sheffield, Sheffield S10 2UQ, UK

JEFFREY M. BLACK, The Wildfowl & Wetlands Trust, Slimbridge, Gloucestershire GL2 7BT, UK

STUART BRADLEY, Biological Sciences, Murdoch University, Perth, Western Australia 6150

SHARMILA CHOUDHURY, The Wildfowl & Wetlands Trust, Slimbridge, Gloucestershire GL2 7BT, UK

ANDRÉ DESROCHERS, Département des sciences du bois et de la forêt, Université Laval, Sainte-Foy, Québec G1K 7P4, Canada

ANDRÉ A. DHONDT, Cornell Laboratory of Ornithology, 159 Sapsucker Woods Road, Ithaca, New York, 14850, USA

STEVEN D. EMSLIE, University of California, Environmental Studies Department, Santa Cruz, California, 95064, USA

BRUNO J. ENS, Institute for Forestry and Nature Research, IBN–DLO, PO Box 167, 1790 AD Den Burg, The Netherlands

JOHN W. FITZPATRICK, Archbold Biological Station, PO Box 2057, Lake Placid, Florida, 33852, USA

PATRICIA ADAIR GOWATY, Institute of Ecology, 711 Biological Sciences Building, University of Georgia, Athens, Georgia 30602-2602, USA

SUSAN HANNON, Department of Biological Sciences, University of Alberta, Edmonton, Alberta T6G 2E9, Canada

PIA LIEVESLEY, Edward Grey Institute for Field Ornithology, Zoology Department, Oxford University, South Parks Road, Oxford OX1 3PS, UK

FRANK McKINNEY, Bell Museum of Natural History and Department of Ecology, Evolution and Behavior, University of Minnesota, 1987 Upper Burford Circle, St. Paul, Minnesota 55105-6097, USA

ELIZABETH B. McLAREN, Point Reyes Bird Observatory, 4990 Shoreline Highway, Stinson Beach, California, 94970, USA

ROBERT D. MAGRATH, Division of Botany and Zoology, Australian National University, Canberra 2064, Australia

KATHY MARTIN, Canadian Wildlife Service, 5421 Robertson Road, RR1, Delta, British Columbia V4K 3N2, Canada

JOHN M. MARZLUFF, Sustainable Ecosystems Institute, 30 East Franklin Road, Suite 50, Meridian, Idaho, 83642, USA

DEBORAH A. MILLS, 5 Skyline Drive, Corning, New York, 14830, USA

JAMES A. MILLS, 5 Skyline Drive, Corning, New York, 14830, USA

DOUGLAS W. MOCK, Department of Zoology, University of Oklahoma, Norman, OK 73019, USA

A. P. MØLLER, Zoological Institute, Copenhagen University, Universitetsparken 15, DK-2100, Copenhagen, Denmark

I. NEWTON, Institute of Terrestrial Ecology, Monks Wood, Abbots Ripton, Huntingdon PE17 2LS, UK

MYRFYN OWEN, The Wildfowl & Wetlands Trust, Slimbridge, Gloucestershire, GL2 7BT, UK

GEOFFREY A. PARKER, Department of Environmental and Evolutionary Biology, University of Liverpool, Liverpool L69 3BX, UK

LAURA PAYNE, Museum of Zoology, University of Michigan, Ann Arbor, Michigan 48109-1079, USA

ROBERT PAYNE, Museum of Zoology, University of Michigan, Ann Arbor, Michigan 48109-1079, USA

CHRISTOPHER PERRINS, Edward Grey Institute for Field Ornithology, Zoology Department, Oxford University, South Parks Road, Oxford, OX1 3PS, UK

RICHARD A. PETTIFOR, The Wildfowl & Wetlands Trust, Slimbridge, Gloucestershire, GL2 7BT, UK

WERNER PLOMPEN, Department of Biology, University of Antwerp, UIA, B-2610 Wilrijk, Belgium.

PETER PYLE, Point Reyes Bird Observatory, 4990 Shoreline Highway, Stinson Beach, California, 94970, USA

EILEEN C. REES, The Wildfowl & Wetlands Trust, Slimbridge, Gloucestershire, GL2 7BT, UK

IAN ROWLEY, CSIRO Division of Wildlife and Ecology, LMB 4, PO, Midland, Western Australia 6056

ELEANOR RUSSELL, CSIRO Division of Wildlife and Ecology, LMB 4, PO, Midland, Western Australia 6056

P. L. SCHWAGMEYER, Department of Zoology, University of Oklahoma, Norman, OK 73019, USA

WILLIAM J. SYDEMAN, Point Reyes Bird Observatory, 4990 Shoreline Highway, Stinson Beach, California, 94970, USA

MURRAY WILLIAMS, Department of Conservation, PO Box 10-420, Wellington, New Zealand

TONY D. WILLIAMS, Department of Biological Sciences, Simon Fraser University, Burnaby, British Columbia, V5A 1S6, Canada

GLEN E. WOOLFENDEN, Archbold Biological Station, PO Box 2057, Lake Pacid, Florida 33852, USA.

RON WOOLLER, Biological Sciences, Murdoch University, Perth, Western Australia 6150

I. WYLLIE, Institute of Terrestrial Ecology, Monks Wood, Abbots Ripton, Huntingdon, PE17 2LS, UK

JOHN W. YARRALL, 1 Vasanta Avenue, Ngaio, Wellington, New Zealand

Initial perspectives

1 Introduction: pair bonds and partnerships

JEFFREY M. BLACK

The study of monogamy

The study of monogamy in birds has been turned on its head. This happened when we learned that sex is often not exclusive to pair bond members. The realization, which is still growing, came when molecular techniques for identifying parentage were included with studies of individual life histories. Many journals continue to provide further examples of socially monogamous birds where 'female choice of a breeding partner is to a large extent uncoupled from her choice of genes for her offspring' (Lifjeld *et al.* 1993). In this book we deal with this change in perception by redefining monogamy and its study, and asking what the social partnership is worth to the individual when sexual activity may not be restricted to the pair bond and when parents may be caring for someone else's offspring (Chapters 2, 3, and 18). Several stimulating hypotheses were tabled between the late 1960s and early 1980s to explain how monogamy came about in the animal kingdom (e.g. Lack 1968; Kleiman 1977; Wittenberger 1981; Gowaty and Mock 1985) and they have been scrutinized by some outstanding studies of monogamous birds that maintain pair bonds during single breeding seasons (e.g. Greenlaw and Post 1985; Arcese 1989; Veiga 1992). In this book we extend the study by asking how and why the ultimate variety of social monogamy, pair bonds that persist for several breeding seasons, is maintained (Chapters 4–17 and 19). The study is one of mate choice, mate fidelity, and divorce following in the wake of the review by Rowley (1983). The first premise that we acknowledge in this endeavour is that individuals behave in a way that promotes their own reproductive success even to the extent that conflicts of interest between the sexes may occur (Davies 1991), and that 'ecological conditions set the stage on which individuals play out their behavioural strategies' (Davies 1992).

Case studies from some of the finest long-term bird studies are included in this book. The foremost challenge in these chapters is to assess the costs and benefits of persistent pair associations and to explain the interplay between the sexes in terms of the birds' reproductive histories. Put simply, we ask, Why monogamy and why persistent mate fidelity? The questions are addressed by identifying the constraints that inhibit other options and assessing reproductive pay-offs to various alternatives, that is, additional mates (monogamy versus polygamy) and/or divorcing a current mate in favour of a different one.

The book has three broad sections, the first providing initial perspectives, then individual case studies, followed by a synthesis section. The middle section, making up the bulk of the book, includes the 14 case studies of birds with perennial monogamy, that is, partnerships that persist from one year to the next, in this case, in at least 50% of the pairs (Table 1.1). The case studies are categorized as either *continuous partnerships* or *part-time partnerships* depending on the amount of time pair members spend together, ranging from Barnacle Geese *Branta leucopsis* that maintain contact throughout the day for the whole year, for life, to Short-tailed Shearwaters *Puffinus tenuirostris* that are together for just a few hours during the year, underground in dark burrows or on the wing. Common themes and patterns for the maintenance of mate fidelity are sought in each chapter using a comparative approach. Within-species comparisons include pairs living in different habitats (e.g. Sparrowhawks *Accipiter nisus* living in poor and good habitats), or in populations with different demographics (e.g. Great Tits *Parus major* at low and high densities). Comparisons between closely related species include migratory versus nonmigratory life histories (e.g. Mute Swans *Cygnus olor* and Bewick's Swans *Cygnus columbianus bewickii*), and species living in different environmental and social situations (e.g. Florida Scrub Jays *Aphelocoma coerulescens* and Pinyon Jays *Gymnorhinus cyanocephalus*). Further comparisons within groups of closely related species were made for ducks, geese, birds of prey, penguins, and gulls/terns. The comparative approach is used to the full in the final chapter in an analysis of mate fidelity and divorce in over 100 bird populations.

The case studies, which span the globe, include a gradient of promiscuous activity outside the pair bond, from Iceland's Whooper Swans *Cygnus cygnus* with exclusive mating within pairs to Australia's Splendid Fairy-wrens *Malurus splendens* whose males father only 30% of the offspring with their social partner.

Table 1.1 Pair bond statistics for 14 case studies indicating the degree of association between pair members, the degree of social monogamy (versus other grouping patterns) and the probability of divorce (i.e. when both pair members survive but are no longer together)

Species	Pair type	Months	Daily	Degree of social monogamy (%)	Probability of divorce	Pair-years
Blue Duck	Continuous	12	3	100	11.3	71
Barnacle Goose	Continuous	12	3	99	2.0	5974
Bewick's Swan	Continuous	12	3	100	0.0	2220
Whooper Swan	Continuous	12	3	100	5.6	305
Mute Swan	Continuous	12	3	100	0.7	603
Florida Scrub Jay	Continuous	12	3	99	2.6	960
Splendid Fairy-wren	Continuous	12	3	—	2.0	346
Pinyon Jay	Continuous	12	2–3	100	1.1	279
European Blackbird	Part-time	9	3	100	31.7	183
Willow Ptarmigan	Part-time	6	3	93	16.4	219
Cassin's Auklet	Part-time	6	2	100	7.3	220
Short-tailed Shearwater	Part-time	6	1	100	17.2	1911
Great Tit	Part-time	5	2	99	25.2	306
Sparrowhawk	Part-time	4 (1–12)	1–2	99	11.3	230
Penguin sp.	Part-time	3.5 (2–5)	2	100	28.2	5850
White-tailed Ptarmigan	Part-time	3	3	92	19.5	41
Red-billed Gull	Part-time	3	1	94	10.5	3903
Indigo Bunting	Part-time	2–4	1	81	51.0	221

The degree of association is indicated by the number of months pair members are together and how closely they associate on a daily basis using an index where 1 = together a little (<2 h), 2 = together some times, and 3 = together all the time.

Whereas previously the function of the pair bond was thought to be centred around exclusive access to a mate, the working hypothesis in this book emphasizes the enhanced ability of the pair members to acquire resources for breeding and survival (see Wickler and Seibt 1983; Mock and Fujioka 1990). The pair bond in birds is an excellent example of co-operative behaviour in action. In addition to the parents' team work while feeding chicks, pairs may alternate alarm calls and flights that repel predators from territories (tits, Regelmann and Curio 1988), join forces in aggressive encounters with flock members for access to food and nest sites (swans, Scott 1980; geese, Black and Owen 1989), alternate vigilance and foraging bouts (jays, McGowen and Woolfenden 1989; geese, Sedinger and Raveling 1990), participate in vocal duets that may signal willingness to remain paired, territory ownership or willingness to fight (parakeets, Arrowood 1988; geese, Hausberger and Black 1990), and synchronize foraging trips during incubation, thus enhancing hatching success (Kittiwakes *Rissa tridactyla*, Coulson 1966; other gulls, Morris 1987; penguins, Davis 1988).

The term pair bond is a literal translation of the German word, Paarbindung, frequently used by Heinroth (1911) and Lorenz (1966). The English word, bond, conjures up concepts like: being tied down; a uniting force; a restraint; a responsibility; a binding agreement; and an emotional attachment (Hawkins and Allen 1991). Welty and Baptista (1988) refer to a psychological bond that holds a pair of birds together, which may result from common attachment to a territory, a comfortable familiarity of each other, or something akin to human affection. A powerful image is painted by Lorenz's (1966) description of his geese grieving after the death of a lifelong mate. Pair bonds have been referred to as strong or weak, implying a measurable adhesive force between pair members (e.g. Nelson 1965; Rasmussen 1981). Studies that attempt to measure the individual's disposition to being attached use terms like attachement strength, companionship, or pair bond quality. In an intriguing experiment with 18 human-reared goslings, Lamprecht (1984) quantified individual differences in the birds' 'internal divorce tendency', meaning strength or degree of attachment to a foster parent. By counting their vocalizations he could predict the degree to which the goslings would stray from the parent when given an option to follow a stranger or semi-stranger. All things being equal, such findings may reflect the birds' predisposition to stay with a partner in the future!

Studies of attachment and companionship need not refer to strength or weakness of a pair bond, but could instead refer directly to the measurements that are taken, like proximity and synchrony of behaviours which may influence fitness (see Wickler 1976). The term partnership, which is used interchangeably with the term pair bond in this book, refers to a male–female 'social' association at any time of year, especially during the breeding season, regardless of their sexual activity (i.e. social monogamy as opposed to genetic monogamy, see Chapters 2 and 3). Using the term partnership, instead of bond, suggests that rather than being disposed to stay with one mate, the relationship between pair members is dynamic, an extension of the mate choice process. In studying the pair bond, therefore, it is helpful to view it as a decision process made by each individual about whether to stay or leave. Pair bonds are not magical marriages but flexible associations that can potentially end any time another option arises. Pair members may continually test the partnership to assess whether they should remain with a current mate.

To develop this continuous mate choice perspective further the case study contributors scrutinize their data sets on mate change and the implication of divorce events (when surviving pair members separate or do not reunite). To identify whether divorce is accidental or proactive several questions are crucial, including determining which behavioural, demographic, and environmental features influence divorce, what mate choice criteria are used, which sex is responsible, and whether pairing with a new mate improves reproductive potential.

Reasons and mechanisms for divorce and mate fidelity

Does divorce occur by choice (birds making optimal decisions) or by some external process (birds split simply due to bad luck)? The potential benefits of mate fidelity are:

(1) improving co-ordination and co-operation with a mate;
(2) prolonged biparental investment; and
(3) reducing the costs involved in mate sampling, like the risk of injury or predation, failing to find a mate in time to initiate breeding, or missing a breeding opportunity altogether

(all of which should enhance reproductive success). At the very least, the costs involve keeping in touch with the chosen mate. For species with continuous partnerships this involves keeping track of and maintaining contact with the mate, whereas, for part-time partnerships, the costs include waiting for the mate to return or searching among the alternatives. When pair members are together contact may be maintained through potentially costly, mutual behaviours including contact calls, duets, beak-touching, bill meeting, allopreening, dancing, greetings, and triumph ceremonies, etc. (see e.g. Butterfield 1970). In some species males defend females against rivals and predators, ultimately investing in their mates' fat/nutrient condition, sometimes at the expense of their own, to ensure successful reproduction (e.g. ducks, Ashcroft 1976; geese, Black and Owen 1988; Black *et al.* 1991; swans, Scott 1980; ptarmigan, Hannon and Martin 1992; and tits, Krebs 1970; Hogstad 1992, 1995). Rowley (1983) predicted that due to these costs mate fidelity should differ for species with different mortality rates and longevity; divorce should be higher in short-lived species with high mortality because previous partners would not be inclined to wait very long for mates to return.

For almost a quarter of a century the foremost hypothesis to explain why, within a species, some pairs divorce between seasons, while other pairs reunite, was the *incompatibility hypothesis* put forward by Coulson (1966, 1972). According to this hypothesis, pairs whose partners are incompatible reproduce poorly and subsequently split, while compatible partners have high success and do not change mates. Throughout this period, the hypothesis was enthusiastically endorsed, especially by students of long-lived seabirds, and often served as the bottom line in papers on mate fidelity. Yet, these papers rarely provided critical tests of the hypothesis. Detailed behavioural observations on Oystercatchers *Haematopus ostralegus* and the notion that pair members may fail to achieve their preferred option when there is a conflict of interest (Davies 1989) led Ens *et al.* (1993) to propose the *better option hypothesis*. This idea can be traced back to Baeyens (1981) who suggested that one member of the pair may initiate the events that lead to divorce because it has a better option, leaving the other as a victim of its choice (also see Slagsvold and Lifjeld 1986; Lindèn 1991).

A persistent interest in the causes of mate fidelity and divorce has led authors to propose no fewer than nine additional hypotheses and/or mechanisms (Table 1.2). The *errors or mate choice hypothesis* proposed by Johnston and Ryder (1987) suggests that divorce may arise from errors made in the original choice of mate. The *extra-pair paternity hypothesis* proposed by Birkhead and Møller (1992) only applies to within-season divorce and suggests that divorce or rapid mate switching within a season may be a male strategy to increase reproductive success through extra-pair paternity. The *keeping company hypothesis* maintains that the likelihood of maintaining contact or reuniting with the same mate will increase with the amount of time pair members associated during the year, an idea that is attributable to Rowley (1983). Butterfield (1970) developed the concept that species living in unpredictable environments maintain their pair bonds to ensure the synchronization of reproductive states, so pairs will be ready when the weather changes (the *climate un-predictability mediated hypothesis*). This view was linked with the idea that pair members stimulate each other's gonadal development through social interactions. According to the *musical chairs hypothesis* divorce is a side effect of differential arrival of the sexes on their territory (Dhondt and Adriaensen 1994; an individual that arrives much later than its mate may find its 'chair' occupied). This '*arrival asynchrony*' concept can be attributed to Coulson and Thomas (1980) who showed that Kittiwake pair members that divorced returned to the colony an average of 22 days apart, compared to only 11 days for pairs that reunited. The *habitat mediated hypothesis* (Desrochers and Magrath Chapter 9; Newton and Wyllie Chapter 14) also maintains that mate

Table 1.2 Hypotheses and mechanisms which have been proposed to explain divorce and mate fidelity in monogamous birds, grouped into three categories of functional explanations as to why and how divorce occurs (updated from Choudhury 1995)

Hypotheses	Background	Source
Choice for mates		
Incompatibility	Pairs that do not reproduce well divorce	Coulson (1972); Rowley (1983)
Better option	Birds divorce when a better mate is available	Ens *et al.* (1993)
Errors of mate choice	Birds divorce when a mistake is discovered	Johnston and Ryder (1987)
Extra-pair paternity	Birds switch mates to increase genetic diversity in brood	Birkhead and Møller (1992); (Chapter 18)
Keeping company	Pair bonds persist more in birds that spend time together	Rowley (1983); Williams (Chapter 15)
Climate un-predictability mediated	Pairs maintain contact ensuring they are ready for the breeding season	Butterfield (1970)
Choice for territories		
Arrival asynchrony or musical chairs	Birds divorce due to asynchronous arrival times	Coulson and Thomas (1980); Dhondt and Adriaensen (1995)
Habitat mediated	Divorce occurs more in poor quality habitats	Desrochers and Magrath (Chapter 9) Newton and Wyllie (Chapter 14)
Salvage strategy		
Accidental loss	Birds divorce if separated on migration or in large flocks	Owen et al. (1988); Owen and Black (1991); Chapter 5
Forced divorce through usurpation	Birds divorce when one member loses a fight	Minton (1968) among others; M. Taborsky and B. Taborsky (personal communication)
Severe weather mediated	Birds divorce when movements occur during bad weather	Minton (1968)

The first two categories of explanations, choice for mates and choice for territories, can be considered as active choice strategies by one or both pair members to improve their reproductive performance by utilizing the opportunity to gain a better mate and/or territory. The third category, salvage strategies, ascribe divorce to some external event, in which divorce is simply a case of making the best of a bad situation and not in the first instance an active choice by either sex. Several of the case studies present a description and empirical evidence in support of these hypotheses and these are reviewed by Ens *et al.* (Chapter 19).

fidelity is linked with site fidelity, but claims, in addition, that the tendency to abandon the territory (and mate) in search of food during the nonbreeding season increases when habitat quality decreases. The *accidental loss hypothesis* was proposed for migratory birds like geese: temporary separation, due to bad weather for instance, coupled with a need to re-pair quickly in order to minimize losses, may lead to divorce (Owen *et al.* 1988). Alternatively, divorce may come about because one of the partners is displaced by a third individual. M. Taborsky and B. Taborsky (personal communication) coined the term *forced divorce* for this hypothesis and others have noted such occurrences (e.g. Minton 1968; Ball *et al.* 1978). Finally, the *severe weather mediated hypothesis* of Minton (1968) states that very severe winters may cause normally sedentary birds to abandon their territories and mates, augmenting the chance that the pair will not reunite the next spring. The evidence for some of these ideas is developed in a number of the case studies and scrutinized in the final chapter (Ens *et al.* Chapter 19).

Prolonged pair bonds in perspective

Although social monogamy is thought to be the mating system of 90% of the 9020+ bird species (Lack 1968), persistent pair bonds are achieved in far fewer species. However, these longer-term partnerships occur in at least 50% of the 28 orders of birds, and in 21% of the 159 bird families, ranging from ancient species like kiwis, penguins, and albatrosses, through slightly more recent groups like cranes and gulls, to numerous families in the most recent order, the passerines.[1]

Prolonged pair bonds or 'mate fidelity' may be most frequent in birds, but the lifestyle also appears in some crustaceans, fish, and mammals. Long-term pair bonds are probably rare in insects, although they may be possible in species like the Horned Beetle *Typhoeus typhoeus* and Burying Beetle *Necrophorus sp.* which are often monogamous and where males defend females (Palmer 1978; Trumbo and Eggert 1994). Singer and Riechert (in press) also describe the monogamous system for the short-lived (49 days) Desert Spider *Agelenopsis aperta*.

The Starfish-eating Shrimp *Hymenocera picta* is an instructive example of an arthropod with prolonged partnerships (Seibt and Wickler 1979). In this species the pair co-operate in catching their prey and male mate guarding limits the female's access to other mates. In fish, seahorse *Hippocampus whitei* pairs, living in Australian seagrasses, perform co-ordinated dances with the same partner each morning for the 6-month breeding season, and this is thought to maintain the pair bond (Vincent 1995). Pairs of butterflyfish, which live in coral reefs, have been recorded together for 6 consecutive years (Fricke 1986; Driscoll and Driscoll 1988). By temporarily removing one partner Fricke (1986) established

[1] These percentages are based on our survey, listed in Appendix 19.1 at the end of the book, augmented by information in Welty and Baptista (1988).

that pair members, as in some birds, work together to defend territory boundaries, sometimes using a dual swimming display.

Only 5% of 4000 mammal species are monogamous (Kleiman 1977; Clutton-Brock 1989). The pairs of those species that are monogamous are apparently together for life; marmosets, tamarins, gibbons, some mice, jackals, and other dogs being common examples. Two generalities as to why monogamy is rare in mammals and common in birds can be gleaned from reviews on monogamy in mammals (Kleiman 1977; Clutton-Brock 1989): mammalian females rarely require male help in rearing offspring; and mammalian males have greater access to more than one female at a time since mammalian females often live in groups. Dunbar (1988) suggests that prolonged pair bonds in large primates are maintained due to the risk of harassment and infanticide by interloping males. A useful set of field studies on monogamous mammals can be found in Clutton-Brock's (1989) review. He ends the subsection stating that explanations for the evolution of monogamy in mammals are not satisfactory and that monogamy remains an enigma.

Guidelines for case studies

Each of the case study contributors in this book provides background information on the birds' ecology, population demographics, a description of the pair formation process, how pair members maintain contact, and parental care. Authors were asked to quantify the degree of social monogamy and polygamy, how long the average pair bond lasts, and how many mates individuals have in a lifetime. Because extra-pair copulations occur in a large variety of birds (reviewed in Westneat *et al.* 1990; Birkhead and Møller 1992) we must devote increasing energy to evaluating the role of extra-pair behaviour in the evolution and maintenance of monogamy and mate fidelity. To do this we need to discover the proximate mechanisms responsible for the behaviour, the fertilization success of the copulations, and the implications of caring for someone else's offspring (Davies 1992; Westneat and Sherman 1993). Contributors with appropriate data sets describe the occurrence of extra-pair behaviour, paternity, and parental care of unrelated offspring, and how such behaviours may influence the social system in their species. This inquiry is the keystone topic developed in the wide ranging chapters by Gowaty (Chapter 2) and Birkhead and Møller (Chapter 18).

Four specific analyses on mate fidelity and divorce were encouraged for those case studies with suitable data sets.

Mate familiarity effect

Does reproductive success really improve with an increasing number of years with the same mate? Pair quality, mate compatibility, and coordination between pair members (which are collectively referred to as the mate familiarity effect) are presumably responsible for patterns in age-related reproductive success in species with long-term partnerships. The suggestion that established pairs reproduce best due to the mate familiarity effect is largely based on the cross-sectional analysis of reproductive output for pairs with new and old mates (see below). It is generally assumed from these analyses that reproductive performance increases with the duration that pair members are together (e.g. Ollason and Dunnet 1978), although this may not be the case. The correlation between pair duration and reproductive performance (a longitudinal analysis) has been claimed for pairs of Florida Scrub Jays (Woolfenden and Fitzpatrick 1984), Little Penguins *Eudyptula minor* (Wienecke and Wooler 1996) and a few seabird species (Coulson 1966; Ollason and Dunnet 1978; Bradley *et al.* 1990; Emslie *et al.* 1992). However, few of these studies could rule out the possibility that age or experience in one or both sexes was responsible for the improvement. So far only one study came close to a clear result after statistically controlling for age (of one pair member) in Cassin's Auklets *Ptychoramphus aleuticus* (Emslie *et al.* 1992; also see Bradley *et al.* 1990 for a less clear example with Short-tailed Shearwaters). One challenge of this book is to see whether reproductive success increases with pair duration after controlling for potentially confounding variables.

The mate familiarity effect idea can be traced to Bryan Nelson's (1965, 1972) classic studies of gannets and boobies. He suggested that familiarity with a mate influenced the pairs' social display quality and ultimately the timing of nesting. New pairs whose display was not synchronized kept practising, whereas the display of pairs that had been together in previous years simply needed brushing up; a similar situation occurs in other species with elaborate displays like Wandering Albatross *Diomedea exulans* (Pickering 1989; see drawing on title page). On a similar vein, Chardine (1987), observing 45 Kittiwakes with the same mate and 24 with new mates, detected some differences in their behavioural repertoires. The behaviours included rate of return to and time spent in attendance on the nest ledges, frequency of greetings and copulations, and copulation success. He concluded that new mates probably had less time to maintain body condition because they spent more time and energy in establishing or maintaining the pair bond than old mates. He also provided some evidence that new pairs spent less time on these 'extra' behaviours as partners became more familiar during the 5-month breeding season (see also Erickson 1973). Numerous other behaviours may require fine tuning between pair members before optimal reproductive rates can be achieved. Besides synchrony in social display, there are several candidate behaviours in this regard: dueting behaviour, allogrooming, courtship feeding, team work in protecting nests against predators, change-over in shared incubation duties, co-ordinating parental care duties, nest defence, alternating vigilance bouts, etc. However, as is indicated in the synthesis (Ens *et al.* Chapter 19), few of these behaviours have actually been adequately measured, so claims that they are a prerequisite to enhanced reproductive success may be premature.

Cost (or benefit) of mate change

Does reproductive success increase, decrease, or stay the same in the year after a mate change or divorce? If the mate familiarity effect is active, then pairs should suffer a cost if they change mates. If the degree of mate fidelity and divorce is caused by the costs of mate change, then species with a high mate fidelity (low divorce rates) should show a decrease in reproductive success after a mate change. Previous studies on seabirds, geese, and tits have shown that birds with new mates do indeed have lower reproductive success than faithful partners (e.g. Coulson 1966, 1972; Mills 1973; Brooke 1978; Ollason and Dunnet 1978, 1988; Cooke *et al.* 1981; Coulson and Thomas 1985; Perrins and McCleery 1985; Owen *et al.* 1988; Wooler *et al.* 1989; Bradley *et al.* 1990; Emslie *et al.* 1992). However, potentially confounding effects of age on reproduction were rarely controlled in these studies, so conclusions about the cost of mate change were tenuous.

Alternatively, reproductive performance should improve if a bird divorces a poor quality mate in favour of a better one (the *better option hypothesis*). Ens *et al.* (1993) predict that divorce should occur more frequently in species with higher mortality rates since more alternative, perhaps better quality, mates would be available. Contributors were asked to compare the reproductive rates between birds that kept and changed (or divorced) mates, while controlling for age and year effects.

Previous reproductive success and divorce

Is divorce more likely to occur after reproductive failure, perhaps due to the incompatibility of pair members? Ens *et al.* (1993) referred to this question as the Kittiwake dilemma after Coulson's (1966, 1972) findings that unsuccessful pairs divorced more than successful pairs. Ens *et al.* point out that the dilemma comes about because divorce as a strategy can be adaptive even though long-term partnerships are meant to be advantageous. However, the original comparison between Kittiwake pairs, and subsequent inquiries in other species (see Rowley 1983; Johnston and Ryder 1987), did not take into account that young birds invariably breed less well than older birds, so divorce may have nothing to do with poor breeding or incompatibility, but everything to do with age. Contributors were, therefore, asked to control for confounding variables in their analyses on the topic.

Pair characteristics and reproduction

Under what circumstances will individuals (of both sexes) be able to breed with the mate that will result in the highest reproductive pay-off (the preferred mate option) despite the conflicting preferences of others? Davies' (1989) question, which was referring to preferred mating systems, can also be attributed to mate choice decisions. By comparing the reproductive pay-offs of males and females with different phenotypic characteristics intra-sexual conflict over preferred mate qualities can be identified. Contributors were asked to check whether there were any preferred options, and which individuals were able to achieve them. The preferred mate is the one with whom a bird could reproduce optimally. Preferred options are achieved when the majority of individuals of a given type (e.g. old males, or large females) are paired with the preferred mate-type. Characteristics that are tested include age, experience, size, natal area, and the degree of relatedness between pair members. Two earlier examples of this approach include Perrins and McCleery (1985) on Great Tits and Marzluff and Balda (1988) on Pinyon Jays.

Terminology

The terminology describing how often pair members reunite and split is quite variable in the literature. To avoid confusion each author has used the same set of terms and these are described below.

Pair duration refers to the length of time, measured in number of years or seasons, that pair members are together. *Mate fidelity* refers to the situation where pair members remain paired or reunite in subsequent years or breeding seasons. *Divorce*, the opposite of mate fidelity, occurs when surviving pair members separate or fail to reunite, usually resulting in eventual re-pairing with a new mate. *Mate change* refers to partnerships that terminate, or fail to reunite for any reason, including death, disappearance, and divorce. The opposite of mate change is *pair fidelity*, a term used in the seabird literature, meaning the proportion of pairs retaining the same partner from one breeding season to the next (used only in Chapters 11, 15, and 16). Whereas *divorce/mate fidelity* occur when pair members from year A are still alive in year B, *mate change/pair fidelity* occur when it is not known whether both pair members are alive in year B. *Probability of divorce*, also referred to as *divorce rate*, is calculated as the number of divorce events/total number of pair-years ×100, where a *pair-year* is when both pair members in year A survive to year B.

Where feasible, terms that are potentially fraught with emotional baggage should be avoided (e.g. divorce). However, the term divorce in this book, meaning the end of a partnership/pair bond, is more manageable and useful than alternative terms that have been used: for example, mate switching, pair splitting, nonretainment, breakage, abandonment, severance, dissolution, etc. No sane reader should misinterpret divorce in birds as implying legal dissolution of a marriage, with alimony and lawyers' fees (D. W. Mock, personal communication)!

Problems in measuring pair bonds and reproductive success

A range of methods are employed to decide whether two birds are 'paired', some of which are more instructive than others. Some assessments are based on catching two birds in the same territory. In these cases no information is collected about the pair members' type or degree of interaction. This kind of method is particularly useful for secretive species that have a dispersed distribution, like forest-dwelling Sparrowhawks (Newton and Wyllie Chapter 14) or seabirds that nest in burrows and only come out at night (see Chapters 11 and 12). At the other extreme, for species that live in large, dense flocks, observers must become familiar with a range of behavioural cues in order to assess which two birds are behaving 'in tune' with each other. Owing to the subjective nature of these assessments, determining the 'paired' status of individuals can be achieved only by checking consecutive records of marked individuals for consistent associations, as is done in the goose and swan studies (see Chapters 5 and 6). For most case study species, however, assessments are more straightforward since pairs tend to associate at a particular nest site or on defined territories where they can be observed.

Authors were encouraged to use the closest reproductive measure to contributing gametes to the next generation; usually chicks that fledge or those that are recruited into the adult population. Authors were encouraged to remove potentially confounding variables when assessing whether pair duration affects reproductive success or if prior reproductive history influences the probability of divorce. Controlling for fluctuating annual variation, due to climate and food conditions, and for the effect of increasing age/experience was of most concern, although the number of helpers and density were also considered in some cases. In most chapters multiple log linear regressions were used in a forward modelling procedure where the variable of interest was the last variable entered and only significant variables were retained (e.g. Crawley 1993). Type III sums of squares were used. Most authors treat reproductive values from the same pairs in successive years as independent events since year and age were controlled for, while others with sufficiently large data sets checked their results with replicate analyses after choosing one record at random per pair (see Chapters 5, 6, and 12). One further problem was that of collinearity in models that contain two or more variables that increase on the same scale (e.g. number of years). This was of particular concern in the analysis of the effect of pair duration on reproductive success while controlling for age. In Chapters 5 and 6 we checked whether this situation was indeed a problem and found that it was not. This was done by repeating the analysis for each age-class separately and assessing whether the pattern for each age was the same as the one in the all-inclusive model. Analyses of long-term data sets may also be confounded by differences in survival which may cause disparate frequencies of superior or inferior phenotypes (see Clutton-Brock 1988, p.4). In most cases, authors still aspire to have large enough data sets to control for this problem.

Traditional measures of reproductive success (e.g. counting the number of chicks) will continue to be open to error in designating correct parentage until the cost of biochemical assessments of actual parentage become less prohibitive. Authors were encouraged to present data on the degree of genetic paternity in their species and Birkhead and Møller (Chapter 18) draw this information together.

Conclusion

This book is about the mate choice process that may persist for an individual's entire lifetime leading to either mate fidelity, divorce, or extra-pair copulation. The focus here is on longitudinal data from long-term bird studies. Authors assess the costs and benefits to each member of the pair in maintaining persistent partnerships and explain the interplay (and conflict) between the sexes in terms of the individuals' reproductive histories. We begin with a review of the study of monogamy to date (including a new approach for the future; Chapter 2) and a review of the intricacies of the monogamous family, between the partners and their offspring (Chapter 3).

References

Arcese, P. (1989). Intrasexual competition and the mating system in primarily monogamous birds: the case of the song sparrow. *Animal Behaviour*, **38**, 96–111.

Arrowood, P. C. (1988). Duetting, pair bonding and agonistic display in parakeet pairs. *Behaviour*, **106**, 129–157.

Ashcroft, R. E. (1976). A function of the pair bond in the Common Eider. *Wildfowl*, **27**, 101–5.

Baeyens, G. (1981). Functional aspects of serial monogamy: the Magpie pair-bond in relation to its territorial system. *Ardea*, **69**, 145–66.

Ball, I. J., Frost, P. G. H., Siegfried, W. R., and McKinney, F. (1978). Territories and local movements of African Black Ducks. *Wildfowl*, **29**, 61–79.

Black, J. M. and Owen, M. (1988). Variation in pair bond and agonistic behaviors in Barnacle Geese on the wintering grounds. In *Waterfowl in winter* (ed. M. Weller), pp. 39–57. University of Minnesota Press, Minneapolis.

Black, J. M. and Owen, M. (1989). Agonistic behaviour in goose flocks: assessment, investment and reproductive success. *Animal Behaviour*, **37**, 199–209.

Black, J. M., Deerenberg, C., and Owen, M. (1991). Foraging behaviour and site selection of Barnacle Geese *Branta leucopsis* in a traditional and newly colonised spring staging habitat. *Ardea*, **79**, 349–358.

Bradley, J. S., Wooller, R. D., Skira, I. J., and Serventy, D. L. (1990). The influence of mate retention and divorce upon reproductive success in short-tailed shearwaters *Puffinus tenuirostris*. *Journal of Animal Ecology*, **59**, 487–96.

Birkhead, T. R. and A. P. Møller (1992). *Sperm competition in birds*. Academic Press, London.

Brooke, M. de L. (1978). Some factors affecting the laying date, incubation and breeding success of the Manx shearwater, *Puffinus puffinus*. *Journal of Animal Ecology*, **47**, 961–7.

Butterfield, P. A. (1970). The pair bond in the zebra finch. In *Social behaviour in birds and mammals* (ed. J. H. Crook), pp. 249–78. Academic Press, London.

Chardine, J. W. (1987). The influence of pair-status on the breeding behaviour of the Kittiwake *Rissa tridactyla* before egg-laying. *Ibis*, **129**, 515–26.

Choudhury, S. (1995). Divorce in birds: a review of hypotheses. *Animal Behaviour*, **50**, 413–29.

Clutton-Brock, T. H. (1988). *Reproductive success*. University Press of Chicago, Chicago.

Clutton-Brock, T. H. (1989). Mammalian mating systems. *Proceedings of the Royal Society*, **235**, 339–72.

Cooke, F., Bousfield, M. A., and Sadura, A. 1981. Mate change and reproductive success in the Lesser Snow Goose. *Condor*, *83*, 322–7.

Coulson, J. C. (1966). The influence of the pair-bond and age on the breeding biology of the kittiwake gull, *Rissa tridactyla*. *Journal of Animal Ecology*, **35**, 269–79.

Coulson, J. C. (1972). The significance of pair bond in the Kittiwake *Rissa tridactyla*. *International Ornithlogical Congress*, **15**, 423–33.

Coulson, J. C. and Thomas, C. S. (1980). A study of the factors influencing the duration of the pair bond in the Kittiwake Gull *Rissa tridactyla*. *International Ornithological Congress*, **25**, 823–33.

Coulson, J. C. and Thomas, C. S. (1985). Differences in the breeding performance of individual kittiwake gulls *Rissa tridactyla*. In *Behavioural ecology* (ed. R. M. Sibley and R. M. Smith), pp. 489–503. Blackwell, Oxford.

Crawley, M. J. (1993). *GLIM for ecologists*. Blackwell, Oxford.

Davies, N. B. (1989). Sexual conflict and the polygamy threshold. *Animal Behaviour*, **38**, 226–34.

Davies, N. B. (1991). Mating systems. In *Behavioural ecology, an evolutionary approach* (ed. J. R. Krebs and N. B. Davies), pp. 263–300. Blackwell, Oxford.

Davies, N. B. (1992). *Dunnock behaviour and social evolution*. Oxford University Press, Oxford.

Davis, L. S. (1988). Coordination of incubation routines and mate choice in Adelie Penguins (*Pygoscelis adeliae*). *Auk*, **105**, 428–32.

Dhondt, A. A. and Adriaensen, F. (1994). Causes and effects of divorce in the blue tit *Parus caeruleus*. *Journal of Animal Ecology*, **63**, 979–87.

Driscoll, J. W. and Driscoll, J. L. (1988). Pair behaviour and spacing in butterflyfishes (Chaetodontidea). *Environmental Biology of Fishes*, **22**, 29–37.

Dunbar, R. I. M. (1988). *Primate social systems*. Croom Helm, London.

Emslie, S. D., Sydeman, W. J., and Pyle, P. 1992. The importance of mate retention and experience on breeding success in Cassin's auklet (*Ptychoramphus aleuticus*). *Behavioral Ecology*, **3**, 189–95.

Ens, B. J., Safriel, U. N., and Harris, M. P. (1993). Divorce in the long-lived and monogamous oystercatcher, *Haematopus ostralegus*: incompatibility or choosing the better option? *Animal Behaviour*, **45**, 1199–1217.

Erickson, C. J. (1973). Mate familiarity and the reproductive behavior of Ringed Turtle Doves. *Auk*, **90**, 780–95.

Fricke, H. W. (1986). Pair swimming and mutual partner guarding in monogamous butterflyfish (Pisces: Chaetodontidae): a joint advertisement for territory. *Ethology*, **73**, 307–33.

Greenlaw, J. S. and Post, W. 1985. Evolution of monogamy in seaside sparrows, *Ammodramus maritimus*: test of hypotheses. *Animal Behaviour*, **33**, 373–83.

Gowaty, P. A. and Mock, D. W. (1985). Avian monogamy. *Ornithological Monographs*, **37**, 1–121.

Hannon, S. J. and Martin, K. (1992). Monogamy in willow ptarmigan: is male vigilance important for reproductive success and survival of females? *Animal Behaviour*, **43**, 747–57.

Hausberger, M. and Black, J. M. (1990). Do females turn males on and off in barnacle goose social display? *Ethology*, **8**, 232–8.

Hawkins, J. M. and Allen, R. (1991). *The Oxford encyclopedic English dictionary*. Clarendon Press, Oxford.

Heinroth, O. (1911). Beitrage zur Biologie, namentlich Ethologie und Psychologie der Anatiden. *Verhhandlung des V Internationalen Ornithologen Kongresses, Berlin*, 589–702.

Hogstad, O. (1992). Mate protection in alpha pairs of wintering willow tits, *Parus montanus, Animal Behaviour*, **40**, 323–8.

Hogstad, O. (1995). Alarm calling by willow tits, *Parus montanus*, as mate investment. *Animal Behaviour*, **49**, 221–5.

Johnston, V. H. and Ryder, J. P. (1987). Divorce in larids: a review. *Colonial Waterbirds*, **10**, 16–26.

Kleiman, D. (1977). Monogamy in mammals. *Quarterly Review of Biology*, **52**, 39–69.

Krebs, J. R. (1970). The efficiency of courtship feeding in the Blue Tit *Parus caeruleus*. *Ibis*, **112**, 108–10.

Lack, D. (1968). *Ecological adaptations for breeding in birds*. Chapman and Hall, London.

Lamprecht, J. (1984). Measuring the strength of social bonds: experiments with hand-reared goslings (*Anser indicus*). *Behaviour*, **91**, 115–27.

Lifjeld, J. T., Dunn, P. O., Robertson, R. J., and Boag, P. T. (1993). Extra-pair paternity in monogamus tree swallows. *Animal Behaviour*, **45**, 213–29.

Lindèn, M. (1991). Divorce in Great Tits: chance or choice? An experimental approach. *American Naturalist*, **138**, 1039–48.

Lorenz, K. (1966) *On aggression*. Harcourt, Brace and Jovanovich, New York.

McGowen, K. J. and Woolfenden, G. E. (1989). A sentinel system in the Florida scrub jay. *Animal Behaviour*, **37**, 1000–6.

Marzluff, J. M. and Balda, R. P. (1988). Pairing patterns and fitness in a free-ranging population of Pinyon Jays: what do they reveal about mate choice? *Condor*, **90**, 201–13.

Mills, J. A. (1973). The influence of age and pair-bond on the breeding biology of the red-billed gull, *Larus novaehollandiae scopulinus*. *Journal of Animal Ecology*, **42**, 147–62.

Minton, C. D. T. (1968). Pairing and breeding in Mute Swans. *Wildfowl*, **19**, 41–60.

Mock, D. W. and Fujioka, M. (1990). Monogamy and long-term pair bonding in vertebrates. *Trends in Ecology and Evolution*, **5**, 39–43.
Morris, R. D. (1987). Time-partitioning of clutch and brood care activities in Herring Gulls: a measure of parental quality? *Studies in Avian Biology*, **10**, 68–74.
Nelson, J. B. (1965). The behaviour of the Gannet. *British Birds*, **58**, 233–88.
Nelson, J. B. (1972). Evolution of the pair bond in Sulidae. *International Ornithological Congress*, **15**, 371–88.
Ollason, J. C. and Dunnet, G. M. (1978). Age, experience and other factors affecting the breeding success of the fulmar *Fulmarus glacialis* in Orkney. *Journal of Animal Ecology*, **47**, 961–76.
Ollason, J. C. and Dunnet, G. M. (1988). Variation in breeding success in fulmars. In *Reproductive success* (ed. T. H. Clutton Brock), pp. 263–78. University of Chicago Press, Chicago.
Owen, M. and Black, J. M. (1991). Geese and their future fortunes. *Ibis*, **133**, S28–S35.
Owen, M., Black, J. M., and Liber, H. (1988). Pair bond duration and the timing of its formation in Barnacle Geese. In *Waterfowl in winter* (ed. M. Weller), pp. 23–38. University of Minnesota Press, Minneapolis.
Palmer, T. S. (1978). A Horned Beetle that fights. *Nature*, **274**, 583–4.
Perrins, C. M. and McCleery, R. H. (1985). The effect of age and pair bond on the breeding success of Great Tits *Parus major*. *Ibis*, **127**, 306–15.
Pickering, S. P. C. (1989). Attendance patterns and behaviour in relation to experience and pair-bond formation in the Wandering Albatross *Diomedea exulans* at South Georgia. *Ibis*, **131**, 183–95.
Rasmussen, D. R. (1981). Pair-bond strength and stability and reproductive success. *Psychological Reviews*, **88**, 274–90.
Regelmann, K. and Curio, E. (1988). How do Great Tit (*Parus major*) pair mates cooperate in brood defence? *Behaviour*, **97**, 10–36.
Rowley, I. (1983). Re-mating in birds. In *Mate choice* (ed. P. Bateson), pp. 331–60. Cambridge University Press, London.
Scott, D. K. (1980). Functional aspects of the pair bond in wintering Bewick's swans (*Cygnus columbianus bewickii*). *Behavioral Ecology and Sociobiology*, **7**, 323–7.
Sedinger, J. S. and Raveling, D. G. (1990). Parental behavior of Cackling Canada Geese during brood rearing: division of labor within pairs. *Condor*, **92**, 174–81.
Seibt, U. and Wickler, W. (1979). The biological significance of monogamy in the shrimp *Hymenocera picta*. *Zeitcshrift für Tierpsychologie*, **50**, 166–79.
Singer, F. and Riechert, S. E. (1995). Mating system and mating success in desert spider, *Agelenopsis aperta*. *Behavioral Ecology and Sociobiology*, **36**, 313–22.
Slagsvold, T. and Lifjeld, J. T. (1986). Mate retention and male polyterritoriality in the pied flycatcher *Ficedula hypoleuca*. *Behavioral Ecology and Sociobiology*, **19**, 25–39.
Trumbo, S. T. and Eggert A. K. (1994). Beyond monogamy: territory quality influences sexual advertisement in male burying beetles. *Animal Behaviour*, **48**, 1043–7.
Veiga, J. P. (1992). Why are house sparrows predominantly monogamous? A test of hypotheses. *Animal Behaviour*, **43**, 361–70.

Vincent, A. C. J. (1995). A role for daily greetings in maintaining seahorse pair bonds. *Animal Behaviour*, **49**, 258–60.

Welty, J. C. and Baptista, L. (1988). *The life of birds*. Saunders College Publishing, New York.

Westneat, D. F. and Sherman, P. W. (1993). Parentage and the evolution of parental care. *Behavioral Ecology*, **4**, 66–77.

Westneat, D. F., Sherman, P. W., and Morton, M. L. (1990). The ecology and evolution of extra-pair copulations in birds. In *Current ornithology* Vol. VII (ed. D. M. Powers), pp. 331–69. Plenum, New York.

Wickler, W. (1976). The ethological analysis of attachment: Sociometric, motivational and sociophysiological aspects. *Zeitschrift für Tierpsychologie*, **42**, 12–28.

Wickler, W. and Seibt, U. (1983). Monogamy: an ambiguous concept. In *Mate choice* (ed. P. Bateson), pp. 33–50. Cambridge University Press, Cambridge.

Wienecke, B. and Wooller, R. D. (1996). Mate and site fidelity in Little Penguins *Eudyptula minor* at Penguin Island, Western Australia. *Emu*, in press.

Wittenberger, J. F. 1981. *Animal social behavior*. Duxbury Press, Boston.

Woolfenden, G. E. and Fitzpatrick, J. W. (1984). *The Florida scrub jay: demography of a cooperative-breeding bird*. Princeton University Press, Princeton.

Wooler, R. D., Bradley, J. S., Skira, I. J., and Serventy, D. L. (1989). Short-tailed shearwater. In *Lifetime reproductive success in birds* (ed. I. Newton), pp. 405–17. Academic Press, London.

2 *Battles of the sexes and origins of monogamy*

PATRICIA ADAIR GOWATY

Introduction

Theories not focused on variation in the lives of females

> Higamous, hogamous, woman's
> monogamous
> Hoggamus higgamus, men are
> polygamous.

Despite the much touted tendency of men to seek polygyny and copulations with more than one woman, as witnessed by William James' famous ditty, monogamy is what most humans do. The fact that humans do what birds do may account for theoretical and empirical interest in the most common of avian mating patterns. Many of our scientific hypotheses about monogamy in birds come from our perceptions about monogamy in people. In fact, empathy probably guides most hypotheses about social behaviour (Hrdy 1981, 1986) even when (or especially when) they are about the naturally selected benefits and costs of monogamy for both sexes of *birds*. Recently, I noticed that William James' little poem provides one clue to our relative lack of success in understanding monogamy: for James, woman was an essentialist concept (notice that he speaks of women in the singular), yet, that was not the case for men; perhaps only a poetic constraint for James, but one that illustrates the trouble with our current theories about monogamy in birds. Our theories still cast females as relatively passive creatures, with so little variation in their lives that the underlying constraints on female behaviour have seldom (never?) been systematically explored. Thus, one of our difficulties in understanding monogamy is our lack of data about females' lives, perpetuated through the lack of adequate theory focused on females (Rosenqvist and Berglund 1992; Berglund *et al.* 1993). A new systematic way to approach understanding monogamy in birds (and perhaps also in people) is in this paper; these ideas are from 'females' perspectives', something that I hope will add balance to our existing theoretical frameworks.

In order to accomplish my goals, I review questions about monogamy and the hypotheses that have been used to explain the evolution and ubiquity of monogamy in birds. I review the hypotheses in terms of what aspects of monogamy the hypotheses focus on to make the point that monogamy hypotheses seldom address the same things. I review two sorts of monogamy hypotheses and briefly explore their relationships: those about current utility and those about origins. Finally, I describe new hypotheses for the origins of social and genetic monogamy and begin evaluation of predictions.

What is monogamy?

Discussions about definitions of monogamy pepper our literature (e.g. see the recent discussion in Ahnesjo *et al.* 1993). Perhaps this is to be expected: operational definitions are necessary for systematic investigations. However, in the monogamy case more seems to be going on than an effort to operationalize definitions.

In the not-too-distant past definitions of monogamy focused on pair bonds (Lack 1968) and the presence of biparental care, co-defence of territories, and assumptions about mating exclusivity (these definitions have been discussed at length by Wickler and Seibt 1982 and Gowaty 1985). Today unmodified 'monogamy' seems inappropriate because it implies only one mate and mating exclusivity, yet multiple mating is quite common (e.g. Smith 1988). The difficult point is that today when we say 'monogamy' we could be referring to social associations or to genetic outcomes. One way around the conundrum is to define the variation in the suite of behaviour of interest in terms of what we can discern readily: the co-operation of one male and one female in breeding, without the attendant assumption that mating between them is exclusive. An operational approach would be to indicate that we are referring to social associations, and to call them 'one female–one male groups' (with emphasis on *group* indicating a social association). This is the meaning of 'social monogamy' and 'sociographic monogamy' as defined originally by Wickler and Seibt (1982) in response to their anticipation of just this conceptual fuzziness. These ways of referring to monogamy leave open the question of the presence or absence of multiple mating by females or males. Why should we maintain the use of 'social monogamy'? One answer is that the variation in social behaviour that we readily observe (social system) remains interesting and compels explanation, whether or not social monogamy is also genetic monogamy, a goal that has readily been achieved by the authors of the 14 case studies in this book.

Social monogamy is often, but not necessarily, associated with some level of biparental care; yet it is also clear that individuals in social monogamy may or may not co-operate in the raising of offspring. Furthermore, socially monogamous individuals may or may not engage in extra-pair fertilizations. In fact, extra-pair fertilizations do occur in a large number of socially monogamous species. From the vantage point of our old definitions of monogamy, the idea that monogamy and extra-pair fertilizations occur simultaneously is an oxymoron; from the perspective of our new definitions, the idea demonstrates some of the possibilities in the question, what is the association between what we readily see (the social mating system) and the pattern of shared gametes (the genetic mating system)? This example serves to illustrate a general observation about monogamy—it's not just one thing. It is multi-faceted. The monogamy of birds is not the monolithic mating system it was once thought to be.

Three main hypotheses for the maintenance of social monogamy

Despite recognition that social monogamy is multi-faceted, existing theories for monogamy sometimes obscure rather than illuminate monogamous themes and variations. Theories for the evolution of mating systems are structured in such a way that some aspects of social behaviour have seemed more crucial than others. For example, almost everyone's (Darwin 1871; Williams 1966; Lack 1968; Orians 1969; Trivers 1972; Emlen and Oring 1977) ideas about selective pressures accounting for the evolution of mating systems pivot around the necessity (or not) of male parental care. For example Lack (1968) explained the origin and maintenance of social monogamy by the necessity of male parental care, and Orians (1969) based his classic model explaining the evolution of social polygyny on the crucial assumption that male parental care was valuable for female reproductive success. As reiterated in Clutton-Brock (1991), when male parental care is nonessential (e.g. Gowaty 1983; see also Wolf *et al.* 1988), and males 'can be emancipated from parental care' (phraseology from Emlen and Oring 1977), a pre-condition for the evolution of social polygny is fulfilled. When this pre-condition for social polygyny is fulfilled and social monogamy still occurs, theorists and empiricists explain social monogamy by reference to constraints (Emlen and Oring 1977; Wittenberger and Tilson 1980). That is, when males 'can be emancipated from parental care', without fitness costs, as they can be in some species all of the time and in other species some of the time, monogamy should be the selective consequence of other constraints on male behaviour rather than the requirement of male parental care.

Three theoretical constraints to social monogamy for males include the temporal distribution of females, the spatial distribution of resources (Emlen and Oring 1977; Oring 1982), and female–female aggression (Wittenberger and Tilson 1980). The temporal distribution argument states that when females breed synchronously, it should be difficult for males to monopolize more than one female at a time thereby constraining males to social monogamy. The resource distribution argument states that when resources necessary for reproduction are distributed so that it is impossible for one male to defend adequate resources for more than one female, males are constrained to social monogamy. Finally, the female–female aggression argument states that males may be constrained to social monogamy because any additional female(s) attracted to a male's territory may be inhibited from settling by the aggression of an already resident female.

The structure of these ideas characterizes monogamy as the default mating system for birds (Gowaty and Mock 1985). By 'default' we meant: if a male fails at polygyny, monogamy is his lot. These theories did not usually take up the equation from female perspectives. For example, if one dared to imagine that multiple mating was a better mating option than single mating for females, theories explaining why females failed at social polyandry would describe social monogamy as the default condition for females. Social or genetic monogamy might be better for males than social or genetic polyandry and better for females than social or genetic polygyny, but our formal and informal theories considered the female side of these equations as the simple obverse of what males do, a tactic that obfuscated the fact that females and males may be (often are) subject to very different selective pressures (importance noted, for example, in Trivers 1972). Constraints inhibiting females from social polyandry may be quite different from those precluding social polygyny to males. These are possibilities that still have not been explored adequately in the mating systems literature.

Male parental care is essential

There have been relatively few attempts to evaluate the three main monogamy hypotheses individually or collectively. Far-and-away, the male-help-is-essential (meaning females always fail without male assistance) hypothesis has received the most attention (e.g. Wright and Cuthill 1992; Wright and Cotton 1994). The striking aspect of the combined data is the relatively large amount of within-species variation for the effects of paternal care (Table 2.1) in socially monogamous species, the between-species variation (Bart and Tornes 1989), and the between-population variation in the occurrence and extent of paternal care in typically polygynous species (e.g. Beletsky and Orians 1990). Some have concluded that requirements for male parental care are adequate for explaining the maintenance of social monogamy in birds. For example,

Table 2.1 Selected male removal studies and the subsequent reproductive success

Species	Timing of male removal (n)	Sample size		Control manipulation	Success		Females compensated?	Source
		Experiment	Control		Experiment	Control		
Reed Warblers	Variable	7	46	Comparison group	73%	96%[1]	Yes	Duckworth (1992)
Acrocephalus scirpaceus								
Black-billed Magpies	Laying–nestling	20	106	Comparison group	0%	52%[1]	No	Dunn and Hannon (1989)
Pica pica								
Snow Buntings	Incubation (7)	14	11	Comparison group				Lyon *et al.* (1987)
Plectrophenax nivalis	Hatch (7)				0–5	3–6[2]	Yes–No	
					2.7	4.5[3]		
Tree Swallows	Laying (7)	47	86	Random experimental			Yes	Dunn and Hannon (1992)
Tachycineta bicolor	Incubation (25)			comparison group	29%	32%[4]		Quinney (1983)
	After hatch (15)				10.7	9.7[3]		
Dark-eyed Juncos		27	37	Comparison group	12%	33%[1]	Yes	Wolf *et al.* (1988); (1990)
Junco hyemalis								
Eastern Bluebirds	1983 SC[6]	13	13	Random, matched	3.8 ±1.5	2.9 ± 1.9[3]	Yes	Gowaty (1983)
Sialia sialis				pairs, experimental				
,,	1987 SC	12	12	,,	3.2 ± 0.4	3.6 ± 0.3[3]	Yes	Gowaty *et al.* (MS*b*)
,,	1987 NY	12	12	,,	1.9 ± 0.4	4.1 ± 0.5[3]	Yes	
,,	1987 ON	12	12	,,	1.2 ± 0.4	2.2 ± 0.4[3]	Yes	
,,	1985 ON	12	8	Comparison group	40 ± 20%	70 ± 40%[5]	Yes	Mackay (1985)

See also Bart and Tornes (1989) for summaries up to 1989. Most male removal studies have used unmanipulated comparison groups for 'controls' so that it remains difficult to separate out whether declines in unaided female nesting success are due entirely to male absence or to experimenter effects.

[1] Percentage of nests with fledglings.
[2] Range of number of fledglings.
[3] Mean number of fledglings.
[4] Percentage of nest failures.
[5] Percentage of nest successes.
[6] SC, South Carolina; NY, New York; ON, Ontario; All Eastern Bluebird males were removed on the last day of incubation or the first day of hatch.

Bart and Tornes (1989) focused on between-species variation in the degree and timing of male help, and noted that removal of male help had the most dramatic effects in species in which males incubated as well as fed nestlings. Their analysis also led them to remark on alternative hypotheses for male parental care (such as male parental care is an aspect of mate guarding, extended courtship, or a mechanism for the resolution of conflict over the sex ratio of offspring (Gowaty and Droge 1991)), none of which has been adequately tested. Their attention to between-species sources of variation in the effects of male parental care seems a productive approach, as is attention to the striking within-species variation in the effects of paternal care on maternal fitness. This within-species and within-population variation in the effects of paternal care on maternal fitness should mean that *some* males at least potentially have other options besides strict social monogamy. Males mated to lucky or highly competent females that are capable of fledging as many young without a male as with one may be under strong selection to attempt social polygyny or extra-pair liaisons, or to desert altogether. Males mated to high-quality females or females in permissive environments may be released from the constraints of 'trade-offs' posited as critical to male options (Westneat *et al.* 1990). Thus, when males mated to essentially self-sufficient females are unsuccessful at attracting secondary partners and yet still do *not* desert their mates (even briefly, say in search of extra-pair copulations), we might ask about alternative selection pressures (Wolf *et al.* 1991) on males that favour their socially monogamous habits. So, even though male parental care is so ubiquitous in birds, the variation in this trait makes it not unreasonable to ask all over again about the selective factors favouring male help (see below). My point here is that selection should favour males' alternative reproductive tactics (seeking extra-pair copulation partners or multiple breeding territories, etc.) whenever they are released from paternal care to whatever extent; thus when we do not observe males engaging in these alternatives (such as seeking multiple breeding territories), our theories should have led us to ask about other advantages for males that stay.

Attention to sources of variation in maternal quality (Wright and Cuthill 1992) and other options during nesting (e.g. Davies 1989) also leads to other questions. What factors affect variation in female reproductive success besides paternal care? How does female quality affect females' options for social monogamy or social polyandry, for genetic monogamy or genetic polyandry? How does environmental quality affect females' options? How does environmental quality interact with female quality to affect females' options? How are females' options constrained by males? How do males affect environmental quality for females? Does paternal care always necessarily have positive effects on maternal fitness? This last question goes against the trend of the last 25 years in that the most crucial assumption of the most influential and published about model for the evolution of mating systems (Orians 1969) assumes that male parental care enhances female fitness. The notion that males' help may not always have positive effects on female reproduction somehow seems difficult to grasp given the theoretical hegemonies of the recent past.

Polygyny threshold not reached

The distribution of resources hypothesis states that social monogamy of males is constrained because the polygyny threshold is not reached (Orians 1969). Social polygyny for males is favoured whenever there is enough habitat or environmental heterogeneity that females choosing to mate with an already mated male may have fitnesses equal to or greater than females choosing to mate with an unmated male.

The idea that social monogamy is constrained by resource distributions has been elaborated to explain other mating systems besides social monogamy and social polygyny (Gowaty 1981; Davies 1991; Bensch and Hasselquist 1991), and tested many times in socially polygynous species (see references in Searcy and Yasukawa 1989 and Björklund and Westman 1986) and at least several times in typically socially monogamous species. Both European Starlings *Sturnus vulgaris* (Smith, in press) and House Sparrows *Passer domesticus* (Veiga 1992), two cavity nesting passerine species, have been experimentally induced to social polygyny by the addition of nesting boxes to territories that males could readily defend.

Veiga (1992) tested the idea that social monogamy prevails in House Sparrows because males are 'unable to defend more than one nest site successfully' by placing new nest boxes close to nest boxes from earlier years. The polygyny threshold hypothesis predicts that this manipulation would facilitate nest defence and increase the likelihood of social polygyny for males. Male House Sparrows did defend more nesting boxes when nesting boxes were close together, but defence of multiple nesting boxes did not affect male pairing success, thus Veiga (1992) rejected this explanation for the maintenance of social monogamy in House Sparrows.

Steve Emlen and I (unpublished data) attempted a similar experiment on Eastern Bluebirds *Sialia sialis* in 1985 (see drawing on title page). We randomly assigned previously successful nest sites to one-box and two-box treatments. We placed nesting boxes in two-box territories about 15–30 m apart, a distance that easily allowed males to defend both nesting cavities. Our experiment, like Veiga's, failed to induce social polygyny, leaving us with the possibility that there was something wrong with the way we did our experiment or that the simple addition of nesting boxes to Eastern Bluebird territories did not introduce enough habitate heterogeneity to allow females to cross the polygyny threshold. Eastern Bluebirds forage closer to their nesting sites than Starlings or House Sparrows usually do, suggesting that in order for Bluebird females to pair socially with an already mated male, we might need to manipulate the availability not just of nest sites, but also of food. Of course, other constraints to social monogamy of males such as female–female aggression might adequately explain why Eastern Bluebirds are so overwhelmingly socially monogamous (Gowaty 1980).

Female–female aggression inhibits settling of additional females

The female–female aggression hypothesis for the maintenance of monogamy has been explicitly tested as a mechanism preventing recruitment of additional females to a male's territory (e.g. Gowaty and Wagner 1988; Veiga 1992). Slagsvold *et al.* (1992) present interesting data that implicate female–female aggression in the maintenance of social monogamy in Pied Flycatchers *Ficedula hypoleuca*. Although male Pied Flycatchers try to attract secondary mates, their usual inability to do so is associated with female aggression to interloper females (measured as aggression to caged females) and increasing distance between nesting boxes simultaneously held by males.

Fewer studies have evaluated patterns of female–female aggression which would allow strong inferences (in the sense of Platt 1964) about the selective factors shaping tendencies of females to be aggressive (e.g. Gowaty and Wagner 1988; Hobson and Sealey 1989). Whether tendencies of females to be aggressive were shaped by selection for nest defence, mate defence, or defence against conspecific nest parasitism, female aggression against intruding females when already settled females are building nests and laying eggs may be sufficient to inhibit recruitment by additional females to males' territories. Thus, female–female aggression for whatever reason during recruitment stages of nest cycles may be sufficient to account for the maintenance of social monogamy in many socially monogamous species. A recent review (Slagsvold and Lifjeld 1994) of female–female aggression indicates that intrasexual aggression by female birds is common. Given that there should almost always be strong selection on females to lay their eggs in the nests of conspecifics (e.g. whenever their own nests are predated or destroyed during egg laying (Gowaty and Wagner 1988; Gowaty and Bridges 1991*a*), female–female aggression shaped by avoidance of conspecific nest parasitism may have as its by-product the maintenance of social monogamy (Gowaty 1980). In fact, female–female aggression may be the most parsimonious explanation for socially monogamous males in many species (Veiga 1992; Gowaty and Wagner 1988; Hobson and Sealey 1989). However, general evaluation of this notion will await rather more explicit tests of variation in timing and degree of female–female aggression. Certainly for Eastern Bluebirds (Gowaty and Wagner 1988) and Yellow Warblers *Dendroica petechia* (Hobson and Sealey 1989), female–female aggression reduces the threat of conspecific nest parasitism, yet probably as effectively limits male Eastern Bluebirds and Yellow Warblers to social monogamy.

Monogamy questions uncoupled

The first observations of extra-pair fertilizations were important because they made clear that assignments of parentage inferred from adult associations at nests were inadequate for tests of hypotheses for the evolution of mating systems. Now understanding the evolution of extra pair fertilizations and their effects on other aspects of social behaviour is a research enterprise itself, but one that has so far left unaddressed the evolution of social monogamy itself. The covariates of social monogamy have been decoupled from the question of the evolution of social monogamy. Understanding the evolution of social monogamy will result from further decoupling of covariates that have been considered important aspects of monogamy. Other questions that deserve independent attention are those in Table 2.2, and there are possibly others.

In some socially monogamous species females do not help males during territorial maintenance, nest building, nest guarding, or incubation (McKitrick 1992, noted also in Wesokowski 1993 as the probable ancestral sort of parental care in birds). In other socially monogamous species males do not help females with territorial maintenance, nest building, incubation or feeding of nestlings or fledglings; however, males stay in monogamous social associations and mating often appears to be exclusive (Verner and Willson 1969). In still other socially monogamous systems males do not stay. And in others, mating is not exclusive. The association of these characteristics with socially monogamous mating systems has nowhere been thoroughly explored. Perhaps a reason for this is that the existing theories for the evolution of social monogamy conflate several separate questions:

1. Why is there only one female?
2. Why is there only one male?
3a. Why are females genetically monogamous (when they are)?
3b. Why are females not genetically monogamous (when they are not)?
4a. Why are males genetically monogamous (when they are)?
4b. Why are males not genetically monogamous (when they are not)?
5. Why do females help?
6. Why do males help?
7. Why do males stay?
8. Why do females stay?

(And several of these questions can be further subdivided.) Table 2.2 is an attempt to identify what aspect of monogamy each of the existing monogamy hypotheses addresses. In fact, mating exclusivity, biparental care, and persistence of adults around nests or on nesting territories are not necessary aspects of social monogamy, whereas the 'pairing' of only

Table 2.2 Hypotheses for social monogamy conflate several questions; some theories leave unaddressed important aspects of monogamy; some aspects of monogamy are unaddressed by any theory

Monogamy hypotheses	Why do males stay?	Why do females stay?	Why 1 male?	Why 1 female?	Why male exclusivity?	Why female exclusivity?	Why do males help?	Why do females help?	Source
Maternal Care essential		X						X	
Paternal Care essential	X						X		Lack (1968); see text
Temporal distribution of breeding females						X			Emlen and Oring (1977)
Resource distribution	X	X	X	X	X				Orians (1969); Emlen and Oring (1977)
Aggression:									
Male–male	X		X		X				Wittenberger and Tilson (1980); Gowaty (1981); Gowaty and Wagner (1988); Slagsvold and Lifjeld (1994)
Female–female		X		X	X				
Male–female		X			X				
Female–male	X				X				
Enhanced health	X	X			X				Sheldon (1994)
Future mateships	X	X			X				Stamps (in press)
RS via EPCs	X								Wagner (1992); Gowaty (in press)
Bet hedging to preserve female condition	X								Lyon *et al.* (1987)

RS = reproductive success; EPCs = extra-pair copulations

one female and one male is (Gowaty 1985; Mock and Fujioka 1990). This decoupling exercise seems especially useful given how often some express surprise that individuals may be genetically monogamous, even when they are not socially monogamous (much less vice versa). The point here is that questions about variants in female and male behaviour may be usefully disassociated into components when asking questions about the likely conditions leading to the origin or maintenance of particular social systems. For example, note that at least two of the hypotheses in Table 2.2 explain why males stay; the male-help-is-essential hypothesis and the extra-pair fertilization hypothesis (Gowaty, in press).

In general, the question, why do males stay, deserves more explicit treatment, separate from consideration of why they help; these two questions are analogous to similar questions about auxiliary individuals in species with 'helpers-at-the-nest' (Brown 1987; Emlen *et al.* 1991). One reason to decouple these questions formally is that it is easy to imagine conditions in which there may have been some positive selective value for male staying, before environmental conditions provided positive selective value for helping by males. Thus there is the possibility that the theoretical assumption that male parental care is pivotal may be misleading and/or unnecessary to the origin, at least, and perhaps, as well, for the maintenance of social monogamy.

When male parental care is essential to female reproductive success what is explained is current utility, why males currently stay and help; and because 'helping' is local, it seems to explain 'staying'. Actually, little may be explained about the conditions leading first to staying by males and second to helping by males. In addition to conflating different questions, the existing theories offer few explanations for why males stay when environmental variations limit males to associating with only one female and when male parental care is nonessential. When males are limited to social monogamy by female–female aggression and when male parental care is nonessential, for example, it would seem advantageous for males to leave permanently to seek additional mating opportunities through defence of additional territories away from their primary territories (Alatalo *et al.* 1981) or to leave temporarily in search of extra-pair fertilizations (Trivers 1972) off of their territories.

Social monogamy and multiple mating

Part of the new interest in social monogamy arises because we now know multiple mating is common and produces offspring. In fact, the worm has turned and today it is far more interesting (and surprising) to report on socially monogamous species (those in which both female and male are simultaneously socially monogamous, in contrast to the case where females only or males only are socially monogamous) that are also genetically monogamous. Of the passerine species studied so far, between

20 and 25% of socially monogamous species also seem to be genetically monogamous (Birkhead and Møller 1992). Because of the widespread existence of extra-pair paternity, social monogamy is no longer considered interesting by default, but rather interesting on its own terms.

Decoupling social monogamy and faithfulness

Now that 'social monogamy' and 'faithfulness' no longer mean the same thing, questions about social monogamy seem mostly to be about who mates multiply and why. The focus of analysis is the individual, so that questions are about monogamous individuals, not monogamous couples. Until very recently, one camp promoted the idea that multiple mating was driven primarily by the interests of males—after all, male reproductive success (and fitness) is a direct function of the number of females with whom a male mates. Because the theoretical answer to the puzzle of why females mate multiply is less readily apparent, proponents of this camp characterized multiple mating as something that males force on relatively passive females (e.g. Wrege and Emlen 1987; Westneat *et al.* 1990). The other camp promoted the idea that advantages to multiple mating exist for females and females are active participants who seek extra-pair partners (Smith 1988; Gowaty and Bridges 1991*b*; Kempenaers *et al.* 1992; Wagner 1992) and 'control' extra-pair fertilizations (Lifjeld and Robertson 1992). However, attention to multiple mating by females is limited by lack of theoretically compelling explanations for it (see below). These polar characterizations of the behaviour of males and females will probably not be reconciled simply by data indicating some happy medium in which both females and males are seen as active participants or in which there is simple symmetry in what females and males do, but by the articulation of theory that will compel belief in advantages for females of multiple mating.

Whether one 'believes' in active or passive female participation in extra-pair fertilizations is irrelevant to the conundrum of females mating with more than one male. Hypotheses (see lists in: Gowaty 1985; Westneat *et al.* 1990; Birkhead and Møller 1992) include: females are directly forced to copulate or are coerced to copulate by the threat of infanticide; females gain fertility insurance; females gain material benefits from nutrients in ejaculates, traded food for copulations, paternal care, or access to extra-territorial food resources; females gain genetic benefits from genetically diverse offspring or from genetically superior sites; females avoid retaliation of males that they may reject; females may gain advantages by reducing sperm supplies available for competing females; females guard against genetic defects caused by 'old' sperm. It is also possible that multiple mating by females will increase their fecundity directly just as extra-pair copulations are expected to increase fecundity of males. In some ways the most curious of the new hypotheses for multiple mating by females is that females mate multiply to avoid close inbreeding, an idea that seems to explain multiple mating by fairy wrens (Brooker *et al.* 1990; Russell and Rowley Chapter 8).

Of the explanations for multiple mating by females, Birkhead and Møller (1992) found little support for any of the hypotheses except those having to do with genetic quality of mates. So far, however, there is neither a synthetic theory from females' perspectives about extra-pair paternity nor enough data to guess at the relative merit of any of these hypotheses for any species except for the curious fairy-wrens (Brooker *et al.* 1990). The challenge is to devise strongly inferential crucial tests (Platt 1964) that will discriminate among explanations for various species. Below is a new way to look at this puzzle, which introduces a female perspective that I later use to explore the origins of social monogamy.

Predicting genetically monogamous females

The solution to the female multiple mating conundrum may be approached by asking when or how females are favoured in genetic monogamy. If one assumes that females are favoured who mate for good genes and that mating is potentially costly, say, because of increased risk of exposure to pathogens and parasites (Sheldon 1993), socially or genetically monogamous females theoretically will be favoured whenever females can choose among available males simultaneously. When there is high variance in male genetic quality (and other nongenetic qualities are equal), females that are unconstrained socially and/or ecologically, so that they are able to choose among all available males simultaneously, theoretically will share their gametes monogamously with the same high quality male (making him genetically polygynous, while most males will not mate at all); the resultant variance in male reproductive success will be high. Conversely, when there is little or no variance in male genetic quality (and other nongenetic qualities are equal), unconstrained females will mate monogamously with any of the males in a population (so that most males are genetically monogamous too); the resultant variance in male reproductive success will be relatively low. However, when ecological and social constraints on female mating decisions are factored in, genetic monogamy may not be so easily achieved for females under selection to mate for good genes. This female perspective facilitates new predictions: for example, in different populations of the same species in which variation in male genetic quality varies and all else is equal, genetic polygyny of males will be higher in populations with high rather than lower variation in male genetic quality (see also Petrie and Lipsitch 1994). The critical variables for females include indicators of genetic variation among males *and* constraints—either ecological (e.g. those suggested in Gyllensten *et al.* 1990) or social (such as those suggested by the battles of the sexes ideas below)—on females preventing them from making their best choices initially (e.g. Slagsvold and Dale 1994).

The assumption that females are favoured who mate for good genes makes intuitive sense and is logically powerful. However, objections to this assumption are common; responses to some follow. In investigations of current benefits to females who mate with more than one male, it may be difficult to observe the effects of selective benefit to females who choose gametic partners for good genes. This may be especially so if selection acts on males to reduce variance in mating success (see below); that is, if evolved social behaviour reflects multiple selection pressures, the force of selection on females for gametic partners with good genes may be obscured (an analogy to finding the costs of reproduction). Potential lack of genetic variation in males would seem unlikely in all but genetically bottle-necked or inbred populations. And, as with other traits, even a very small amount of variation will be enough to have selective effects. Thus, it seems that a theory of female genetic monogamy requires examination of the sources of ecological and social constraint on females. The conditions under which female birds are able to make comparisons among all available males without temporal, spatial, or social constraint are probably relatively rare (see below), thus, ecological and social constraint on females may be such that females mating for good genes end up mating multiply—even when there are potential costs from pathogen and parasite exposure. All of which brings us to the hypothesis that multiple mating by females may indicate females making 'the best of a bad job' mating for good genes under ecological or social constraint, i.e. the Constrained Female Hypothesis.

Origins of social monogamy through battles of the sexes

Most theories for the evolution of mating systems assume that the current costs and benefits to females and males of particular mating systems reflect selective factors that contributed to the origin of particular patterns. For example, one might posit that male parental care arose as a random event, but was so selectively advantageous that individuals exhibiting parental care were favoured over those that did not. Alternatively, an origins hypothesis that does not depend on capitalizing on random advantage from one male–one female groups follows from a consideration of selection pressures on resource-users and their limiting resources (Gowaty 1992, in press).

Resource-users should be under strong selection to manipulate or control the reproduction of their limiting resources (Gowaty 1992, in press). When resources are living, selection should act on them to resist any manipulation or control that is deleterious to their own fitnesses. Thus, whenever access to females limits male reproduction, males should be under strong selection to manipulate or control female reproduction, and whenever female fitness is less when under the control of others than when under own control, females should be under strong selection to resist. Whenever access to females limits males' reproduction more than access to males limits females' reproduction, what is pivotal to understanding the variation in mating patterns is females' abilities to resist manipulation or control of their own reproduction by males (Gowaty, in press). Thus, males should attempt to manipulate or control females by acting on females' bodies *directly* through conditioning or force as well as indirectly through brokering females' access to crucial resources (Gowaty 1992, in press). Whether males are successful depends critically on females' intrinsic abilities to resist males' efforts and on environmental potential for resistance by females; therefore, females' abilities to resist and retain control of their own reproduction is of central importance to the evolution of social behaviour in sexual species (Gowaty 1992, in press).

Thus, social monogamy may arise via the following pathway. When females who share their gametes with males with superior genes are favoured by selection, some males will be chosen and females will freely mate with them. Some males, however, will not be chosen and selection will act on these males to manipulate or control females' behaviour such that the probability that females will mate with them increases. In this scenario there is always some cost to females in terms of the reproductive success component of their fitnesses whenever they are unable to mate with males they freely choose. A heuristic categorization from females' perspectives of mechanisms males use is in terms of their effects on the survival component of females' fitnesses (Gowaty, in press). Smuts and Smuts (1993) focus on aggressive conditioning of females' behaviour and forced copulation, mechanisms at one end of this continuum (Gowaty 1992, in press). Males can use mechanisms that theoretically:

(1) increase females' survival probabilities (e.g. guard her against predation, help her raise her offspring, facilitate her foraging for herself or her dependent offspring, or feed her directly);

(2) have no effect on females' survival probabilities (e.g. exploit pre-existing aesthetic biases);

(3) decrease females' survival probabilities (e.g. aggressive conditioning of female behaviour, forced copulations, or infanticide resisted by females).

All else being equal, those control mechanisms that do not decrease females' survival probabilities should be favoured over those that do (Gowaty, in press). Obviously, a range of options exists for males non freely chosen by females to affect the likelihood that females will mate with them. The continued development of this scenario depends on whether males use mechanisms that increase or decrease females' survival probabilities. Mechanisms available to males depend on females' abilities to resist males' efforts at control. Note that this discussion and the discussion in the following two sections is based on females' *survival* probabilities as distinct from the *reproductive success* components of females' fitnesses, because of the assumption that usually female reproductive success will be less when she mates with males not freely chosen by her than when she does. Discussions of alternatives based on both survival and reproductive success components of females' fitnesses are much more complex; this discussion and the following take the simpler route as a heuristic that focuses attention on the dynamics of male/female interactions. Whether these ideas match existing variation in nature is an empirical point to be investigated.

Social monogamy via helpful coercion

'Helpfully coercive' males are those who manipulate females into mating with them by helping. Factors affecting the outcome of such intrasexual selection among males *on females* depend on the variation in male genetic quality and whether fitness effects on females through helpful coercion (through her enhanced survival) are enough to over-ride the disadvantage of being unable to produce offspring with a male of higher genetic quality. If mating with a male of high genetic quality is more advantageous than mating with a helpful male, females would be selected to attempt to be fertilized with the best genetic quality male (and possibly also to exploit the labour of helpfully coercive males). There would be variation among females within populations in their abilities to resist males' efforts, that is, in how and how much they interact with males. Some females would copulate with males and continue breeding activities alone. Some females' survival would be enhanced enough to overcome the difference between their helpful mate's genetic quality and the best male's genetic quality (note that this situation provides the conditions for the further evolution of help, say with incubation and/or the care of nestlings and fledglings). Some females will not be helped enough to nullify the difference between their helpful mate's genetic quality and the best male's genetic quality, that is, either help and/or genes will be low quality, so that females' reproductive success will be higher without 'help' than with it. Such females may be selected to resist help by males and to be fertilized by a male of high genetic quality and to continue breeding activities unassisted, or to exploit the low quality help, yet seek to be fertilized by males with better genetic quality. A scenario like this may explain much of the extra-pair paternity in birds that has recently been reported (Fig. 2.1, discussed further below).

If female quality is equal, the behaviour of females will depend on the variation in genetic quality of males and on males' abilities to manipulate them into mating through 'helpful coercion'. However, when females' quality varies too, the behaviour of females will depend on the interaction of variation in females (and the environments they find themselves in) and variation in males.

Female vulnerability to control by males

least	→	most
♀♀	sex that arrives first	♂♂
♀	larger sex	♂
♀	philopatric sex	♂
♀♀	same-sex kin coalitions	♂♂
NO	feeding niche overlap	YES
NO	territory overlap	YES
NO	'co-operation' in parental care	YES
♀♀	adult sex ratio bias	♂♂

Fig. 2.1 Demographic and social factors theoretically associated with female ability to resist manipulation and control of their reproduction by males.

This is an origins hypothesis. In the male–female groups in this scenario there is no requirement for males to stay beyond the period of initial helpful coercion that manipulates females into mating with them; there is no requirement for biparental care or co-defence of territories. Males compete in two ways: through variation in both freely expressed female choice and in males' abilities to 'helpfully coerce' females. Note, however, that this backdrop provides opportunity for further selection on males to 'helpfully coerce' females into mating with them; that is for the further elaboration of male staying and helping in the ways that are often associated with social monogamy.

Social monogamy via aggressive coercion

When males not freely chosen by females fail to manipulate females into mating with them using mechanisms that increase females' survival probabilities (as might happen when lack of male genetic quality is not made up by males 'helping'), selection should act on males to manipulate or control females using more drastic measures, such as mechanisms that potentially decrease females' survival probabilities (Gowaty 1992, in press; Smuts and Smuts 1993) such as aggressive conditioning of female behaviour and forced copulation. In birds, so-called 'mate guarding' may be a type of aggressive conditioning of females' behaviour. At least, this might profitably be explored in further evaluations of male behaviour towards females.

Agressive conditioning will facilitate social monogamy whenever it trains females to remain away from other males, limits females' movements, or females' opportunities to sample multiple males. Selection on females to resist will be even more likely because the decrement in female fitness due to declines in survival probabilities from forced coercion will be relatively more difficult to make up than in the scenarios above where males did enhance females' survival probabilities. Despite resistance, females will sometimes be coerced aggressively into social monogamy, yet this route to social monogamy should be less successful and less frequently observed than the others.

Social monogamy via resource brokering

When females are able to resist direct control mechanisms, males should be selected to attempt to coerce females into mating for access to male-brokering resources (Gowaty 1992, in press). This may be especially common in birds because female birds seem able to resist many mechanisms of direct male control, and because so many aspects of avian reproduction are vulnerable to 'take over' by males. Thus, social monogamy in birds may be enforced by males who manipulate female behaviour by brokering their access to essentials.

Crucial resources that limit reproduction by female birds include territories, which may contain essential nutrients and nest sites. Male territoriality may be one of the most reliable and efficient ways that male birds use to manipulate females into mating with them, an idea that puts the spring arrival of males before females on territories in a new light. The ecological potential for females to resist male resource brokering is pivotal to this pathway to social monogamy (Gowaty, in press). Social monogamy via male brokering of females' access to nesting sites is discussed further below.

Variation in female resistance and female genetic monogamy

As the previous discussion makes clear, females' abilities to resist males' manipulation or control attempts is theoretically pivotal to the evolution of social monogamy. It is also the theoretical key to understanding genetic monogamy of females, no matter whether the social mating system is polygynous, monogamous, polyandrous, polygynandrous, or promiscuous.

Until empirical evidence of females' resistance mechanisms or abilities are available, they can be logically inferred. For example, Smuts (1995) argues that when male primates are more likely than females to form kin coalitions, females may be less able to resist aggressive conditioning or forced copulations. By analogy, female birds may be most able to resist when females are larger than males, when females are more philopatric than males (so that female kin or same-sex coalitions are favoured or female knowledge and experience with local resources are enhanced relative to males), when females arrive at essential resources before males do, and in situations where extended interactions between females and males are inhibited (e.g. when there is no feeding niche or little territorial overlap for females and males or when males and females do not co-operate in parental care). In addition, females are more likely to resist successfully when the adult sex ratio favours females, as it does in some bird species (Breitwich 1990).

Figure 2.1 shows traits associated with females' abilities to resist. In fact these traits may be the phylogenetic vestiges of selection for male control and female resistance. Species of grouse and other lekking species and shorebirds would seem to represent avian species in which females are most able to resist male efforts, while typical helper-at-the-nest species, in which helpers are mostly males, would seem to represent birds in which females are least able to resist successfully. This logic says that genetic monogamy in female birds is most likely for females at each end of the control continuum, that is, when females have the most and the least abilities to resist males' control efforts.

One can rank theoretically females' abilities to resist in phylogenetically related species as well, and use this variation to predict the likelihood of extra-pair paternity. For example, among female Turdinae, the open-cup nesting thrushes like American Robins *Turdus migratorius* should be better able to resist than females in cavity nesting thrushes like bluebirds (*Sialia*). This is theoretically so because cavity nest sites are rarer than open cut sites, making it easier for males to broker females' access to them. Therefore, American Robins will have more extra-pair paternity than Eastern Bluebirds.

A model of social monogamy and genetic polyandry: the Constrained Female Hypothesis

Figure 2.2 describes theoretical effects on female fitness of the interaction between female quality and environmental quality, when females are 'helped' by males and when they are not. The *x*-axis represents variation in the combined interactive effects of female intrinsic quality *and* the quality of the environments that females find themselves in. The two curves on the graph converge on the right hand side, because of the assumption that some combinations of females and environments allow females to reach their intrinsically limited reproductive success without male help; that is when females are of high quality and/or environments are permissive, thus facilitating females' resistance efforts, male efforts will be unable to increase female fitness further. This is a notion in keeping with the idea that female reproductive success is limited by intrinsic variation in females (Bateman 1948; Williams 1966; Trivers 1972).

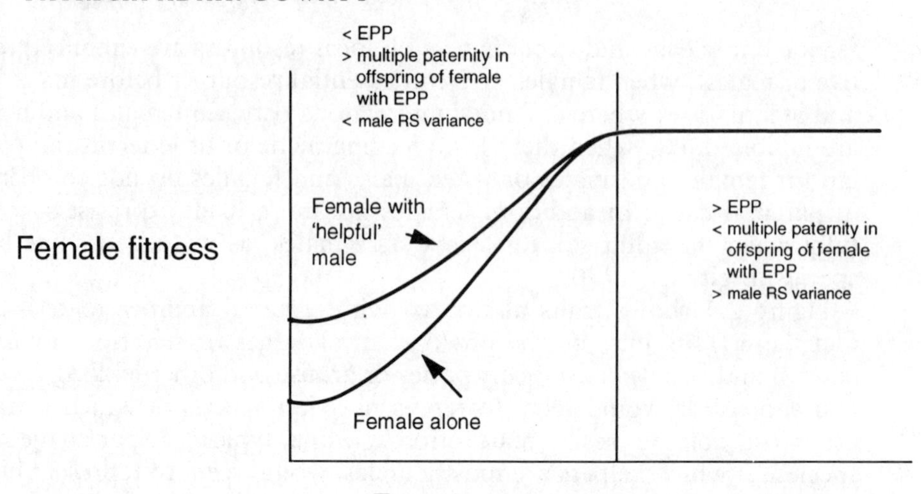

Fig. 2.2 The Constrained Female Hypothesis predicts genetic mating patterns of socially monogamous females. The model describes the relationship between female fitness as a function of interactions between female quality and environmental quality. The upper curve describes how female fitness varies when females are 'helped' by males and when they are not. The curves converge for high female × environmental quality because female fitness is limited by intrinsic variation in females. This version of the model says that females of poorer quality or in less permissive environments are more vulnerable to manipulation by male 'helpful coercion' than females of higher quality or those in highly permissive environments. Predictions of the model include relative rates of extra-pair paternity (EPP) and multiple paternity for females × environments below and above the point at which the curves converge. RS = reproductive success.

This hypothesis focuses on how variation in female intrinsic quality and the environments they are in fosters female abilities to resist male *helpful coercion*. When all else is equal, for some females—low intrinsic quality females and/or females in poor environments—vulnerability to male helpful coercion will be high. One of the potential meanings of this is that these low intrinsic quality females or those in poor environments have 'more to lose' if male help is withdrawn, for any reason including retaliation through withholding of help for lack of female 'faithfulness', for example. For other females—high intrinsic quality females and/or females in permissive environments—vulnerability to male helpful coercion will be much lower; that is, these females have less to lose from withdrawal of male 'help'.

Variation in paternal provisioning of nestlings as a function of their genetic paternity (Møller and Birkhead 1993; Dixon *et al.* 1994, Gowaty *et al.* MS*a*) is obviously germane to this idea. The hypothesis says that what underlies these patterns is variation in female abilities to make up for any withdrawal of male provisioning that might be correlated with extra-pair paternity.

The variation in reproductive success of experimental females in male removal experiments (Table 2.1) is quite interesting in relation to this hypothesis, and supports the view that what is pivotal to understanding the origins and maintenance of social monogamy is not the average value *per se* of male parental care to females, but the underlying interaction between intrinsic quality of females and their environments that makes them needy or not of male help in the first place (fostering social constraint of females who depend on males). Thus, we might imagine that social monogamy arose because males defended all available territories and brokered females' access to them—in effect requiring females to 'trade' mating for access to crucial resources. Females with the most to lose from withdrawal of male help should be the most highly constrained and, thus, coerced by males; in other words when the difference in fitness for a female of reproducing alone after fertilization with a superior male is much less than reproducing with the help (and fertilizations) of an inferior male, such females should be expected to remain 'faithful' to their social mates. However, when the fitness differences between these options are less dramatic, we should observe unconstrained females, able to mate with males of superior genetic quality when they are available, while perhaps simultaneously exploiting the helpful coercion of males on females' territories.

This *Constrained Female Hypothesis* is testable. If the assumptions of the hypothesis are met, it predicts that high quality females or females in permissive environments are more likely to produce extra-pair offspring than low quality females or females in less permissive environments. Tests of this prediction require assessment—independent of fitness measures—of females' intrinsic qualities and environmental quality. However, even at this point, this prediction is especially interesting in relation to reports of higher fecundity or higher offspring survival or quality for females with broods from extra-pair paternity versus those without extra-pair paternity (Gowaty *et al.* unpublished data). These effects may be due as much to high quality females as they are to good genes in males. A second prediction is that lower quality females or females in poor environments who do have extra-pair young will be genetically polyandrous, while higher quality females or females in better environments will be more likely to be genetically monogamous (meaning all of a females' offspring are sired by one extra-pair male). See Chapter 16 for an example of females in low quality situations (i.e. with a poor quality male) that seek extra-pair copulations, and females in good quality situations that avoid extra-pair copulations.

Genetically monogamous females and male reproductive success variance: origins of social mating patterns

If females sharing gametes with males with good genes are selectively favoured, and if exposure to pathogens and parasites is costly to female

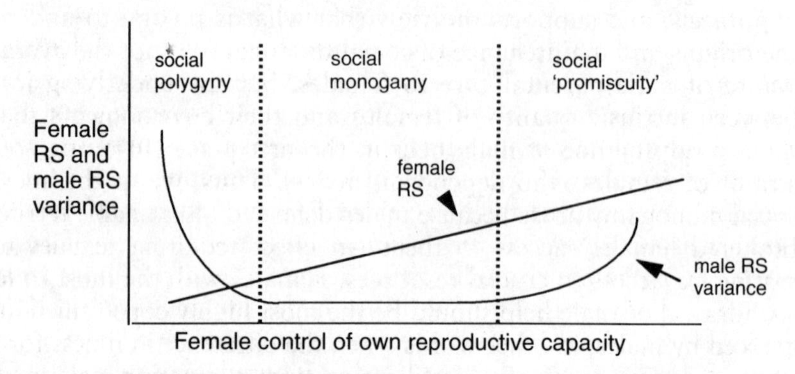

Fig. 2.3 A graphical model of the relationship between how much control of their own reproductive capacities females have and (1) female reproductive success and (2) variance in male reproductive success. This model says that at each end of the continuum of female control of own reproduction variance in male reproductive success is relatively high. On the right hand side of the graph, variance in male reproductive success is high because of free female choice for gametic partners of superior genetic quality. On the left hand side of the graph, variance in male reproductive success is high because of mechanisms that are highly successful in the control of female reproduction. Male variance is drawn higher on the left than the right hand side because of the assumption that male efforts to control females and correlated traits associated with other male/male competitive interactions will lead to higher reproductive success variance than female choice for superior genes in males (see text for additional explanation).

fitness, I hypothesize that females within species should be genetically monogamous whenever (1) they are free of ecological and social constraints, so that they can choose among available males simultaneously, or (2) they are completely controlled by males. When either of these two conditions is met, the mating system of females will be genetically monogamous, though perhaps seldom socially monogamous (Fig. 2.3). And similarly, if these ideas have merit, social monogamy in birds will seldom indicate genetic monogamy; rather social monogamy and genetic polyandry of females will be strongly associated. Predictions include the following. In species such as lekking grouse, in which females are able to make simultaneous choices among males and in which females' abilities to resist males are theoretically large (Fig. 2.1), females are genetically monogamous. Females in typical helpers-at-the-nest species such as Florida Scrub Jays *Aphelcoma coerulescens* are genetically monogamous (Woolfenden and Fitzpatrick Chapter 7), not because of females' freedom to choose among available males but theoretically because of females' inabilities to resist male control (Fig. 2.2). New questions seem relevant. For example, is the well-known, supposedly anti-predator *sentinel* behaviour (McGowan and Woolfenden 1989) of Scrub Jays a mechanism whereby males manipulate females' movements?

Differences in the variance in reproductive success of males may also be associated with these two contrasting routes to female genetic monogamy (Fig. 2.3). When females are free, variance in male mating success will be relatively high. When females' mating choices are controlled by males, variance in male mating success may be even greater (because of runaway processes of contests among males for access to females). In response to high variance, some males may join coalitions against especially successful resource brokers (despots) or free females. Social monogamy through male resource control may be favoured because of advantages among those males not favoured in either social polygyny or social promiscuity. Such variance in relative reproductive success among males not otherwise favoured may also select for male coalitions in preventing or retarding female access to resources not brokered by males.

Variance in male reproductive success under freely expressed female choice should seldom be as great as under extreme male control of females, because there probably exist limits on the ability of the most superior male to fertilize all females. If a female is unable to be fertilized by the best male, she will seek the next best, and so on. Selection may then act on females to seek not just best genes, but best genes given their own complement of genes. So, males and females will assort roughly by quality. The ratchet of sexual selection on males would lead to extraordinary abilities of males to inseminate multiple females, while the ratchet of sexual selection on females would lead to superior abilities to differentiate genetic quality of males.

Females mate multiply under imperfect social constraint and 'pair bondage'; predictions and potential tests

In general it should be difficult for male birds to control females totally, because of female intrinsic quality and environmental potentials favouring successful female resistance (Buschhaus and Gowaty, unpublished data). Most bird species lack intromittent organs so that females must co-operate with males in order for sperm transfer to be successful; therefore forced copulation should seldom be a successful alternative reproductive tactic for males (Fitch and Shugart 1984). In socially monogamous species of birds males and females are often size monomorphic or females are larger (Dunning 1993), so males should seldom have any size-related advantages in between-sex aggression. Therefore, aggressive conditioning of female behaviour by males should be easily resisted or reciprocated by female birds. Because most birds fly, females can readily escape many copulation attempts by males, and when birds live in complex habitats such as forests, it should be relatively easy for females to escape males' influence by evasive flying and hiding. Mate guarding in fact may be male attempts to corral females, to keep them on home territories, and to constrain them from seeking fertilizations with better-quality, extra-pair partners (Gowaty and Bridges 1991*b*). Thus, in general among birds males should seldom be able to control females completely.

However, when females' reproduction is limited by male-brokered access to resources, females may accept a 'pair bond' with a given male in order to have access to the resources that limit her. For birds, for the reasons listed in the preceding paragraphs, such social constraint, such 'pair bondage', should seldom represent complete control, because male resource brokering should seldom be completely successful. For example, there are very few cases of complete dependence of female birds on males for food (hornbills, perhaps, excepted).

When females 'pair bond' with males who broker resources in order to gain access to essential resources, females will sometimes be coerced in this way into mating with males of relatively low genetic quality. These females will be under selection to seek extra-pair fertilizations with better genetic quality males, whenever females can escape their social mates' control. This hypothesis says that most socially monogamous avian species will have some genetically polyandrous females. Genetically polyandrous females in socially monogamous mateships may be females making the best of a bad job under social constraint.

Between-species comparisons

The Constrained Female Hypothesis makes several predictions.

1. In comparisons of related species in which males' abilities to control females' access to limiting resources vary (e.g. say, in comparisons of related species in which females build open-cup nests and those that use nesting cavities), extra-pair paternity is less prevalent in the species with the more resource constrained females (e.g. among cavity nesting rather than open-cup nesting species) (Fig. 2.4(a)).
2. Variance in male reproductive success is a positive function of the degree of freely expressed female choice; so that variance in male reproductive success is higher among those species with less rather than more constrained females (Fig. 2.4(b)).
3. Among females with extra-pair offspring, more females are likely to be genetically monogamous in the relatively unconstrained species (Fig. 2.4(c)). This might explain the interesting variation in Red-winged Blackbirds *Agelaius phoenicens*, in which the observed frequency of extra-pair broods sired by only one male was significantly higher than the expected frequency (Westneat 1992).
4. Mixed paternity of broods is a negative function of the degree of freely expressed female choice; so that broods of more constrained

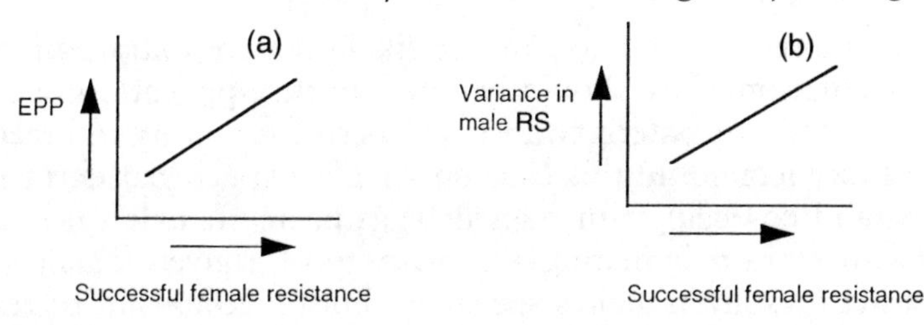

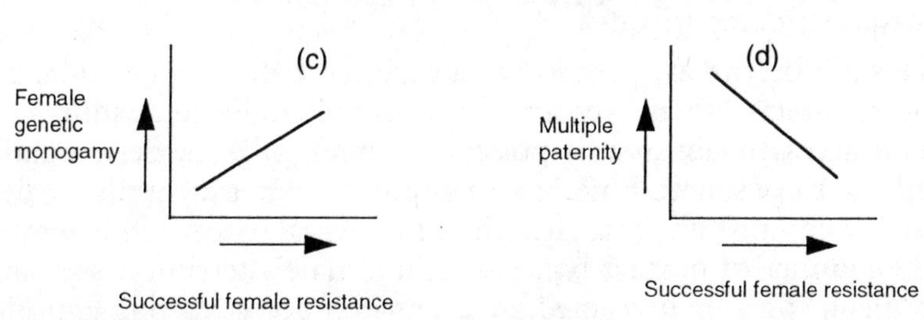

Fig. 2.4 Some predictions of the model of female genetic mating patterns under imperfect social and ecological constraint, modified for interspecific comparisons. When males are unable to control females completely and, yet, when females are not completely free to make choices of gametic partners for good genes, in species with better female resistance abilities, extra-pair paternity (EPP) will be more common (a) and variance in male reproductive success (RS) will be higher (b) than in species with females with less ability to resist. Among females with extra-pair offspring, genetic monogamy will be less common (c) and multiply sired broods more common (d) when females have less rather than more control of their own reproduction.

females are more likely to have higher levels of mixed paternity than the broods of less constrained females (Fig. 2.4(d)).

5. In addition, females may control copulations and solicit males more (Sheldon 1994) in freely expressed female choice mating systems, while males may solicit females more in resource brokering systems. Thus, I expect that American Robin females solicit males more than Eastern Bluebird females.
6. This perspective also invites speculation about other mechanisms of female resistance to male control such as hostility of females' reproductive tracts to sperm. Thus, I predict that in mating systems characterized by free female choice, hostility of females' reproductive tracts to certain males' sperm (perhaps measured as percentage of infertile eggs) will be less than in male control systems; that is, hatchability will be higher in American Robins than Eastern Bluebirds. This might be known as 'facultative infertility'. Said another way, fecundity will

be a positive function of the likelihood of extra-pair mating in the population. Care should be taken in the application of this prediction to extra-pair paternity and evidence of fecundity in breeding attempts of *one* female. In this case one might expect that extra-pair offspring would co-occur with infertile eggs in one female's nesting attempt, if both extra-pair mating and hostility of a given female's reproductive tract to a given male's sperm are both mechanisms of resistance. This is an alternative to the sperm depletion explanation used to explain the positive association between extra-pair offspring and infertility within broods of House Sparrows (Wetton and Parkin 1991).

Within-species comparisons

This perspective also invites within-species, between-individual comparisons of the effects of male resource control on female behaviour. For example, in Eastern Bluebirds it is possible to manipulate experimentally females' access to resources (i.e. to manipulate experimentally males' abilities to broker resources to females) in several ways. One way is to vary the number of nesting boxes within a male's territory, so that it is more difficult for a male to mediate a female's access to this limiting resource. If females' control of this limiting resource is increased in two-cavity versus one-cavity territories, females should be less constrained to limit their matings to the territorial male when better genetic quality males are available. A prediction of this idea is that females in two-cavity territories will be more likely to seek extra-pair paternity of their broods than females in one-cavity territories. Similar predictions follow from consideration of variation in the defensibility of food resources by males.

These are interesting predictions not only because of their novelty, but because they can be used to construct crucial tests of alternative hypotheses (Platt 1964). For example, in socially monogamous birds, if males, rather than females (as suggested here), are the primary seekers of extra-pair fertilizations as suggested by other perspectives (e.g. Westneat *et al.* 1990), one would expect more extra-pair paternity as female control of their own reproduction decreased. Variance in male reproductive success would be a function of male–male behavioral contests and would show either no relationship to female control of own reproduction or a decline as female control increased. For females with extra-pair offspring, no genetic monogamy would be expected under any circumstances; all broods with extra-pair offspring would be sired by more than one male. There would be a negative association between female solicitations for copulations and extra-pair paternity, rather than a positive one. Lastly, fecundability would have no relationship or be negatively correlated with extra-pair mating rather than the positive correlation predicted by the Constrained Female Hypothesis to explain multiple mating.

Whichever predictions hold up to empirical testing, it is my hope that this admittedly female centred view, with its emphasis on variation in females, will encourage more complete explorations of the determinants of reproductive behaviour in birds and perhaps other monogamous creatures—like ourselves—as well. Perhaps a caricature of James' ditty will express our surprise:

Higamous, hogamous, most men are
monogamous.
Whambamdrous, bedamndrous, many women are
polyandrous!

Summary

Of the three main hypotheses for the evolution of social monogamy only the male parental care hypothesis has been tested adequately. The effect of male parental care on female reproductive success is variable both within most species and between species, thus this hypothesis is probably not the most parsimonious explanation for social monogamy. None of the three main hypotheses, male parental care essential, polygyny threshold not reached, or female–female aggression, explains the existence, prevalence, or distribution of extra-pair young in socially monogamous birds. A new theory of coevolutionary selective pressures acting on males for control of female reproductive capacities and on females for resistance to males' efforts to control them says that social monogamy will often be genetic polyandry. This Constrained Female Hypothesis says that within species females' options for extra-pair mating depend on female quality and the environments females are in. It says that high quality females or females in good environments are able to resist males' control efforts better than low quality females or females in poor environments and predicts that within species the broods of high quality females will have more extra-pair paternity represents females under social and ecological constraint 'making the best of a bad job' in their quest to share their gametes with males of high genetic quality. The Constrained Female Hypothesis also predicts the conditions for association of social monogamy in birds (and humans) with extra-pair offspring, extra-pair paternity, genetic monogamy, and genetic polyandry of females with extra-pair paternity, and variance in male reproductive success. The hypothesis also addresses the co-occurrence of genetic monogamy and social polygyny and social 'promiscuity'.

Acknowledgements

I thank the US National Science Foundation (Grant Number IBN–9222005) and the National Institute of Mental Health ADAMHA (Research Scientist Development Award, Grant Number 7 KO2 MH00706–04) for partial support during the preparation of this chapter. I thank Monique Bolgerhoff Molder, Nancy Buschhaus, Judy Guinan,

Jon Plissner, Gunilla Rosenqvist, Steve Shuster, and Judy Stamps for comments on a draft. I thank the participants in the Vienna (November 1993) Female Control Meeting for discussions about evolution of social and genetic monogamy. I thank Doug Mock, Tim Birkhead, and Jeff Black for providing formal comments during the review process. For the 'whambams' and 'bedamns' I thank Judy Guinan.

References

Ahnesjo, I., Vincent A., Alatalo, R., Halliday T., and Sutherland, W. J. (1993). The role of females in influencing mating patterns. *Behavioral Ecology*, **4**, 187–9.

Alatalo, R. V., Carlson A., Lundberg, A., and Ulfstrand, S. (1981). The conflict between male polygamy and female monogamy: The case of the pied flycatcher, *Ficedula hypoleuca. American Naturalist*, **117**, 738–53.

Bart, J. and Tornes, A. (1989). Importance of monogamous male birds in determining reproductive success: evidence for house wrens and a review of male-removal experiments. *Behavioral Ecology and Sociobiology*, **24**, 109–16.

Bateman, A. J. (1948). Intrasexual selection in *Drosophila. Heredity*, **2**, 349–68.

Bletsky, L. D. and Orians, G. H. (1990). Male parental care in a population of Red-winged Blackbirds, 1983–1988. *Canadian Journal of Zoology*, **68**, 606–9.

Bensch, S. and Hasselquist, D. (1991). Nest predation lowers the polygyny threshold: A new compensation model. *American Naturalist*, **138**, 1297–306.

Berglund, A., Magnhagen, C., Bisazza, A., Konig, B., and Huntingford, F. (1993). Female–female competition over reproduction. *Behavioral Ecology*, **4**, 184–7.

Birkhead, T. R. and Møller, A. P. (1992). *Sperm competition in birds: evolutionary causes and consequences*. Academic Press, San Diego.

Björklund, M. and Westman, B. (1986). Adaptive advantages of monogamy in the great tit (*Parus major*): an experimental test of the polygyny threshold model. *Animal Behaviour*, **34**, 1436–40.

Breitwich, R. (1989). Mortality patterns, sex ratio, and parental investment in monogamous birds. In *Current ornithology*, Vol. 6, (ed. D. M. Power), pp. 1–50. Plenum Press, New York.

Brooker, M. G., Rowley, I. Adams, M., and Baverstock, P. R. (1990). Promiscuity: an inbreeding avoidance mechanism in a socially monogamous species? *Behavioral Ecology and Sociobiology*, **26**, 243–58.

Brown, J. L. (1987). *Helping and communal breeding in birds*. Princeton University Press, Princeton.

Clutton-Brock, T. H. (1991). *The evolution of parental care*. Princeton University Press, Princeton.

Darwin, D. (1871). *The descent of man, and selection in relation to sex*. Murray, London.

Davies, N. B. (1989). Sexual conflict and the polygamy threshold. *Animal Behaviour 38*, 226–34.

Davies, N. B. (1991). Mating systems. In *Behavioural ecology: an evolutionary approach, (3rd edn)*, (ed. J. R. Krebs and N. B. Davies), pp. 263–94, Blackwell, Oxford.

Dixon, A., Ross, D., O'Malley, S. L. C., and Burke, T. (1994). Paternal investment inversely related to degree of extra-pair paternity in the reed bunting (*Emberiza schoeniclus*). *Nature*, 371, 698–700.

Duckworth, J. W. (1992). Effects of mate removal on the behaviour and reproductive success of Reed Warblers *Acrocephalus scirpaceus*. *Ibis*, **34**, 164–70.

Dunn, P. O. and Hannon, S. J. (1989). Evidence for obligate male parental care in Black-billed Magpies. *Auk*, **106**, 635–44.

Dunn, P. O. and Hannon, S. J. (1992). Effects of food abundance and male parental care on reproductive success and monogamy in tree swallows. *Auk*, **109**, 488–99.

Dunning, B. (1993). *CRC handbook of avian body masses*. CRC Press, Boca Raton.

Emlen, S. T. and Oring, L. W. (1977). Ecology, sexual selection, and evolution of mating systems. *Science*, **197**, 215–23.

Emlen, S. T., Reeve, H. K., Sherman, P. W., Wrege, P. W., Ratnieks, F. L. W., and Shellman-Reeve, J. (1991). Adaptive versus nonadaptive explanations of behavior: the case of alloparental helping. *American Naturalist*, **138**, 259–70.

Fitch, M. A. and Shugart, G. W. (1984). Requirements for a mixed reproductive strategy in avian species. *American Naturalist*, **124**, 116–26.

Gowaty, P. A. (1980). Origin of mating system variability and behavioral and demographic correlates of mating system variation of Eastern Bluebirds (*Sialia sialis*. Unpublished Ph.D. thesis. Clemson University.

Gowaty, P. A. (1981). An extension of the Orians–Verner–Willson model to account for mating systems besides polygyny. *American Naturalist*, **118**, 851–9.

Gowaty, P. A. (1983). Male parental care and apparent monogamy in eastern bluebirds (*Sialia sialis*). *American Naturalist*, **121**, 149–57.

Gowaty, P. A. (1985). Multiple parentage and apparent monogamy in birds. In *Avian monogamy* (ed. P. A. Gowaty and D. W. Mock), pp. 11–21. American Ornithologists' Union, Washington, D.C.

Gowaty, P. A. (1992). Evolutionary Biology and Feminism. *Human Nature*, **3**, 217–49.

Gowaty, P. A. (in press). Sexual dialectics, sexual selection, and variation in mating behavior. In *Feminism and evolutionary biology* (ed. P. A. Gowaty) Chapman and Hall, New York.

Gowaty, P. A. and Bridges, W. C. (1991*a*). Nest box availability affects extra-pair fertilization and conspecific nest parasitism in eastern bluebirds, *Sialia sialis*. *Animal Behaviour*, **41**, 661–76.

Gowaty, P. A. and Bridges, W. C. (1991*b*). Behavioral, demographic, and environmental correlates of extra-pair fertilizations in eastern bluebirds, *Sialia sialis*. *Behavioral Ecology*, **2**, 339–50.

Gowaty, P. A. and Droge, D. L. (1991). Sex ratio conflict and the evolution of sex-biased provisioning. In *Acta XX Congressus Internationalis Ornithologici, Vol. II* (ed. B. D. Bell, R. O. Cossee, J. E. C. Flux, B. D. Heather, R. A. Hitchmough, C. J. R. Robertson, and M. J. Williams) pp. 932–45. New Zealand Ornithological Congress Trust Board, Wellington.

Gowaty, P. A. and Mock, D. W. (1985). *Avian monogamy*. American Ornithologists' Union, Washington, D.C.

Gowaty, P. A. and Wagner, S. J. (1988). Breeding season aggression of female and male eastern bluebirds (*Sialia sialis*) to models of potential conspecific and interspecific egg dumpers. *Ethology*, **78**, 238–50.

Gowaty, P. A., Richardson, D. and Burke, T. (MS*a*). Paternal provisioning varies with genetic paternity in eastern bluebirds *Sialia sialis*

Gowaty, P. A., Robertson, R. J., Dufty, A., and Ball, G. (MS*b*). Geographic variation in the necessity of male parental care in eastern bluebirds: a multinational, experimental study.

Gyllensten, U. B., Jakobsson, S., and Temrin, H. (1990). No evidence for illegitimate young in monogamous and polygynous warblers. *Nature*, **343**, 168–70.

Hobson, K. A. and Sealey, S. G. (1989). Female–female aggression in polygynously mating yellow warblers. *Wilson Bulletin*, **101**, 84–6.

Hrdy, S. B. (1981). *The woman that never evolved*. Harvard University Press, Cambridge.

Hrdy, S. B. (1986). Empathy, polyandry, and the myth of the coy female. In: *Feminist approaches to science* (ed. R. Bleier), pp. 119–46. Pergamon Press, New York.

Kempenaers, B., Verheyen, G. R., VandenBroeck, M., Burke, T., Banbroeckhoven, C., and Dhondt, A. D. (1992). Extra-pair paternity results from female preference for high-quality males in the blue tit. *Nature*, **357**, 494–6.

Lack, D. (1968). *Ecological adaptations for breeding in birds*. Methuen London.

Lifjeld, J. T. and Robertson, R. J. (1992). Female control of extra-pair fertilization in Tree Swallows. *Behavioral Ecology and Sociobiology*, **31**, 89–96.

Lyon, B. E., Montgomerie, R. D., and Hamilton, L. D. (1987) Male parental care and monogamy in snow buntings. *Behavioral Ecology and Sociobiology*, **20**, 377–82.

McGowan, K. and Woolfenden, G. E. (1989). A sentinel system in the Florida scrub jay. *Animal Behaviour*, **37**, 1000–6.

Mackay, K. (1985). Adoption, indifference, or infanticide in replacement male eastern bluebirds *Sialia sialis* and how females cope with male help. Unpublished M.Sc. thesis. Queen's University, Kingston, Ontario.

McKitrick, M. C. (1992). Phylogenetic analysis of avian parental care. *Auk*, **109**, 828–46.

Mock, D. and Fujioka, M. (1990). Monogamy and long-term pair bonding in vertebrates. *Trends in Ecology and Evolution*, **5**, 39–43.

Møller, A. P. and Birkhead, T. R. (1993). Certainty of paternity covaries with paternal care in birds. *Behavioral Ecology and Sociobiology*, **33**, 261–8.

Orians, G. H. (1969). On the evolution of mating systems in birds and mammals. *American Naturalist*, **103**, 589–603.

Oring, L. W. (1982). Avian mating systems. In *Avian biology* Vol. 6 (ed. D. S. Farner, J. R. King, and K. C. Parkes), pp. 1–92. Academic Press, New York.

Petrie, M. and Lipsitch, M. (1994). Avian polygyny is most likely in populations with high variability in heritable male fitness. *Proceedings of the Royal Society of London B*, **256**, 275–80.

Platt, J. R. (1964). Strong inference. *Science*, **146**, 347–53.

Quinney, T. E. (1983). Tree swallows cross a polygyny threshold. *Auk*, **100**, 750–4.

Rosenqvist, G. and Berglund, A. (1992). Is female sexual behaviour a neglected topic? *Trends in Ecology and Evolution*, **7**, 174–6.

Searcy, W. A. and Yasukawa, K. (1989). Alternative models of territorial polygyny in birds. *American Naturalist*, **134**, 323–43.

Sheldon, B. C. (1993). Sexually transmitted disease in birds: occurrence and evolutionary significance. *Philosophical Transactions of the Royal Society London B.*, **339,** 491–7.

Sheldon, B. C. (1994). Sperm competition in the chaffinch: the role of the female. *Animal Behaviour*, **47,** 163–73.

Slagsvold, T. and Dale, S. (1994). Why do female pied flycatchers mate with already mated males: deception or restricted mate sampling? *Behavioral Ecology and Sociobiology*, **34,** 239–50.

Slagsvold, T. and Lifjeld, J. T. (1994). Polygyny in birds: the role of competition between females for male parental care. *American Naturalist*, **143,** 59–94.

Slagsvold, T., Amundsen, T., Dale, S., and Lampe, H. (1992). Female-female aggression explains polyterritoriality in male pied flycatchers. *Animal Behaviour*, **43,** 397–407.

Smith, H. G. (In press). Experimental demonstration of a trade-off between mate attraction and parental care. *Proceedings of the Royal Society of London B.*

Smith, S. M. (1988). Extra-pair copulations in black-capped chickadees: the role of the female. *Behaviour*, **197,** 15–23.

Smuts, B. B. (1995). The evolutionary origins of patriarchy. *Human Nature*, **6,** 1–32.

Smuts, B. B. and Smuts, R. W. (1993). Male aggression and sexual coercion of females in nonhuman primates and other mammals: evidence and theoretical implications. In *Advances in the study of behavior*, Vol. 22. (ed. P. J. B. Slater, M. Milinski, J. S. Rosenblatt, and C. T. Snowdon), pp. 1–63.

Stamps, J. (In press). The role of females in extra-pair copulations in socially monogamous territorial animals. In *Feminism and evolutionary biology* (ed. P. A. Gowaty) Chapman and Hall, New York.

Trivers, R. L. (1972). Parental investment and sexual selection. In *Sexual selection and the descent of man* (ed. B. G. Campbell), pp. 136–79. Aldine, Chicago.

Veiga, J. P. (1992). Why are house sparrows predominantly monogamous: a test of hypotheses. *Animal Behaviour*, **43,** 361–70.

Verner, J. and Willson, M. F. (1969). Mating systems, sexual dimorphism and the role of male North American passerine birds in the nesting cycle. *Ornithological Monographs*, **9,** 1–76.

Wagner, R. (1992). The pursuit of extra-pair copulations by monogamous female razorbills: how do females benefit? *Behavioral Ecology and Sociobiology*, **29,** 455–64.

Wesokowski, T. (1993). On the origin of parental care and the early evolution of male and female parental roles in birds. *American Naturalist*, **143,** 39–58.

Westneat, D. F. (1992). Do female red-winged blackbirds engage in a mixed mating strategy? *Ethology*, **92,** 7–28.

Westneat, D. F., Sherman, P. W., and Morton, M. L. (1990). The ecology and evolution of extra-pair copulations in birds. In *Current Ornithology*, Vol. 7 (ed. D. M. Power). pp. 331–69. Plenum, New York.

Wetton, J. H. and Parkin, D. T. (1991). An association between fertility and cuckoldry in the house sparrow, *Passer domesticus. Proceedings of the Royal Society London B.* **245,** 227–33.

Wickler, W. and Seibt, U. (1982). Monogamy, an ambiguous concept. In *Mate choice*, (ed. P. Bateson), pp. 33–52. Cambridge University Press, Cambridge.

Williams, G. C. (1966). *Adaptation and natural selection*. Princeton University Press, Princeton.

Wittenberger, J. F. and Tilson, R. L. (1980). The evolution of monogamy: hypotheses and evidence. *Annual Review of Ecology and Systematics*, **11**, 50–68.

Wolf, L., Ketterson, E. D., and Nolan, V. Jr. (1988). Paternal influence on growth and survival of dark-eyed junco young: do parental males benefit? *Animal Behaviour*, **36**, 1601–8.

Wolf, L., Ketterson, E. D., and Nolan, V. Jr. (1990). Behavioural response of female dark-eyed juncos to the experimental removal of their mates: implications for the evolution of male parental care. *Animal Behaviour*, **39**, 125–34.

Wolf, L., Ketterson, E. D., and Nolan, V. Jr. (1991). Female condition and delayed benefits to males that provide parental care: A removal study. *Auk*, **108**, 371–80.

Wrege, P. H. and Emlen, S. T. (1987). Biochemical determination of parental uncertainty in white-fronted bee-eaters. *Behavioral Ecology and Sociobiology*, **20**, 153–60.

Wright, J. and Cotton, P. A. (1994). Experimentally induced sex differences in parental care: an effect of certainty of paternity? *Animal Behaviour*, **47**, 1311–22.

Wright, J. and Cuthill, I. (1992). Monogamy in the European starling. *Behaviour*, **120**, 262–85.

3 The model family

DOUGLAS W. MOCK, P. L. SCHWAGMEYER, AND GEOFFREY A. PARKER

Introduction

From a comparative perspective, monogamy is a truly anomalous mating system. Its relative scarcity makes intuitive sense when one recognizes that the anisogamy[1] that formally defines *males* (as the purveyors of small gametes) typically leaves them less constrained by parenting tasks (e.g. Bateman 1948; Parker *et al.* 1972; Trivers 1972). Most males manifest their relative irresponsibility by departing right after insemination, often to seek further mating opportunities. What makes monogamy special as a mating system, then, is the spectacle of males remaining in the picture after fertilization and continuing to invest in their offspring. While it is possible for males to be monogamous for relatively ignoble reasons (e.g. inability to attract additional mating partners), the phenomenon in need of special explanation is the pattern of sizeable male investment that is not met with immediate desertion by the female (as occurs in many polyandrous taxa): in short, we come quickly to the matter of biparental care.

There are many aspects of a given species' life history that apparently contribute to the evolutionary stability of biparental care, including how much total investment is needed for the successful development of offspring, the operational sex ratio of the local population, and the existence (or lack) of any morphological specializations pre-adapting one sex more than the other for parenting. Many of these topics have been aired

[1] Literally 'unequal-sized gametes'.

recently in considerable detail (Clutton-Brock 1991). Our discussion here takes off from the premises of genetic monogamy (sexual fidelity) and biparental care, surely the bedrock of the much-heralded Traditional Family, and explores some of the theory developed over the past two decades to see what social complexities are likely to be present and perhaps under-appreciated. This exercise seems particularly appropriate for a book comprised principally of field data collected in many cases over the full span of those same decades. In most, if not all, of these impressive empirical studies the questions posed at their outsets probably differed sharply from what seems most pressing today. Such shifts of focus are a measure of scientific progress and deserve to be celebrated. In a parallel manner, theory is incrementally enriched by the discoveries of new phenomena needing tentative explanations and by the failures of previous hypotheses to account for growing piles of facts.

Our discussion will embrace the whole monogamous family, the unit that materializes fully as an emergent property of biparental care. We have three general objectives for this chapter: to advocate the study of *family structure* in multiple dimensions (not just the pair bond); to encourage interest in and greater use of simple mathematical modelling techniques with the topic; and to highlight some short-term experimental approaches that complement the impressive long-term studies featured in much of this book. Our basic take-home message is that even the very simplest—perhaps idealized—form of monogamous sexual union can generate families of remarkable social complexity. That stems inevitably from the genetic diversity produced by sex itself. As such, the monogamous family is quite a useful unit in which to study the balances of co-operation and conflict, being populated with parties of varying genetic harmony and discord. That is, each family member that is not a monozygotic twin contains unshared genetic material potentially capable of influencing its phenotype in selfish directions. And if our strict assumption of genetic monogamy is relaxed—as when close scrutiny of real-life systems reveals various imperfections (e.g. extra-pair fertilizations, intraspecific brood parasitism, etc.)—predictions of social dissonance ascend even further. From Hamilton's rule[1] we see clearly that any factors tending to lower r automatically tend to promote the genetic incentives for selfishness/conflict. And because we humans like to think of ourselves as 'monogamous' (at least most of the time), scrutiny of monogamy in birds and other taxa may enlighten us about our own aspirations, that is, about the goal or model we set for ourselves. This morass of intrafamily co-operation and conflict can quickly become fatiguing to contemplate so, as a conceptual and organizational framework, we tentatively atomize

[1] $rb-c>0$, where b is the fitness benefit to the recipient of a social act, c is the fitness cost to its performer, and r is the coefficient-of-relatedness (the probability of a rare allele being shared because of common ancestry) between the two parties (Hamilton 1964; for a lucid summary see Grafen 1984).

the family into three 'social dimensions' (parent–parent, parent–offspring, and sibling–sibling), recognizing that while each may be considered in isolation all are inexorably intertwined.

For each of these social dimensions, we offer a quick, and hopefully painless, peek at some quantitative models that have been developed for exploring various aspects of monogamy's mysteries. Our message here is a simple sales-pitch: there is an increasingly rich literature of theoretical models that can be used to generate provocative, frequently counterintuitive, and most importantly *testable* predictions. We believe theoreticians and field empiricists really have much to learn from each other, if only they would learn to communicate better. And we believe that such interaction has *not* been a major feature of most studies of monogamy.

Finally, for each of our three dimensions we review and highlight a bit of the recent experimental work that addresses these layers of conflict and/or co-operation in family structure. Because this book focuses overwhelmingly on commendably long-term data sets for monogamous birds, it may be worth noting that one does not always need decades of effort to make exciting progress. In these days of shrinking and stochastic research support, especially, the knack for identifying a discrete problem and illuminating some of its secrets experimentally in a few breeding seasons is likely to become increasingly important. Needless to say, there are many issues that cannot be addressed with such an expeditious method.

For taxa with substantial biparental care, the simplest *family* worthy of the label consists of the mother, the father, and a single offspring, wherein the two putative parents are indeed the genetic parents. Most of these parents will have additional offspring in the future so, in the abstract, we can consider the evolutionary interests of one or more siblings as well. For explanatory simplicity, let us focus mainly on a contemporaneous sibship of two, reserving the option of discussing present–future trade-offs as a special case. In such a family facing the possibility of brood reduction (loss of one offspring to fatal sibling rivalry), O'Connor (1978) identified three categories of players: Parent, Surviving Offspring, and Victim. We modify his approach (Fig. 3.1) to stress the *relationships* between

(1) the male and female adult sexual partners (the so-called 'pair bond');
(2) the parent–offspring dynamic (wherein we lump the two parents so as to accentuate intergenerational processes); and
(3) the siblings' own 'love–hate' interaction.

On the assumption of out-breeding, the parents are expected to share no rare alleles with each other because of common ancestry ($r = 0$), but each parent–offspring coefficient is exactly 0.50 (we assume diploidy), and the sib–sib coefficient has a mean of 0.50 (Hamilton 1964). From this

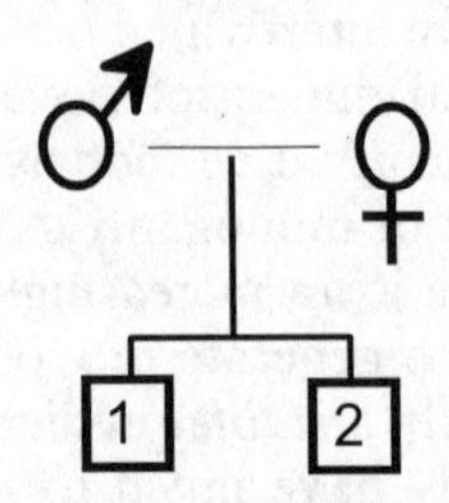

Fig. 3.1 The three social dimensions of monogamous family structure. Because parents are typically unrelated to each other in outbred taxa (shown as a thin line), they potentially have genetic incentives for selfishness. The parent–offspring and sib–sib relationships, on the other hand, feature a high level of genetic relatedness (thick lines), but may none the less contain significant conflict because of the basic relatedness asymmetries (sibling rivalry and parent–offspring conflict) that may be aggravated by a mismatch between ecological supply-and-demand.

information alone, the potential for both co-operation and conflict is apparent on each social dimension, though numerous other factors can (and frequently do) displace the fulcrum between those two realms in dramatic ways.

Dimension I: monogamous parents and sexual conflict

Though biparental care obviously can offer mutualistic advantages to each parent's fitness, the potential for sexual conflict between the parents may arise over the contributions each makes to the offspring (Parker *et al.* 1972; Trivers 1972; Parker 1979). Without genetic relatedness, the balance between co-operation and conflict here is set solely by extrinsic factors, including the remaining needs of the current dependent brood (Dawkins and Carlisle 1976), the relative resource-harvesting abilities of each parent, the supply of alternative sexual partners that may be available asymmetrically to each parent (Maynard Smith 1977), the perceived social compatibility between the mates, and anything else likely to affect whether the current partnership will persist into the future. At one rare if not absurd heuristic extreme, which we call 'True Monogamy' (following Parker 1985), neither partner has any sexual interests or opportunities beyond the pair bond—each becomes permanently celibate if and when its initial partner dies. Under True Monogamy, there is no incentive possible for selfishly foisting more of the parental burden onto one's mate: the two adults' lifetime interests are exactly congruent, by definition. It goes without saying that in the great majority of monogamous relationships these onerous stipulations will not be met. Extra-pair matings are obviously quite common in many species and dissolution of 'pair bonds' can even be a routine operating procedure (a very recent set of optimization models explores the conditions under which *divorce* pays higher fitness dividends than mate retention: McNamara and Forslund, in press). Thus one must be on the alert for signs of sexual conflict, specifically for attempts at mutual exploitation, within most real-world monogamous relationships.

There have been quite a few models for what we might call 'antagonistic' biparental care in recent years (e.g. Maynard Smith 1977; Chase 1980; Houston and Davies 1985; Parker 1985; Winkler 1987; Winkler and Wallin 1987; Hussell 1988), of which we will discuss one evolutionarily stable strategy (ESS) model in a bit of verbal and graphical detail (Fig. 3.2). Of interest is the compromise that unfolds when two unrelated sexual partners negotiate with each other (*sensu* Maynard Smith 1982) over the amounts of care each will provide to the current brood. If either mate could wave a magic wand and have whatever balance of effort it dearly wanted, presumably it would choose selfishly to have its spouse do *all* the work for this brood, thereby saving itself from parental burnout and enjoying enhanced future success (with a fresh partner). Such extreme selfishness is checked by two stark realities: (1) the requisite supply of replacement partners may not exist and, in any case, (2) the original mate is likely to view the system in exactly the same way, hence is presumably unwilling to take up all the damned slack! Thus neither parent gains as much as it might if it were free to parasitize completely; instead each gains the most available after assessing how much its current partner will invest. Chase (1980) and Houston and Davies (1985) viewed such adjustments as an iterated 'bargaining' process, whereby each individual responds to the current workload of its mate with its own 'best reply,' until the ESS compromise levels are reached (see also Winkler 1987). These best replies can be plotted across a range of partner efforts, yielding a reaction curve with an intercept equal to the individual's optimal effort when its mate makes *no* contribution, and a slope representing its optimal decrease in effort as the mate's contribution rises (Fig. 3.2, upper graph). When both mates' reaction curves are superimposed on each other (Fig. 3.2, lower graph), it can be seen that the only ESS exists at the intersection (see explanation in legend).

ESS levels of effort may change, however, with changes in the reaction curves of either or both parents. A 'best reply' specifies the level of parental effort that optimizes an individual's current productivity plus its residual reproductive value (given its partner's behaviour), so it can be influenced by *any* factor that would normally affect current/future tradeoffs. Chase (1980), for instance, suggested that changes in the unit costs of parenting (such as might accompany declines in food availability) should precipitate adjustments to the equilibrium. And he predicted similar modifications to the ESS if the unit costs of non-parental activities (e.g. self-maintenance) were to change for either or both partners;

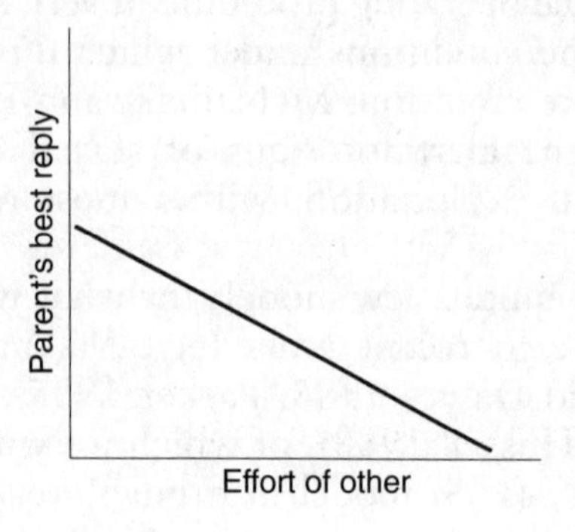

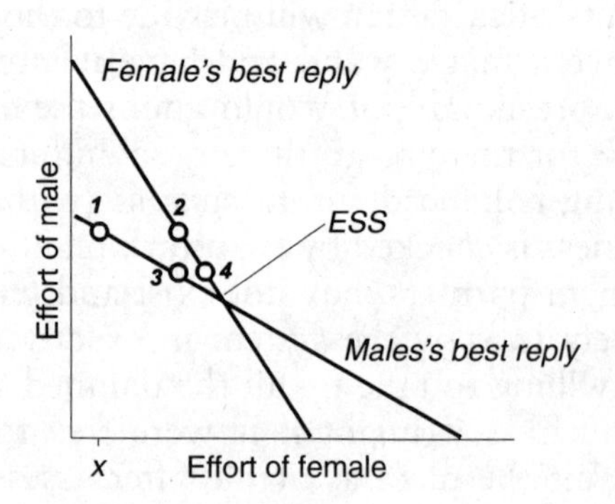

Fig. 3.2 A graphical presentation of sexual conflict in biparental care (from Chase 1980; Houston and Davies 1985). In the *upper figure*, a stylized 'reaction curve' for an individual parent whose interests differ from those of its mating partner ('Other'). The slope is less than –1, indicating *partial compensation* (our focal parent should react to a step decrease in Other's effort by making a smaller increase of its own). Any savings potentially available from exploiting the partner will be invested with a different ('fresh') partner in the future. The catch, of course, is that the original mate has a similar perspective. In the *lower figure* we find a pair of such curves (each faithful to its owner's axis) and their intersection represents the evolutionarily stable strategy (ESS). A hypothetical negotiated series of bids and counter-bids ('best replies') leads inevitably to this spot. For example, if the female makes a very low initial offer (only level x), then the male's best reply is circle '1' on his best-reply line. However, when he provides only that amount of effort, the female's best second offer is '2', after which he bids '3', which she follows with '4' and so on till the ESS intersection is reached by successive approximation. (NB These reaction curves are conventionally drawn as straight lines for simplicity: they need not be monotonic.)

indeed, ESS levels of investment can be affected by virtually any variable that impinges on even one parent's parental investment optimum (e.g. Carlisle 1982; Nur 1984*a*; Winkler 1987; Beauchamp *et al.* 1991; Westneat and Sherman 1993).

That avian parental workloads are flexible has been demonstrated in a variety of ways. Clutch or brood manipulations sometimes reveal abridgement of parental feeding to broods of reduced size (e.g. Mock and Lamey 1991) and augmentations in parental feeding to broods that have been enlarged (Nur 1984*b*; Gustafsson and Sutherland 1988; Gustafsson and Pärt 1990; Mock and Lamey 1991; Martins and Wright 1993) or to broods with artificially synchronized hatching (e.g. Fujioka 1985; Gibbons 1987; Mock and Ploger 1987). Male removal studies (Gowaty 1983; Greenlaw and Post 1985; reviewed in Wolf *et al.* 1988, 1990; and Bart and Tornes 1989) have commonly shown that the effort of widowed females increases. These studies demonstrate that: (1) parents often adjust their effort facultatively to the prevailing effort of their partners; and (2) in many species, parents are demonstrably *not* working at their maximum levels. Yet we are still a long way from *quantitatively* predicting the relative workloads of males and females belonging to monogamous species. This short-coming is partially due to the fact that rarely has there been any attempt to predict the degree to which individuals ought to adjust their effort in response to changes in their partner's effort. The exceptional case, on both counts, has been the Dunnock *Prunella modularis* study, where the observed feeding effort by monogamous partners closely matched those predicted from an ESS model (Houston and Davies 1985).

In short, an ESS approach to biparental care considers the partner's contribution when evaluating the optimum behaviour of each parent. Moreover, it leads to testable predictions about biparental care, since the relative contribution of each parent should be affected by that of its partner, which in turn may be shaped by any number of natural factors. For example, if one's mate is slowed by a temporary ailment (e.g. minor injury, mild infection, etc.) or suffers a string of bad luck in finding or capturing food for the offspring, what should be done? This model makes the clear prediction that the unimpaired parent should either decline to compensate for its mate's affliction or to do so only partially: specifically, it should either simply maintain its current effort level or increase its effort enough to make up no more than part of the shortfall[1]. Without full compensation, of course, the brood will take a pay-cut overall. For a responding parent to do more than this (escalating its efforts so much that the brood does not suffer at all) could lead to two bad consequences. First, its own future reproduction (technically, its 'residual reproductive value') might be eroded, costing it more fitness in the long run than it salvages from the current family. Second, such largess invites further reductions in partner effort, essentially by making laziness pay.

[1]Alternatively, recent models of honest signalling (Grafen 1990; Godfray 1991) may allow a temporarily disabled parent to convey an honest signal of its disability (provided that the signal is costly) and for the unimpaired parent to compensate fully. Such extensions of honest signalling models to sexual conflict over parental investment remain to be studied.

An innovative recent approach to the study of compensations by monogamous partners has involved the experimental application of a physical handicap (a simulated 'ailment') to one member of a pair. This can be done quite simply by attaching small lead fishing weights to the rectrices with super-glue (Wright and Cuthill 1989, 1990*a*, *b*), such that the burdened individual reduces the mean rate at which it provisions the brood. In the first of these studies, European Starling *Sturnus vulgaris* parents showed the predicted partial compensation response when their partners were encumbered and it made no difference which sex was being tested (Wright and Cuthill 1989). Surprisingly, though, compensations in subsequent studies (Wright and Cuthill 1990*a, b*) appeared to be complete. It is unclear at this point what might account for these discrepant results, though interseasonal variation and limited statistical power seem likely contributors to the confusion. Interestingly, weighting of individual Antarctic Petrel *Thalassoica antarctica* parents led to no compensatory responses by the mates at all (Sæther *et al.* 1993), which the authors suggest may relate to a high mortality cost of reproduction in that long-lived species.

In a sense, the empirical studies of 'compensatory' responses provide a barometer of the co-operation and conflict inherent in biparental care systems. Consider once more True Monogamy, in which sexual conflict is absent. In contrast to the predicted 'partial' or no compensatory response expected with sexual conflict, True Monogamy parents should compensate *completely* for a partner's shortfall (Mock and Parker, in press). Also, it is not hard to imagine intermediate levels of conflict, where increased or decreased expenditures by one's partner have some impact on the focal individual's future. The option of taking a new partner after their current reproductive episode may not be feasible for all monogamous animals, so future reproduction may be linked to the condition of one's current mate (for theoretical treatment, see McNamara and Forslund, in press; for empirical support, see Ens *et al.* Chapter 19). When pairs typically remain together for successive nesting cycles, a male's selfishness should be mitigated by a trade-off between his *partner*'s current expenditure level and the *pair*'s future reproduction together. Thus, the cost to future fitness borne by females is likely often to impinge also on their males, even if only partially (given that a new mate might be located). 'Intermediate conflict' seems most likely to occur at detectable levels when: (1) current effort influences future reproductive performance per se (rather than affecting only survival); and (2) alternative mates are limited, costly to obtain, or are not of sufficiently higher value to make re-pairing worthwhile (see Breitwisch *et al.* 1986; Freed 1987). This line of reasoning may have important implications for the relative division of labour across species, as well as how much total care a given brood receives. The Parker (1985) model's dichotomous contrast between True Monogamy (no conflict) and full-conflict monogamy indicated that increasing levels of sexual conflict over biparental care should push ESS levels of total parental investment downward (as each partner negotiates by successive mild decrements in its contribution), hence should lead to production of greater numbers of lower quality offspring. Extrapolating from both that model and the Houston and Davies (1985) model suggests intuitively that intermediate conflict is likely to favour total parental investment levels somewhere between the parental investment level of True Monogamy and that of full conflict.

Sexual conflict over parental investment between labour-sharing mates clearly has the power to influence the other social dimensions of the family. If conflict is moderately strong, the lowered total amount of parental care is likely to exacerbate the supply:demand balance for competing siblings, while broadening the potential zone of parent–offspring conflict (Parker and Macnair 1979; Parker 1985).

Dimension II: parent–offspring relationships

Although many aspects of parenting are appropriately viewed as positive, nurturing, and co-operative, Trivers (1974) recognized that parents and offspring are also likely to disagree over how parental investment ought to be allocated, a logic that, in hindsight, can be seen to flow naturally from the prevalence of sibling rivalry in closed sibships (e.g. Mock and Parker, in press). Parents are genetically related to all their own offspring equally ($r = 0.5$ in each case), but an offspring is asymmetrically more related to itself ($r = 1.0$) than to its nestmate. All else being equal, then, parents should favour a meticulously egalitarian distribution of parental investment, while each of their progeny should try to skew resources towards itself (as in Fig. 3.3). In short, offspring should be selected to be more selfish than parents should allow. Superimposed on this framework, however, is an obvious *power* asymmetry between most parents and their offspring, such that the latter may seldom be in a position to impose their will upon the parents. Nevertheless, Trivers suggested that the incongruence of genetic interests should generate a relentless selection pressure that would find some means of expression: in particular, he proposed that feeble offspring should use 'psychological weaponry' (especially deceptive mis-reporting on their true level of need), thereby conning parents into over-investment errors (Trivers 1974).

This argument proved to be very seductive and quickly gained a substantial following. What is surprising, though, is the remarkable absence of unambiguous evidence that parent–offspring conflict theory actually accounts for the evolution of behavioural phenotypes (Stamps and Metcalf 1980; Mock and Forbes 1992, 1993). That is, we find a remarkable array of behaviours 'consistent with' the Trivers view—for example, all manner of neonatal begging, whining, screaming, and throwing of tantrums *might* have evolved as tricks for cajoling parental investment

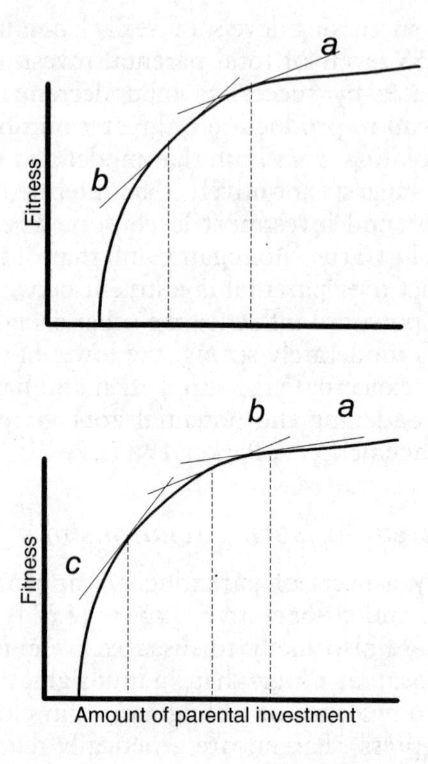

Fig. 3.3 A chick that (hypothetically) enjoys total control over parentally delivered resources is imagined to take the 'next' unit of parental investment (e.g. the next item of food) whenever so doing boosts its own viability by *more than half* of what its full-sibling would derive from the same item. That is, it should direct that investment towards the 'direct' component of its inclusive fitness (over the 'indirect' component) until the alternative of sharing pays sufficiently to compensate for the relatedness difference. This is because the despot is twice as related to itself as to its sibling. In a graphical representation plotting amount of parental investment against contribution to inclusive fitness, which is assumed to be a decelerating function of consumption, the despot should continue to be selfish until its own *marginal gain* from consumption (the slope) drops to one-half that available to its only full-sibling *upper figure*; below that point, its fitness fares better from letting the item go to its sibling (from Parker *et al.* 1989). Things get considerably more complicated when a third sibling is admitted to the model *lower figure*. Then a fully despotic *a*-sibling must consider not only how its own marginal gain compares with that of its two nestmates, but also how the next-most-powerful sibling is likely to rank its priorities. Because the *b*-sibling is likely to consume items that the *a*-sibling might otherwise prefer to see ingested by the *c*-sibling, a game-within-a-game is set up between the two senior-most siblings that leads each to be rather more selfish (because of the threat each poses to the other) than would otherwise be the case. As a result, the *c*-sibling is left with an even smaller share of the total parental investment, as the other two carry off over-sized shares (from Parker *et al.* 1989).

from resistant parents—but virtually no evidence that these traits *did* evolve under the proposed selection pressures. There is no evidence in particular showing that parents end up investing in such manipulative offspring *to the detriment* of their ability to invest in others, as required by the theory. At this point, we can only acknowledge the soundness of the genetic asymmetry logic, which Godfray and Parker (1992) called the 'battleground' of parent–offspring conflict, but we cannot yet judge whether that asymmetry has played an important evolutionary role in shaping the 'resolutions' of that conflict.

This distinction between battleground and resolutions allows alternative explanations for some of the interesting behavioural traits, such as intense begging, siblicidal aggression, and mammalian 'weaning conflict' that heretofore have usually been ceded to parent–offspring conflict (Godfray 1991; Mock and Forbes 1992; Forbes 1993). For example, a controversy currently exists over begging in the context of parent–offspring conflict and sibling competition. Stamps *et al.* (1978) and Parker and Macnair (1979) viewed costly begging as the product of sibling competition over parental resources: siblings that beg more *manipulate* their parents in ways enabling them to get more. An alternative to the idea of manipulating the parent is that begging represents an honest (but costly) signal of true need by an offspring to its parent (Godfray 1991, 1995). Ultimately, manipulation and signalling may represent two ends of a continuum, with the degree of parental control acting as the determining factor. Relatively ineffective (or totally abdicated) parental control over parental investment allocation may well result in the offspring seizing control ('manipulation'); conversely, tight parental control may be more likely to result in honest signalling (Mock and Parker, in press).

As with any evolutionary conflict, experimental demonstration of parent–offspring conflict ought to show that when each party gets its way (when effective countermeasures by the putative 'opponent' are artificially nullified), its fitness is enhanced and opponent fitness is diminished. This has not been done for the great majority of cases held forth as parent–offspring conflict (Stamps and Metcalf 1980, Mock and Forbes 1992). With monogamous birds, it is not even clear where the true intergenerational conflicts lie, though some have been proposed. For example, siblicidal brood reduction was cast in this light (O'Connor 1978) on the logical grounds that it involves certain offspring actively destroying other offspring, ostensibly cutting into parental fitness for their own selfish gains. But this view assumes that parents do best by raising all of their hatchlings, that clutch size is set to coincide exactly with the parental optimum for family size. Once one entertains that possibility that parents benefit from a downward adjustment of family size, this conflict may vanish (Mock and Forbes 1992). To date, the logistical problems associated with arranging for each party (parents and senior offspring) to 'win' have not been solved: the key experiments remain to be performed.

Dimension III: sibling–sibling

Although sibling relationships are powerfully influenced by the gene-sharing (especially under our assumptions of parental sexual fidelity), they are still very much coloured by extrinsic factors. Indeed, because parents generally set clutch and brood size at an 'optimistic' level (i.e. somewhat in excess of what they ordinarily can—or will—support: reviewed in Mock and Parker 1986; Kozlowski and Stearns 1989; Forbes 1991; Mock and Forbes 1995), avian nestlings of many species routinely commence life in a situation so highly competitive as to require brood reduction. What makes that rivalry so interesting is that the players are passing through one of the narrowest resource bottlenecks they will ever face, yet can gain relief only by sacrificing one or more of their closest genetic relatives (Mock and Parker, in press). And, of course, because the critical limiting resources being contested (typically food) are controlled to a degree by parental effort, the overall economy of the system is under strong parental influence (see Dimension II).

The fitness pay-offs for sib-sacrificing selfishness can be expressed mathematically (and verbally): each of these individuals is expected to regard a given unit of parental investment received by Self to be twice as valuable, *ceteris paribus*, as a unit received by its (full-sibling) nestmate. The upshot is that, given the choice, each nestling with any say in the matter should favour taking that 'next' unit of parental investment again and again, relentlessly, until it is so lavishly provisioned that its own phenotype obtains a marginal boost only half as great as its sibling would gain if the item were allowed to pass along (Fig. 3.3). That is, when passing up a unit of parental investment enhances the 'indirect component' of a despotic sibling's inclusive fitness (see Brown and Brown 1981) by a value at least twofold what would be available from its own consumption of that parental investment (and padding the 'direct component' of its own inclusive fitness), then generosity is the better option (Hamilton 1964; Dawkins 1976; Parker *et al.* 1989). Clearly there must be many circumstances when selfish consumption is favoured, even at the expense of a nestmate.[1]

The number of empirical studies of avian sibling rivalry has grown sharply in recent years, largely stimulated by the models of O'Connor (1978) and by the discovery of siblicidal aggression in several colonially nesting species (for reviews see Mock *et al.* 1990; Forbes and Mock 1994; Mock and Parker, in press). Quite a few variables have been explored experimentally under field conditions, including food amount (how it influences sibling aggression (Drummond *et al.* 1986; Drummond

[1]Of course, if the nestmate is only a half-sibling, the threshold for compensation jumps to a four-fold difference; if it is unrelated altogether, there is usually no point at which a despotic resident offspring's fitness benefits from sharing.

and Garcia Chavelas 1989) or fails to do so (Mock *et al.* 1987)), hatching asynchrony (how it affects brood reduction: e.g. Fujioka 1985; Mock and Ploger 1987; Magrath 1989), brood size (how it shapes fighting: Mock and Lamey 1991), relative hunger among nestmates (how it affects begging intensity: Smith and Montgomerie 1991), prey monopolizability (Mock 1984, 1985), and so on. For example, if the Cattle Egrets' *Bubulcus ibis* normal 3-day hatching span (which parents create by starting incubation during laying) is obliterated such that all three nestlings commence life as equals, the chicks respond by tripling their aggressiveness and parents end up with a lower efficiency in their conversion of delivered prey into offspring (Fujioka 1985; Mock and Ploger 1987). This swapping of hatching eggs also illustrates another point, that the sibship is often the most logistically accessible of the monogamous family's three social dimensions, simply because nestling birds are conveniently sessile. Nevertheless, several key questions remain almost totally open. For example, we have little appreciation for how temporary food deprivations during the nestling period might curtail lifetime reproductive performance (Mock and Forbes 1994; but see Perrins 1965; Magrath 1991 for some relevant data), despite the fact that sublethal starvation is a conspicuous feature in the lives of many nestling birds (Ricklefs 1968).

Concluding remarks

Parents make the initial determination of maximum family size at the time of oviposition, presumably as an evolved life history trait shaped by its influence on lifetime reproductive success. In taxa that practise monogamy and provide biparental care, this early clutch-size decision sets in motion a host of consequences and processes for all members of the family, at least during their association period and sometimes well beyond. As the season's ecological realities become better known, the level of parental support for the brood can be assessed and altered, the maternal and paternal contributions re-negotiated. Each parent may work at levels considerably below its capacity and there is reason to suspect that each is also judging, as best it can, the commitment and performance of its partner along the way. The level of parental investment actually delivered should usually be of keen interest to each dependent offspring, which has its own decisions to make about whether to try extracting more from the parents and/or to attempt various ploys for skewing the parental investment unfairly towards itself. At present, though we must concede that the costs and benefits envisioned here for monogamous family members have not been measured empirically for any species, piecemeal bits of encouragement are apparent across the broad taxonomic sweep of relatively well-studied birds. And we express optimism that the general effort to understand the diversity of monogamous avian partnerships will be well served in the long run by considering the evolutionary interests of all family members in a relatively comprehensive way.

Acknowledgements

For research support during the preparation of this chapter, we thank the US National Science Foundation (DEB 9107246 to DWM and BNS-9021182 to PLS) and the UK Science and Engineering Research Council (to GAP). For feedback on an earlier draft of this chapter, we thank P. A. Gowaty, J. M. McNamara, I. Cuthill, and J. M. Black.

References

Bart, J. and A. Tornes. (1989). Importance of monogamous male birds in determining reproductive success. *Behavioral Ecology and Sociobiology*, **24**, 109–16.

Bateman, A. J. (1948). Intra-sexual selection in *Drosophila*. *Heredity*, **2**, 349–68.

Beauchamp, G., Ens, B. J., and Kacelnik, A. (1991). A dynamic model of food allocation to starling (*Sturnus vulgaris*) nestlings. *Behavioral Ecology*, **2**, 21–37.

Breitwisch, R., Merritt, P. G., and Whitesides, G. (1986). Parental investment by the Northern Mockingbird: male and female roles in feeding nestlings. *Auk*, **103**, 152–9.

Brown, J. L. and Brown, E. R. (1981) Kin selection and individual fitness in babblers. In *Natural selection and social behavior: recent results and new theory* (ed. R. D. Alexander and D. W. Tinkle), pp. 244–56. Chiron Press, New York.

Carlisle, T. R. (1982). Brood success in variable environments: implications for parental care allocation. *Animal Behaviour*, **30**, 824–36.

Chase, I. D. (1980). Cooperative and noncooperative behaviour in animals. *American Naturalist*, **115**, 827–57.

Clutton-Brock, T. H. (1991). *The evolution of parental care*. Princeton University Press, Princeton.

Dawkins, R. (1976). *The selfish gene*. Oxford University Press, Oxford.

Dawkins, R. and Carlisle, T. R. (1976). Parental investment, mate desertion, and a fallacy. *Nature*, **262**, 131–3.

Drummond, H. and Garcia Chavelas, C. (1989). Food shortage influences sibling aggression in the blue-footed booby. *Animal Behaviour*, **37**, 806–19.

Drummond, H., Gonzalez, E., and Osorno, J. L. (1986). Parent–offspring cooperation in the blue-footed booby (*Sula nebouxii*): social roles in infanticidal brood reduction. *Behavioral Ecology and Sociobiology*, **19**, 365–73.

Forbes, L. S. (1991). Insurance offspring and brood reduction in a variable environment: the costs and benefits of pessimism. *Oikos*, **62**, 325–32.

Forbes, L. S. (1993). Avian brood reduction and parent–offspring 'conflict'. *American Naturalist*, **142**, 82–117.

Forbes, L. S. and Mock, D. W. (1994). Proximate and ultimate determinants of avian brood reduction. In *Protection and abuse of young in animals and man* (ed. S. Parmigiani and F. vom Saal), pp. 237–56 Harrwood Academic, London.

Freed, L. A. (1987). The long-term pair bond of tropical house wrens: advantage or constraint? *American Naturalist*, **130**, 507–25.

Fujioka, M. (1985). Food delivery and sibling competition in experimentally even-aged broods of the cattle egret. *Behavioral Ecology and Sociobiology*, **17**, 67–74.

Gibbons, D. (1987). Hatching asynchrony reduces parental investment in the jackdaw. *Journal of Animal Ecology*, **56**, 403–14.

Godfray, H. C. J. (1991). Signalling of need by offspring to their parents. *Nature*, **352**, 328–30.

Godfray, H. C. J. 1995. Signalling of need between parents and young: Parent–offspring conflict and sibling rivalry. *American Naturalist*, **146**, 1–24.

Godfray, H. C. J. and Parker, G. A. (1992). Sibling competition, parent–offspring conflict and clutch size. *Animal Behaviour*, **43**, 473–90.

Gowaty, P. A. (1983) Male parental care and apparent monogamy among eastern bluebirds (*Sialia sialis*). *American Naturlist*, **121**, 149–57.

Grafen, A. (1984). Natural selection, kin selection and group selection. In *Behavioural ecology: an evolutionary approach* (ed. J. R. Krebs and N. B. Davies), pp. 62–84. Sinauer Associates, Sunderland, Massachusetts.

Grafen, A. (1990). Biological signals as handicaps. *Journal of theoretical Biology*, **144**, 517–46.

Greenlaw, J. S. and Post, W. (1985). Evolution of monogamy in seaside sparrows, *Ammodramus maritimus*: tests of hypotheses. *Animal Behaviour*, **33**, 373–83.

Gustafsson, L. and Pärt, T. (1990). Acceleration of senescence in the collared flycatcher *Ficedula albicollis* by reproductive costs. *Nature*, **347**, 279–81.

Gustafsson, L. and Sutherland, W. J. (1988). The costs of reproduction in the collared flycatcher *Ficedula albicollis. Nature*, **335**, 813–5.

Hamilton, W. D. (1964). The genetical theory of social behaviour. I & II. *Journal of theoretical Biology*, **7**, 1–52.

Houston, A. I. and Davies, N. B. (1985). The evolution of co-operation and life history in the dunnock, *Prunella modularis*. In *Behavioural ecology: ecological consequences of adaptive behaviour* (ed. R. M. Sibly and R. H. Smith), pp. 471–87. Blackwell, Oxford.

Hussell, D. J. T. (1988). Supply and demand in tree swallow broods: a model of parent–offspring food-provisioning interactions in birds. *American Naturalist*, **131**, 175–202.

Kozlowski, J. and Stearns, S. C. (1989). Hypotheses for the production of excess zygotes: models of bet-hedging and selective abortion. *Evolution*, **43**, 1369–77.

McNamara, J. M. and P. Forslund. Divorce rates in birds: predictions from an optimization model. *American Naturalist*. (In press).

Magrath, R. D. (1989). Hatching asynchrony and reproductive success in the blackbird. *Nature*, **339**, 536–8.

Magrath, R. D. (1991). Nestling weight and juvenile survival in the blackbird, *Turdus merula. Journal of Animal Ecology*, **60**, 335–51.

Martins, T. L. F. and Wright, J. (1993). Brood reduction in response to manipulated brood sizes in the common swift *Apus apus. Behavioral Ecology and Sociobiology*, **32**, 61–70.

Maynard Smith, J. (1977). Parental investment—a prospective analysis. *Animal Behaviour*, **25**, 1–9.

Maynard Smith, J. (1982). *Evolution and the theory of games*. Cambridge University Press, Cambridge.

Mock, D. W. (1984). Siblicidal aggression and resource monopolization in birds. *Science*, **225**, 731–3.

Mock, D. W. (1985). Siblicidal brood reduction: the prey-size hypothesis. *American Naturalist*, **125**, 327–43.

Mock, D. W. and Forbes, L. S. (1992). Parent–offspring conflict: a case of arrested development. *Trends in Ecology and Evolution*, 7, 409–13.

Mock, D. W. and Forbes, L. S. (1993). Reply to Gomendio. *Trends in Ecology and Evolution*, **8**, 218.

Mock, D. W. and Forbes, L. S. (1994). Life history consequences of avian brood reduction. *Auk*, **111**, 115–25.

Mock, D. W. and Forbes, L. S. The evolution of parental optimism. *Trends in Ecology and Evolution*, **10**, 130–4.

Mock, D. W. and Lamey, T. C. (1991). The role of brood size in regulating egret sibling aggression. *American Naturalist*, **138**, 1015–26.

Mock, D. W. and Parker, G. A. (1986). Advantages and disadvantages of egret and heron brood reduction. *Evolution*, **40**, 459–70.

Mock, D. W. and Parker, G. A. (In press). *The evolution of sibling rivalry*. Oxford University Press, Oxford.

Mock, D. W. and Ploger, B. J. (1987). Parental manipulation of optimal hatch asynchrony in cattle egrets: an experimental study. *Animal Behaviour*, **35**, 150–60.

Mock, D. W., Lamey, T. C., and Ploger, B. J. (1987). Proximate and ultimate roles of food amount in regulating egret sibling aggression. *Ecology*, **68**, 1760–72.

Mock, D. W., Drummond, H., and Stinson, C. H. (1990). Avian siblicide. *American Scientist*, **78**, 438–49.

Nur, N. (1984*a*). The consequences of brood size for breeding blue tits II. Nestling weight, offspring survival and optimal brood size. *Journal of Animal Ecology*, **53**, 497–517.

Nur, N. (1984*b*). The consequences of brood size for breeding blue tits I. Adult survival, weight change, and the cost of reproduction. *Journal of Animal Ecology*, **54**, 479–96.

O'Connor, R. J. (1978). Brood reduction in birds: selection for fratricide, infanticide and suicide. *Animal Behaviour*, **26**, 79–96.

Parker, G. A. (1979). Sexual selection and sexual conflict. In *Sexual selection and reproductive competition in insects* (ed. M. S. Blum and N. A. Blum), pp. 123–66. Academic Press, New York.

Parker, G. A. (1985). Models of parent–offspring conflict. V. Effects of the behaviour of the two parents. *Animal Behaviour*, **33**, 511–8.

Parker, G. A. and Macnair, M. (1979). Models of parent–offspring conflict. IV. Suppression: evolutionary retaliation by the parent. *Animal Behaviour*, **27**, 1210–35.

Parker, G. A., Baker, R. R., and Smith, V. G. F. (1972). The origin and evolution of gamete dimorphism and the male–female phenomenon. *Journal of theoretical Biology*, **36**, 529–53.

Parker, G. A., D. W. Mock, and T. C. Lamey. (1989). How selfish should stronger sibs be? *American Naturalist*, **133**, 846–68.

Perrins, C. M. (1965). Population fluctuations and clutch size in the great tit, *Parus major* L. *Journal of Animal Ecology*, **34,** 601–47.

Ricklefs, R. E. (1968). On the limitation of brood size in passerine birds by the ability of adults to nourish their young. *Proceedings of the National Academy of Sciences, USA*, **61,** 847–51.

Sæther, B-E., Andersen, R., and Pedersen, H. C. (1993). Regulation of parental effort in a long-lived seabird: an experimental manipulation of the cost of reproduction in the antarctic petrel, *Thalassoica antarctica. Behavioral Ecology and Sociobiology*, **33,** 147–50.

Smith, H. G. and Montgomerie, R. (1991). Nestling robins compete with siblings by begging. *Behavioral Ecology and Sociobiology*, **28,** 307–12.

Stamps, J. A. and Metcalf, R. (1980). Parent–offspring conflict. In *Sociobiology: beyond nature–nurture?* (ed. G. A. Barlow and J. Silverberg), pp. 598–618. Westview Press, Boulder, Colorado.

Stamps, J., R. A. Metcalf, and V. V. Krishnan. (1978). A genetic analysis of parent-offspring conflict. *Behavioral Ecology and Sociobiology*, 3, 369–92.

Trivers, R. L. (1972). Parental investment and sexual selection. In *Sexual selection and the descent of man* (ed. B. Campbell), pp. 136–79. Aldine Press, Chicago.

Trivers, R. L. (1974). Parent–offspring conflict. *American Zoologist*, **14,** 249–64.

Westneat, D. and Sherman, P. W. (1993). Parentage and the evolution of parental behavior. *Behavioral Ecology*, **4,** 66–77.

Winkler, D. W. (1987). A general model for parental care. *American Naturalist*, **130,** 526–43.

Winkler, D. W. and Wallin, K. (1987). Offspring size and number: a life history model linking effort per offspring and total effort. *American Naturalist*, **129**,708–20.

Wolf, L., Ketterson, E. D., and Nolan, V. (1988). Paternal influence of growth and survival of dark-eyed junco young: do parental males benefit? *Animal Behaviour*, **36,** 1601–18.

Wolf, L., Ketterson, E. D. and Nolan, V. (1990). Behavioural response of female dark-eyed juncos to the experimental removal of their mates: implications for the evolution of male parental care. *Animal Behaviour*, **39,** 125–34.

Wright, J. and Cuthill, I. (1989). Manipulation of sex differences in parental care. *Behavioral Ecology and Sociobiology*, **25,** 171–81.

Wright, J. and Cuthill, I. (1990*a*). Manipulation of sex differences in parental care: the effect of brood size. *Animal Behaviour*, **40,** 462–71.

Wright, J. and Cuthill, I. (1990*b*). Biparental care: short-term manipulation of partner contribution and brood size in the starling, *Sturnus vulgaris. Behavioral Ecology*, **1,** 116–24.

Continuous partnerships

The term *continuous partnership* is used to describe a relationship where pair members associate together for the whole year, even during the non-breeding season. Five case studies, including eight species with *continuous partnerships*, are presented in this section. Five species live in or near their breeding grounds the whole year and three others migrate long distances between breeding and wintering areas. In three of the species, pairs receive help from young nonbreeders.

Clive Minton (1968) was one of the first authors to present a detailed account of a species with *continuous partnerships*. He studied about 100 colour-banded pairs of Mute Swans *Cygnus olor* in the ponds, lakes, and rivers of the Black Country in Staffordshire, England, over a 7-year period. In addition to useful data on first breeding and pairing age, his paper provides the first estimates of divorce and mate fidelity in water-fowl; 52 divorce events in 492 pair-years giving a 10.6% probability of divorce or a value of 89.4% mate fidelity. He suggested weather conditions may have increased the occurrence of divorce (i.e. *severe weather mediated hypothesis*), since 42% of the divorce events occurred in an unusually severe winter. Minton also documented a case where a usurper drove the territory-holding male away after a prolonged battle while the female, attending a brood of four, 2-month-old cygnets, did not take part (i.e. the *forced divorce through usurpation hypothesis*). The female 'sailed off down the stream with the new mate and family as if nothing happened'.

Only a few experiments testing relevant mating system hypotheses have been performed with species whose pair bonds last a lifetime. Martin *et al.* (1985) provide one test of Lack's (1968) notion that monogamy evolved because both sexes were necessary to produce offspring. By removing male Lesser Snow Geese *Anser caerulescens* at different times during the nesting period (Martin *et al.* 1985) it was found that beyond a certain stage of incubation, females were able to hatch the eggs on their own, although they finished incubation in poorer condition. Males were thought to be useful during territory establishment, as eggs were being laid, and during brood rearing; goslings with single parents survived less well in some years. From other studies on geese it is known that single females, without male assistance, rarely achieve breeding condition in the first place (see Chapter 5).

This section begins with Blue Ducks *Hymenolaimus malacorhynchos* from the mountainous ravines of New Zealand and proceeds to the wet-lands of Britain for Barnacle Geese *Branta leucopsis* and three species of swans (*Cygnus columbianus bewickii*, *Cygnus cygnus*, and *Cygnus olor*), over to scrublands in Florida and Arizona for Florida Scrub Jays *Aphelocoma coerulescens* and Pinyon Jays *Gymnorhinus cyanocephalus*, and ends with Splendid Fairy-wrens *Malurus splendens* that live in the heathlands of south-western Australia.

Comparison between the five chapters shows that a significant life-history feature in species with *continuous partnerships*, in terms of reproductive success, is the amount of time pair members spend together on a daily basis. In most of these species partners maintain close proximity throughout the day, every day: in the Blue Duck, Barnacle Goose, Bewick's, Whooper and Mute Swan, and the Florida Scrub Jay. In these species, reproductive performance improves the longer a pair is together (*the mate familiarity effect*) and a temporary decline in reproductive output is experienced when partners change mates (*the cost of mate change*). The exceptions to this pattern are found in the Pinyon Jay and Splendid Fairy-wren. Although Pinyon Jay pairs live in the same wintering flock, they may not always be each other's closest associate when travelling over a large home range. Splendid Fairy-wren partners sometimes leave their mates to visit neighbouring territories when extra-pair copulations occur. These slight but notable exceptions may explain why reproductive success does not improve with pair duration and why there was no reduction in reproductive success after changing mates in these species. Overall, the degree of '*keeping company*' through the year seems to make a difference (cf. Rowley 1983).

References

Lack, D. (1968). *Ecological adaptations for breeding birds*. Chapman and Hall, London.

Martin, K., Cooch, E. G., Rockwell, R. F., and Cooke, F. (1985). Reproductive performance in lesser snow geese: are two parents essential? *Behavioral Ecology and Sociobiology*, **17**, 257–63.

Minton, C. D. T. (1968). Pairing and breeding in Mute Swans. *Wildfowl*, **19**, 41–60.

Rowley, I. (1983). Re-mating in birds. In *Mate choice* (ed. P. Bateson), pp. 331–60. Cambridge University Press, London.

4 *Long-term monogamy in a river specialist—the Blue Duck*

MURRAY WILLIAMS AND FRANK McKINNEY

Introduction

The mallard-sized Blue Duck *Hymenolaimus malacorhynchos*, endemic to New Zealand, is found mostly on rivers of steep gradient, a distinctive and linear habitat subjected to brief and dramatic changes in flow at any time of year. Adult ducks live year-round as pairs confined to exclusive 0.5–1.5 km territories along the river bed. Only territory-holders attempt breeding. Pairs keep close company at all times, males regularly patrol the territory boundaries, and territory (and mate) ownership are subjected to frequent challenges. Field and laboratory studies have yielded no evidence of mate infidelity nor of any breeding relationship other than strict monogamy (Triggs *et al.* 1991; Williams 1991). Males and females both care for the ducklings until fledging, that parental care consisting primarily of vigilance by the male and close guarding of the ducklings by the female. Both sexes are highly philopatric and birds resident in one catchment are effectively separate from those adjacent.

This appraisal of the Blue Duck mating system concentrates on the reproductive histories of 46 pairings over 14 breeding seasons (95 pair-years) during which 119 ducklings were fledged. From an analysis of pair characteristics and partner histories and reproductive success, we seek to appraise why long-term monogamy is characteristic of this species and other river-dwelling ducks, identify some of the costs and benefits of that system to individuals, and briefly summarize the nature of pair relationships in other ducks.

Background, study site, and procedures

The Manganuiateao River in central North Island, New Zealand (see Fig. 4.1) is probably more typical of the bird's former and prime habitat than most areas in which it presently occurs. The study area extends over 9.3 km of river at the lower extremity of the bird's range on the river and at an elevation of 240–360 m. Blue Ducks occur on a further 30 km of the river up to an elevation of about 1000 m. Comparative demographic data from four more elevated and generally colder locations elsewhere in New Zealand indicate that within the Manganuiateao River study area Blue Ducks live longer, have smaller territories, are consistently more productive, and are more diurnally active.

In 1975 a volcanic lahar travelled down much of the Manganuiateao River rendering the entire study area (as well as a considerable area above it) uninhabitable by Blue Ducks. Although some of the displaced birds lived for a time in side streams, most then resident simply disappeared. This study commenced in the austral summer of 1980 with four pairs beginning the process of recolonization, and data up to and including the 1993 breeding season are included in the analyses. Most data were obtained during brief 3–4 day visits to the area approximately eight times annually. All resident adults and almost all of their progeny were caught and banded with unique combinations of colour bands. See Williams (1991) for details.

The mean clutch size was six eggs. Annual productivity after a 70–82 day fledging period was 1.3 young per breeding pair. Annual survival of breeders was 0.86 (male 0.92, female 0.80) and that of juveniles in their first year 0.44. Mortality, particularly of nesting females, resulted mostly from mammalian predation. Juveniles dispersed and established permanent residency within a few months of fledging and although some wandered extensively throughout, and possibly beyond, the natal catchment, both sexes were highly philopatric and sought to establish territories close to or within their natal range.

Although most territory-holding Blue Ducks nested each year, only about half successfully raised ducklings. Hence, the reproductive values are heavily skewed with zeros. A logistic regression analysis was performed, thus controlling for the Poisson distribution of the data. The dependent variables, when significant, such as male age and female age, were included in the models. Variation in reproductive success due to year (between 1980 and 1993) were also included in the analyses.

Results

The pair bond

The mating system in Blue Ducks was one of strict social monogamy. In all 87 confirmed breeding attempts territories and nests were defended

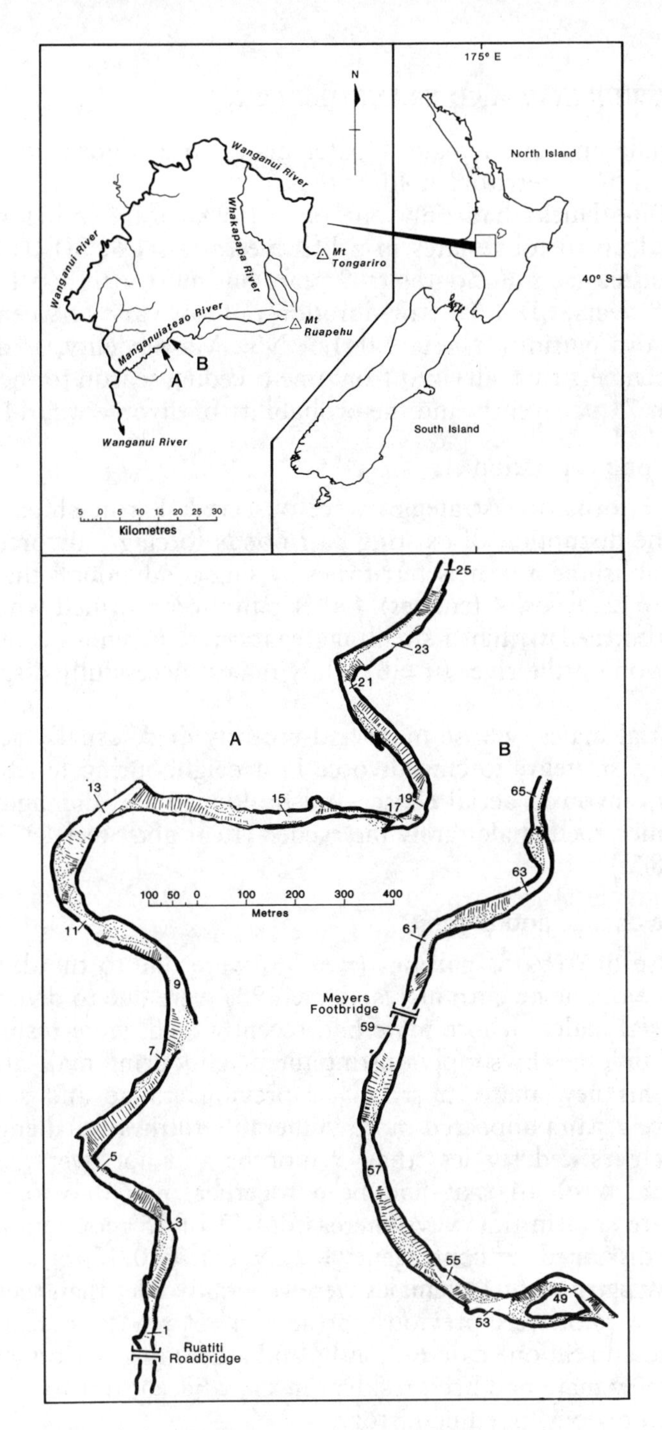

Fig. 4.1 Location of the Manganuiateao River study site within New Zealand. Throughout each day of each year of the study Blue Duck pairs occupied and defended territories that included several pools and rapids as indicated on two stretches of river (A and B). For example, two territories, occupied in all 14 years, encompassed pools 1 to 9, and 55 to 61. The smallest two territories included only pool 11, occupied for 15 months, and pools 49 to 55 for 3 years.

by one male and one female. Floater males were vigorously challenged and evicted by all territory-holding males.

Most Blue Ducks had only one mate (59% of 54 individuals) while others had up to four mates in a lifetime (mean 1.6, SD 0.8). Partnerships persisted from 6 months to 8 years but most included 1 or 2 years (mean 1.8 years, SD 1.7). Pair duration did not vary between first, second, or third pairings (*t*-tests 1.0–1.5, NS). Mate fidelity, as determined by pair members that survived from one breeding season to the next, was 88.7% (in 71 pair-years) and the probability of divorce was 11.3%.

Mechanism of pair formation

Three pair formation strategies were observed. Partnerships were initiated by the disruption of existing pair bonds forcing a divorce (n = 13), or the establishment of new territories by single individuals that attracted mates (n = 12 males, 4 females). Other pairs were formed when a dominant pair formed within a small aggregation of juveniles at an unoccupied location on the river (n = 6). Only males successfully disrupted pair bonds.

Territorial males, whose mate had recently died, usually adopted the most active strategy, forcing divorce in a neighbouring territory. These encounters involved aerial chases and prolonged and damaging fighting during which the female rarely intervened (Kear and Steel 1971; Eldridge 1985, 1986*a*).

Causes of mate change and divorce

The demise of 61% of pairings (n = 36) were due to the death or disappearance of one or both mates, while 39% were due to divorce.

Territorial males, whose mate had recently died, were responsible for 11 forced divorces by simply ousting the neighbouring male and occupying with his new mate most of her previous range and some of his former. New pairs appeared more vulnerable to these challenges; 5 of 7 pairs that persisted for less than 9 months and not over the breeding season, and 5 of 14 that had been together over only one breeding season were split in this way whereas only 3 of 15 more enduring pairings were disrupted (χ^2 contingency = 2.89, $0.1>P>0.05$).

The new partners of 14 females were generally older than their previous partner (n = 7) or were previous partners (n =3) whose return challenges were successful (in one case 6 months and a breeding season later). In 12 cases the new mate had been resident in the adjacent territory and all but one had a previous breeding history.

There were two instances where female choice was apparent in the divorce event. One mate switch involved two yearlings who, for 4 months at the peak of the breeding season, occupied a vacant section of river at the lower extremity of the range occupied by Blue Ducks on the river and who made no detectable breeding attempt. While the male was in wing moult, the female moved approximately 4 km upriver, 2 km above her natal range, to consort immediately with a territorial male whose mate had just died. Possibly in similar vein was the demise of a sibling pair which occupied a small range between two long-held territories for 15 months and over one breeding season in which they made no detectable breeding attempt. When the female on the adjacent territory died, her mate entered the young pair's range and was seen chasing the male and consorting with the female; she then moved to his territory. These are the only instances in which a female changed her range.

Correlates of reproductive success

Blue Duck breeding success varies annually depending on the severity of nest predation and winter/spring river flows (Williams 1991) and total reproductive failure across the population occurred in one of the 14 years of this study. In addition to year effects, we controlled for male age in the following analyses, as reproductive success increased throughout the age-classes for this sex (logistic regression $\Delta D = 5.93$, $\Delta df = 1$, $P < 0.02$). Young males produced fewer fledglings than middle-aged males which produced fewer than older males; for example, yearling males produced 0.5 fledglings per year (adjusted for year effects), 2–4-year-old males 1.1 fledglings, and 5+-year-old males 1.7 fledglings. Female reproductive success also increased with age; for example, yearling females produced 0.4 fledglings, 2–4-year-old females 1.3 fledglings, and 5+-year-old females 1.8 fledglings. Female age did not contribute significantly to variance in reproductive success of the pair when male age was included in the analyses ($\Delta D = 1.66$, $\Delta df = 1$, NS).

Familiarity with a partner

After controlling for year and age effects, the variation in reproductive success was not influenced by pair duration ($\chi^2 = 1.93$, $df = 1$, NS). However, the fitted values clearly increased between the first and eighth year pairs were together (Fig. 4.2), meeting the nonparametric criteria for a significant correlation (Spearman rank correlation, $r_s = 0.93$, $n = 8$, $P < 0.01$). Therefore, we suspect that pair familiarity in Blue Duck does influence a pair's reproductive potential.

Cost of mate change

In order to determine whether there was a cost to changing partners, the confounding effects of male age and year were removed in an analysis of fledging success in pairs with new or the same partners (i.e. the mate status). Reproductive success varied significantly with mate status ($\Delta D = 4.27$, $\Delta df = 1$, $P < 0.05$). For example, faithful mates produced a mean of 1.6 fledglings compared to a mean of 0.74 fledglings for new partnerships.

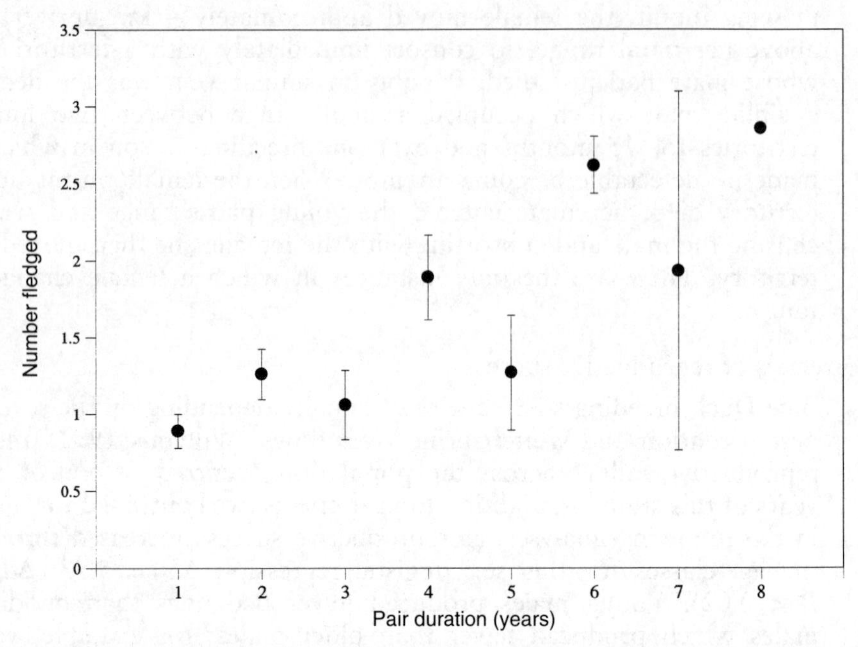

Fig. 4.2 Pair duration and reproductive success in Blue Ducks. The fitted values and standard errors are shown (Spearman rank correlation $r_s = 0.93$, $n = 8$ years, $P < 0.01$). Sample sizes for each point were 35, 20, 15, 9, 7, 5, 3, 1 pairs.

Correlates of divorce

There was little indication that prior reproductive histories influenced the probability of divorce in Blue Ducks. Considering pair members that survived over two consecutive breeding seasons, 11% of 28 divorced after reproductive failure and 14% of 37 separated after reproductive success. This difference was not significant.

Partner characteristics and reproduction

Yearling males tended to pair with females of the same age, but thereafter, and because of their longer lives, re-paired with younger females. Overall, in 19 (44%) of 43 different pairings males were older than females, in six (14%) females were older, in 11 (26%) partners were of similar ages (relative ages not known in seven (16%)). The most extreme example was a male at least 11 years old with a consort a mere 3 months on the wing!

The reproductive consequences of pairing with different aged partners is shown in Fig. 4.3. Any pair that contained a yearling of either sex reproduced poorly (mean 0.5 fledglings, SD 0.2) compared to when both partners were older (mean 1.7 fledglings, SD 0.5). In some pairings it was obvious that an older mate was advantageous. For both sexes the

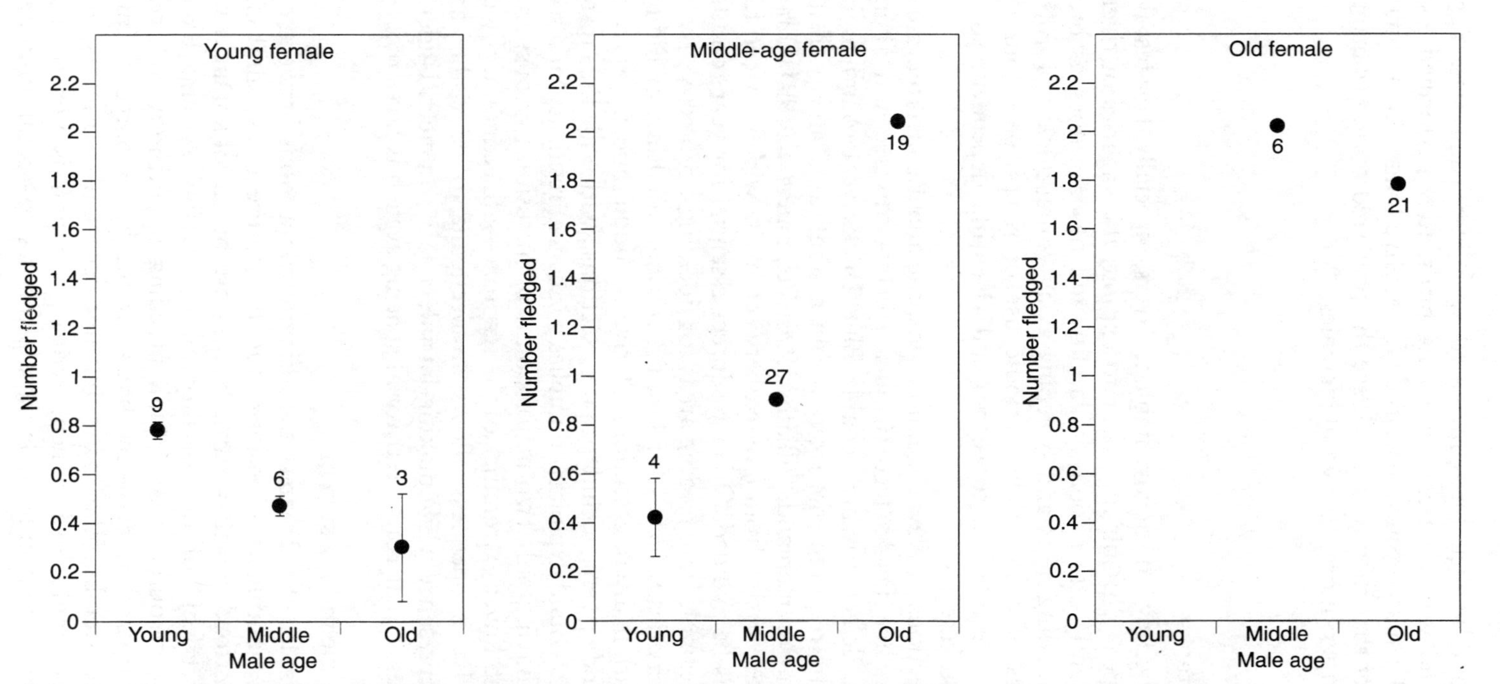

Fig. 4.3 The combined effect of male and female age on annual number of ducklings fledged, after controlling for year variation. Both male and female age effects were significant (logistic regression χ^2 males = 4.7, $df = 1$, $P < 0.05$; females = 5.6, $df = 1$, $P < 0.02$). Sample sizes accompany each point.

success of middle-aged birds increased with increasing mate age. The opposite trend was apparent for young birds; they performed less well with the older mates. Since male Blue Ducks live longer than females we would expect that they would achieve the preferred option (pairing with an older partner) more than would females.

Discussion

The riverine habitat

Blue Ducks occupy a distinctive habitat. Rivers, especially their fast-flowing sections, appear difficult for ducks to exploit; the vagaries of river flow may significantly affect food availability and be hazardous as breeding habitat by flooding nests and sweeping away young ducklings. Although many species of ducks make periodic use of rivers, only four species worldwide have adapted to year-round river life dependent solely on riverine resources.

In some sections of New Zealand rivers, the aquatic invertebrate fauna upon which Blue Ducks feed (Collier 1991) is diverse and abundant, occurring at densities which enable Blue Ducks year-round to obtain their energetic requirements from within a small area and with limited feeding effort (Veltman and Williams 1990; Veltman *et al.* 1991). Energetic costs and risks arising from moving between many widely spaced feeding sites can thus be avoided. Other resources essential for year-round habitation, for example, safe feeding areas for ducklings, protective riverbank vegetation, cavities or caves for nesting, resting, moulting, or hiding can be found within discrete sections of river. Middle and upper zones of New Zealand rivers include many sections with all the resources required. Year-round exploitation implies permanent residency, and sections of river, by having only two major points of intrusion by conspecifics—the upper and lower extremities of the territory—are economically defensible. The river's openness allows an occupant to be visible and to advertise its presence while making intruders easy to detect. Under these circumstances, a year-round territorial spacing system is favoured.

Territoriality and the mating system

Oring and Sayler (1992) stress the intimate connections between spatial organization, mating relationships, and the nature of parental care in shaping waterfowl social systems. In the case of the Blue Duck, we see three major effects of territoriality on the mating system. Exclusive ownership of resources would limit breeding opportunity to territory-holders. Co-operation between mates would be favoured to prevent territory take-over by other pairs, but not necessarily so in response to intrusions by single birds when territory eviction is not threatened. Prolonged investment in a single mate promotes co-operation in defence and enhances both breeding opportunity and performance, especially important during egg formation and laying, and brood rearing.

Restriction of breeding opportunity

The subdivision of suitable riverine habitat into a series of exclusively occupied and defended segments restricts non-territorial individuals to an itinerant and furtive existence that denies breeding opportunity. Intense competition for that exclusive space, and obvious attempts at obtaining it as soon as possible, are indicated by the strong philopatry of both sexes (the area best known is the area of birth), instances of just-fledged juveniles consorting with and eventually pairing with solitary territory-holders, and examples (given above) of young pairs holding territories before being physiologically capable of breeding. The importance of having a partner with whom to co-operate in defence of the territory against intruding pairs and ensure certainty of breeding opportunity is emphasized by territorial males, upon the death of their mate, immediately attempting to steal a neighbour's irrespective of time of year (including one case while the female was incubating, another during care of a young brood, and two immediately prior to laying), and their ready acceptance of newly fledged juveniles as potential partners. Temporary liaisons with neighbours when existing partners are vulnerable to eviction (e.g. during wing moult) may also serve this purpose (Eldridge 1986*a*).

Costs and benefits of divorce

If an individual's only chance to breed is as a member of a territory-holding pair, then those with the opportunity should vigorously defend their territory. Co-operation between partners in challenging intruders is obviously advantageous when challenges come from a pair. However, although partners should stay alert to opportunities to switch mates when the pair bond is threatened by a single intruder (which may explain why in many instances they do not show strong aggressiveness to intruders of the opposite sex), there are obvious reproductive costs in doing so. For example, we have identified that the initial breeding effort with a new partner is less productive than a further attempt with the existing mate, and pair familiarity increases reproductive success. Significantly, there was a tendency for new pairings and those comprising young birds to be more readily disrupted than longer-established ones comprising older birds.

For young females, however, the reproductive costs of divorce may be minor. A partnership with an older, established territory-holder (and possibly proven breeder) is likely to be more enduring, and thus more productive, than one with a young male. The relative shortage of females within the study population provides most females with the opportunity to pair with older males. Old females, on the other hand, have much to lose from a change of mate and are predicted to be more co-operative in territorial defence. Enduring pair bonds and enduring residency on a particular territory separately and/or together may enhance reproductive success through familiarity with territorial resources. For example, the feeding locations and ranges of experienced parents with their broods are more consistent day by day than those of new parents or new territory occupants (M. Williams, unpublished data). A more precise knowledge of a territory's resources may be especially important during prolonged river spate when food is scarce.

Prolonged breeding effort

The breeding effort demands a long annual contribution from both sexes. In contrast to most other ducks, but in common with other river specialists, the reproductive effort of Blue Ducks extends over about 45% of the year (Williams 1991). Prior to and during egg laying, the pair reduce their range within the territory and the male is especially belligerent towards intruders of either sex and ducks of other species as he guards his mate and his breeding opportunity. Further support of the reproductive effort is demanded by the ever-present hazard of the river current to duckling survival; separated ducklings rarely survive, and if they do, fledge considerably lighter than those raised with biparental care (Williams 1991). In response to this hazard, selection has favoured a significant male contribution to care of the brood (Veltman and Williams 1990).

Comparisons with other riverine waterfowl

Three other ducks, the Torrent Duck *Merganetta armata*, African Black Duck *Anas sparsa*, and Salvadori's Duck *Salvadorina waigiuensis*, are also river specialists. Torrent Ducks share similar spatial and mating systems, evidence of a common response to common ecological influences. For example, Torrent Ducks are dispersed as pairs on discrete territories (Johnsgard 1966) and although both birds co-operate in territorial defence, Eldridge (1986*b*) noted the tendency for the individual partners to react more strongly to intruders of the same sex. Males contribute fully to brood care (Moffett 1970). Salvadori's Duck from the highlands of New Guinea has been little studied but is dispersed as discrete territorial pairs along the rivers, males contribute to care of the young, and pairs occupy and defend territories year-round (Kear 1975). African Black Ducks, as pairs, defend discrete, nonoverlapping territories year-round (Ball *et al.* 1978; McKinney *et al.* 1978). Pair bonds, some of which persist over several years, are frequently tested and disrupted (Ball *et al.* 1978). At times, pair bonds appear to weaken (e.g. during moult or when the partner is injured) and individuals may form temporary liaisons with neighbours or other birds. Such liaisons may provide opportunities to assess alternative mates or they may be useful in preventing eviction from a territory. Breeding is restricted to territory-holders. In contrast to other river specialists, male African Black Ducks do not contribute parental care to the brood although they remain in contact with the female throughout (McKinney *et al.* 1978). The riverine habitat of this species is considerably more tranquil than that of the other three river ducks thus posing less of a hazard to brood survival and the basic *Anas* pattern of female-only brood care persists. This suggests male parental care is a by-product of year-round territoriality and mate fidelity and occurs only in those species inhabiting environments hazardous to ducklings where male participation confers a clear selective advantage (McKinney 1991).

Why strict monogamy?

Forced extra-pair copulations (FEPCs) occur in many duck species and mixed male reproductive strategies (monogamy + FEPC) appear to be common in Anatidae (McKinney *et al.* 1983; Afton 1985; Evarts and Williams 1987; Sorenson 1994). There are no observations of FEPC attempts in African Black Ducks (McKinney *et al.* 1978) or Blue Ducks (Williams 1991), and no evidence of extra-pair fertilizations in Blue Ducks (Triggs *et al.* 1991). We believe that FEPC is incompatible with the social system of river specialists. Paired territory-holding males give priority to guarding mate plus territory, especially during the breeding season, and the extra-pair excursions that would be needed to monitor the breeding status of females on neighbouring territories and to attempt FEPCs on fertile females apparently entail costs and risks that males cannot afford.

Long-term pair bonds in other ducks

Monogamy, involving seasonal or long-term pair bonds, is the primary mating system in most waterfowl (Oring and Sayler 1992). Seasonal pair bonds are characteristic of migratory Holarctic ducks (*Anas, Aythya, Mergus, Aix*) in which males desert their mates during the breeding season, leaving the females to care for the ducklings alone, and new bonds form while the birds are in flocks on the wintering areas. As well as being usual in geese and swans, long-term pair bonds are believed to be characteristic of whistling ducks (Dendrocygnini), shelducks and sheldgeese (Tadornini), and certain tropical and Southern Hemisphere ducks, although this belief is based on data from wild populations of only a few species (Table 4.1). Long-term partnerships are suspected to occur also in steamer ducks (Tachyerini), Torrent Ducks and in certain southern ducks with biparental care (Brazilian Merganser *Mergus octosetaceus,* Silver Teal *Anas versicolor,* Chiloe Wigeon *A. sibilatrix,* Bronze-winged Duck *A. specularis,* Chestnut Teal *A. castanea,* Brown Teal *A. chlorotis,* Crested Duck *Lophonetta specularioides,* Salvadori's Duck). In most ducks, pair formation occurs away from the breeding

Table 4.1 Statistics for long-term pair bonds and probability of divorce in 19 ducks and sheldgeese

Species	No. pairs	Pair-years	Number of divorce events	Probability of divorce	Biparental care?	Source
Continuous partnerships						
Black-bellied Whistling Duck	19	19	2	10.5	Yes	1
Dendrocygna autumnalis	106	123+	2	—		2
Magellan Goose						
Chloephaga picta	13	13	2	15.4	Yes	3
Maned Duck						
Chenonetta jubata	6	6	0	0	Yes	4
African Black Duck						
Anas sparsa	7	10	1	10.0	No	5
Blue Duck (sedentary)						
Hymenolaimus malacorhynchos	28	71	8	11.3	Yes	6
Part-time partnerhips; mates reunite on breeding area						
Shelduck	71	130	—	—	Yes	7
Tadorna tadorna	41	41	8	19.5		8
Australian Shelduck						
Tadorna tadornoides	21	63+	—	—	Yes	9
White-checked Pintail (sedentary)						
Anas bahamensis	36	44	15	34.1	No	10/11
Laysan Duck (sedentary)						
Anas laysanensis	11	11	3	27.3	No	12
Mallard (urban, sedentary)						
Anas platyrhynchos	8	11	1	9.1	No	13
Part-time partnerships; mates reunite on wintering area						
Bufflehead						
Bucephala albeola	2	5	—	—	No	14
Barrow`s Goldeneneye						
Bucephala islandica	6	6	1	16.7	No	15

European Wigeon						
Anas penelope	3	4	—	—	No	16
Long-term bonds frequent; mechanism unknown						
Speckled Teal						
Anas flavirostris	28	47	4	8.5	Yes	17
Cape Teal						
Anas capensis	4	14	—	—	Yes	18
Grey Teal						
Anas gibberifrons gracilis	7	17	—	—	Yes	19

Note: the terms pair-years and probability of divorce are defined in Chapter 1.

Sources: 1. Bolen (1971); 2. Delnicki (1983); 3. Summers (1983); 4. Kingsford (1990); 5. Ball *et al.* (1978); 6. this study; 7. Young (1864); 8. Williams (1973); 9 Riggert (1977); 10. Sorenson (1992); 11. sorenson *et al.* (unpublished); 12. Moulton and Weller (1984); 13. Mjelstad and Saetersdal (1990); 14. Gauthier (1987); 15. Savard (1985); 16. C. R. Mitchell (unpublished); 17. J. Port (unpublished); 18. Siegfried *et al.* (1976); 19. Marchant and Higgins (1990).

grounds, females are strongly philopatric, and each male follows his mate back to her familiar nesting area (Rohwer and Anderson 1988; Anderson *et al.* 1992; McKinney 1992). In species where males desert their mates, pair bonds apparently can persist only if breeding partners are able to rendezvous and reunite at some point before the next breeding season, for example at the breeding area before birds leave, at wintering or migratory stop-over sites, or at the breeding area after birds return. Reuniting of mates on the wintering grounds has been recorded in two migratory species of sea ducks (Mergini: genus *Bucephala*; Savard 1985; Gauthier 1987) and this pattern may be more widespread in this tribe. C. R. Mitchell (in litt.) recorded a few cases of European Wigeon *Anas penelope* pair members reuniting in a subsequent year at super-rich wintering sites, where supplemental feeds were provided. Although mate fidelity has been recorded in some other migratory *Anas* species (Dwyer *et al.* 1973; Fedynich and Godfrey 1989; Seymour 1991), it appears to be rare because pair formation takes place in large flocks (where mates may not encounter each other) and competition between males for mates is strong because of male-biased sex ratios. Mate fidelity can be more frequent in sedentary *Anas* populations (Moulton and Weller 1984; Sorenson 1992; Mjelstad and Saetersdal 1990) (Table 4.1). Reuniting of mates after return to the breeding area occurs in three shelduck species (Tadornini: genus *Tadorna*; Young 1964; Williams 1973, 1979; Riggert 1977) and this pattern may be more widespread in this tribe. Banding studies on breeding adults of seven migratory North American *Anas* and *Aythya* species showed that males returned to the same areas at lower rates (1.0–10.5%) than females (14.7–74.5%), and nearly all returning males were unpaired (Anderson *et al.* 1992).

Persistent pair bonds occur regularly in certain *Anas* species living in tropical and Southern Hemisphere regions where breeding seasons are often extended and/or irregular and many populations are nonmigratory (McKinney 1991; Sorenson 1991). Three studies of *Anas* species with strongly developed biparental brood care (*capensis, gibberifrons gracilis, flavirostris*) indicated a high incidence of long-term partnerships (Siegfried *et al.* 1976; Marchant and Higgins 1990; J. Port unpublished data), while data for two sedentary species with female-only brood care (*laysanensis, bahamensis*) showed a much higher proportion of pairs that divorce (Moulton and Weller 1984; Sorenson 1992).

In Table 4.1, biparental care is recorded in four of five species with continuous partnerships, all three species with long-term but uncertain pair bonds, but only two of eight species in which bonds break annually. Of these species, Tadornini (2), Dendrocygnini (1) and Cairinini (1) exhibit biparental care, one of the Mergini species does not, and only about half of the *Anas* species do so. The probability of divorce ranged between 0% and 34.1%, mean of 14.0% (SE 2.7). Species with biparental care were generally less likely to divorce (mean 9.4%) than those with female-only care (mean 19.4%). Thus, there are preliminary indications that mate fidelity and phylogeny may be associated to some degree with patterns of parental care. Analyses of the relative importance of these factors will not be possible without evidence for many more species.

Sorenson's (1992) study has addressed the possibility of a relationship between divorce and breeding success but no such relationship was found. Probability of divorce rates were similar among pairs of White-cheeked Pintails *Anas bahamensis* that were successful in raising ducklings (7 of 11 pairs) and those that were unsuccessful (6 of 12 pairs).

The early stages of pair formation in young, inexperienced ducks (less than 1 year old in *Anas* and *Aythya*, less than 2 years old in Tadornini) involve temporary bonds as described for Barnacle Geese *Branta leucopsis* (Choudhury and Black 1993), and divorce is preceded by extra-pair courtship and liaisons with the new mate (McKinney 1992). Many duck populations have male-biased adult sex ratios and males tend to be the most active sex in courtship. Chiloe Wigeon *Anas sibilatrix* are an extreme example: unpaired males begin to court females while they are still ducklings, apparently because adult females which are already paired are no longer available (Brewer 1991). The usurping of mates by male take-over, as described for Blue Duck and African Black Duck, is an unlikely strategy for nonterritorial ducks and has not been described. Several shelduck species are unusual in having female-biased sex ratios and females compete actively with one another for mates.

The comparative evidence for long-term pair bonds in ducks is still scarce and studies of Southern Hemisphere species are especially needed. The strong linkage that has been assumed between the incidence of long-term pair bonds (see Chapter 3) and the incidence of biparental care should be examined carefully, and attention should be given to the behavioural mechanisms involved in mate acquisition and mate fidelity.

Summary

Most Blue Ducks had one mate in a lifetime, usually lasting for 2, but sometimes up to 8, years. Reproductive success of pairs tended to increase with pair duration after controlling for year and age effects. A common cause of mate change (39%) was usurpation by neighbouring males that had lost their mates. These lone males were usually older than victim males. In addition to losing their mates, victim males lost their territory to the usurper. Divorce was not related to the pairs' prior reproductive failure/success. Ducks that changed mates had fewer fledglings the next year. Pairs with one young member bred less well than pairs with two experienced mates. Considering data from 11 duck and sheldgeese species the probability of divorce ranged between 0% and

34.1%, mean of 14.0%. Species with biparental care were apparently less likely to divorce (probability of 9.4%) than those with female-only care (probability of 19.4%).

Acknowledgements

We gratefully acknowledge Jeff Black's very considerable help with statistical analyses. We thank Jeff Port and Lisa Sorenson for permission to cite unpublished data and for their comments on the manuscript. We also thank the Editor, Mike Anderson, Carl R. Mitchell, and Clare Veltman for helpful comments on earlier drafts.

References

Afton, A. D. (1985). Forced copulation as a reproductive strategy of male Lesser Scaup: a field test of some predictions. *Behaviour*, **92,** 146–67.

Anderson, M. G., Rhymer, J. M., and Rohwer, F. C. (1992). Philopatry, dispersal, and the genetic structure of waterfowl populations. In *Ecology and management of breeding waterfowl* (ed. B. D. J. Batt *et al.*), pp. 365–95. University of Minnesota Press, Minneapolis.

Ball, I. J., Frost, P. G. H., Siegfried, W. R., and McKinney, F. (1978). Territories and local movements of African Black Ducks. *Wildfowl*, **29,** 61–79.

Brewer, G. (1991). Courtship of ducklings by adult male Chiloe Wigeon *Anas sibilatrix*. *Auk*, **108,** 969–73.

Choudhury, S. and Black, J. M. (1993). Mate-selection behaviour and sampling strategies of geese. *Animal Behaviour*, **46,** 747–57.

Collier, K. J. (1991). Invertebrate food supplies and diet of Blue Duck *Hymenolaimus malacorhynchos* on rivers in two regions of the North Island, New Zealand. *New Zealand Journal of Ecology*, **15,** 131–8.

Delnicki, D. (1983). Mate changes by Black-bellied Whistling Ducks. *Auk*, **100,** 728–9.

Dwyer, T. J., Derrickson, S. R., and Gilmer, D. S. (1973). Migrational homing by a pair of Mallards. *Auk*, **90,** 687.

Eldridge, J. L. (1985). Display inventory of the Blue Duck. *Wildfowl*, **36,** 109–21.

Eldridge, J. L. (1986*a*). Territoriality in a river specialist: the Blue Duck. *Wildfowl*, **37,** 123–35.

Eldridge, J. L. (1986*b*). Observations on a pair of Torrent Ducks. *Wildfowl*, **37,** 113–22.

Evarts, S. and Williams, C. J. (1987). Multiple paternity in a wild population of Mallards. *Auk*, **104,** 597–602.

Fedynich, A. M. and Godfrey Jr. R. D. (1989). Gadwall pair recaptured in successive winters on the southern high plains of Texas. *Journal of Ornithology*, **60,** 168–70.

Gauthier, G. (1987). Further evidence of long-term pair bonds in ducks of the genus *Bucephala*. *Auk*, **104,** 521–2.

Johnsgard, P. A. (1966). The biology and relationships of the Torrent Duck. *Wildfowl*, **30,** 5–15.
Kear, J. (1975). Salvadori's Duck of New Guinea. *Wildfowl*, **26,** 104–11.
Kear, J. and Steel, T. H. (1971). Aspects of social behaviour in the Blue Duck. *Notornis*, **18,** 187–98.
Kingsford, R. T. (1990). Flock structure and pair bonds of the Australian Wood Duck *Chenonetta jubata. Wildfowl*, **41,** 75–82.
McKinney, F. (1991). Male parental care in Southern Hemisphere dabbling ducks. *Proceedings International Ornithological Congress*, **20,** 868–75.
McKinney, F. (1992). Courtship, pair formation and signal systems of waterfowl. In *Ecology and management of breeding waterfowl* (ed. B. D. J. Batt *et al.*), pp. 214–49. University of Minnesota Press, Minneapolis.
McKinney, F., Siegfried, W. R., Ball, I. J., and Frost, P. G. H. (1978). Behavioral specialisations for river life in the African Black Duck (*Anas sparsa* Eyton). *Zeitschrift für Tierpsychologie*, **48,** 349–400.
McKinney, F., Derrickson, S. R., and Mineau, P. (1983). Forced copulation in waterfowl. *Behaviour*, **86,** 250–94.
Marchant, S. and Higgins, P. (1990). *Handbook of Australian, New Zealand and Antarctic birds.* Oxford University Press, Oxford.
Mjelstad, H. and Saetersdal, M. (1990). Reforming of resident Mallard pairs *Anas platyrhynchos*, rule rather than exception? *Wildfowl*, **41,** 150–1.
Moffett, G. M. (1970). A study of nesting Torrent Ducks in the Andes. *Living Bird*, **9,** 5–27.
Moulton, D. W. and Weller, M. W. (1984). Biology and conservation of the Laysan Duck *Anas laysanensis. Condor*, **86,** 105–17.
Oring, L. W. and Sayler, R. D. (1992). The mating systems of waterfowl. In *Ecology and management of breeding waterfowl* (ed. B. D. J. Batt *et al.*), pp. 190–213. University of Minnesota Press, Minneapolis.
Riggert, T. L. (1977). The biology of the Mountain Duck on Rottnest Island, Western Australia. *Wildlife Monographs* No. 52, 1–67.
Rohwer, F. C. and Anderson, M. G. (1988). Female-biased philopatry, monogamy, and the timing of pair formation in migratory waterfowl. *Current ornithology*, **5,** 187–221.
Savard, J-P. L. (1985). Evidence of long-term pair bonds in Barrow's Goldeneye *Bucephala islandica. Auk*, **102,** 389–91.
Seymour, N. R. (1991). Philopatry in male and female American Black Ducks. *Condor*, **93,** 189–91.
Siegfried, R. W., Frost, P. G. H., and Heyl, C. W. (1976). Long-standing pair bonds in Cape Teal. *Ostrich*, **47,** 130–1.
Sorenson, L. G. (1992). Variable mating system of a sedentary tropical duck; the White-cheeked Pintail *Anas bahamensis bahamensis. Auk*, **109,** 277–92.
Sorenson, L. G. (1994). Forced extra-pair copulation and mate guarding in the white-cheeked pintail: timing and trade-offs in an asynchronously breeding duck. *Animal Behaviour*, **48,** 519–33.
Summers, R. W. (1983). The life cycle of the Upland Goose *Chloephaga picta* in the Falkland Islands. *Ibis*, **125,** 524–44.
Triggs, S., Williams, M., Marshall, S., and Chambers, G. (1991). Genetic relationships within a population of Blue Duck *Hymenolaimus malacorhynchos. Wildfowl*, **42,** 87–93.

Veltman, C. J. and Williams, M. (1990). Diurnal use of time and space by breeding Blue Duck *Hymenolaimus malacorhynchos*. *Wildfowl*, **41**, 62–74.

Veltman, C. J., Triggs, S., Williams, M., Collier, K. G., McNab, B. K., Newton, L., Haskell, M., and Henderson, I. M. (1991). The Blue Duck mating system—are river specialists any different? *Proceedings International Ornithological Congress*, **20**, 860–7.

Williams, M. (1973). Dispersionary behaviour and breeding of Shelduck *Tadorna tadorna* on the River Ythan Estuary. Unpublished Ph.D. thesis. Aberdeen University.

Williams, M. (1979). Social structure, breeding and population dynamics of Paradise Shelduck in the Gisborne–East Coast district. *Notornis*, **26**, 213–72.

Williams, M. (1991). Social and demographic characteristics of Blue Duck *Hymenolaimus malacorhynchos*. Wildfowl, **42**, 65–86.

Young, C. M. (1964). An ecological study of the Common Shelduck *Tadorna tadorna, L.* with special reference to the regulation of the Ythan population. Unpublished Ph.D thesis. Aberdeen University.

5 *Do Barnacle Geese benefit from lifelong monogamy?*

JEFFREY M. BLACK, SHARMILA CHOUDHURY, AND MYRFYN OWEN

Introduction

Barnacle Geese *Branta leucopsis* are renowned long-distance migrants, flying from the high arctic in summer to winter on coastal marshes and pastures in temperate areas. After the journey, accompanied by goslings, each pair spends the winter months in dense flocks of thousands of birds. In spite of the potential for losing contact in these flocks, each male–female pair remains together throughout the day, each season, often for life. The pair's offspring also stay close for much of the first year, benefiting from parental vigilance and assistance in encounters with flock members (Black and Owen 1989*a,b*). Parents may care for unrelated goslings that join families mainly through adoption and intraspecific nest parasitism. The pairing process in geese happens long before the nesting season, so pairs generally arrive together on the breeding grounds to compete for food patches and nest sites. During the breeding season the male and female work as a team, protecting the clutch and subsequent brood from predation by gulls and Arctic Foxes *Alopex lagopus* (Prop *et al.* 1984). Exceptions to this monogamous strategy are rare.

Drawing from a sample of over 6000 individually marked birds with known life histories, we assess the significance of the pair bond by asking why certain pairs stay together and why others switch mates. We quantify whether geese reproduce better by keeping the same partner and whether one type of partner is better than another. Some of the behavioural changes that evolve with increasing years with the same partner are described. We also review some ideas regarding parental care for unrelated goslings.

Background, study sites, and procedures

Barnacle Goose males are larger (2.1 kg) than females (1.9 kg), but both have similar black and white patterned plumage. They are specialist vegetarians, feeding up to 200 bites per min, still taking only the most nutritious items. Like other goose species, Barnacle Geese tend to return to the same locations year after year, making use of familiar breeding and foraging sites. The population that we study breeds mostly on small islands in Svalbard and winters on the coast of Scotland and northern England. The geese nest in colonies of a few to some hundreds of pairs, with nest sites being approximately 7 m apart. The geese only have a short period of time in which to initiate and complete breeding during the brief arctic summer, so each pair only attempts one nest per year, laying 4–5 eggs. By the time the geese depart for the wintering grounds, predation reduces broods to a mean of about three goslings, and after migration a brood of two is the average (Prop *et al.* 1984; Owen and Black 1989*a*). Family groups monopolize the best feeding areas, as dominance rank in goose flocks is related to the size of the social unit (Black and Owen 1989*b*; Black *et al.* 1992).

During the study period, from 1973 to 1991, between 10 and 25% of the population were marked with coded plastic leg bands, readable from 250 m. During the ringing process, birds were sexed, aged, and measured. Birds that were not ringed as goslings or yearlings were assigned estimated minimum ages of 1 or 2 years depending on the season and location of the catch (Owen 1980). In winter, the birds used managed fields bordered by numerous blinds and observation towers at The Wildfowl and Wetlands Trust's reserve at Caerlaverock, Scotland. In some years the geese were also observed in Norway on the staging area during migration (n = 13 years) and on the breeding grounds (n = 11 years). Over 95% of the marked birds were recorded between 3 and 25 times (mean 8 records) each year, and birds that were not seen in two consecutive years were assumed to have died. Only 0.15% of birds assumed to be dead by this criterion appeared subsequently (Owen and Black 1989*a*).

At each sighting, we also recorded the identity of the mate and goslings; a number of behavioural cues were used to determine the identity of pairs, including synchronized behaviours in vigilance, feeding, defence of foraging space, routes taken while foraging and in flight, as well as maintenance of proximity, social display, and preflight signalling. The date of pair formation with a particular partner was taken as the first of multiple sightings when the pair was recorded together. The end of their partnership was attributed to the last date they were recorded together (Owen *et al.* 1988). To reduce the possibility of accepting trial partners (Choudhury and Black 1993) as true partners, we excluded associations that lasted less than 2 months, and only included pairs that spent at least one breeding season together. In order to obtain population estimates for the mean number of mates and pair durations, we used a sample of birds hatched before 1984. These birds had the opportunity to attain at least the average age of a goose (8 years) within our data set (Black and Owen 1995), and were therefore effectively followed over a lifetime. To determine the causes and consequences of mate change, we used data from pairs where both members were ringed and whose fate was known. Mate change was assumed to be caused by death if one mate was recorded as dead or missing in the next year.

Our main measure of reproductive success of a pair was the number of goslings in the family unit at arrival on the wintering grounds; we also used brood success, meaning the occurrence of at least one gosling in the brood. Most brood loss occurred prior to or during migration, so that once goslings reached the wintering areas, they had an 83% chance of entering the breeding population (Owen and Black 1989*b*). Since the majority of the geese arrived on the wintering grounds without any young, our measure of breeding success followed a Poisson distribution. Therefore, log-linear and logistic models (GLIM, NAG 1993) were used for most analyses. Since reproductive success is known to vary with breeding seasons (Owen and Black 1989*b*, 1991*b*, 1991*a*) and age (Black and Owen 1995), we controlled for both factors by entering them prior to the measure of interest (e.g. divorce rate, pair duration). Since reproductive performance follows a quadratic pattern, where appropriate we included the square of age (and pair duration) in the models. However, when the age of both pair members was included, one or the other sex dropped out of the model (i.e. P–values > 0.5, owing to a high collinearity between the ages of partners; Black and Owen 1995), so we controlled for male and female age separately or present results for just one sex. The data presented here generally include multiple observations of the same individuals, although from different years. To test whether our models were reliable, we repeated the analyses using one randomly selected sample from each bird. In all cases, the results were similar to those obtained from the complete data set.

Demographics

During the study period, the population increased from 4400 to 13 300 birds, resulting in several demographic changes. These included an expansion in the range of nesting and foraging sites, an older age of first breeding, a large increase in the number of nonbreeders, and an increase in gosling and adult mortality during migration (Owen and Black 1989*a*, 1991*b*). During this study, recruitment rate (percentage of young on arrival on the wintering grounds) fluctuated between 2 and 28% of the population depending largely on the severity of the arctic summer (Prop and de Vires 1993). On average, about one in six birds was an unpaired yearling or 2 year old looking for its first mate (Black and Owen 1995).

Pair formation process

First-time pairing in Barnacle Geese usually occurs between 1 and 5 years of age, with the majority (75%) pairing at the age of 1 or 2 years. Pair formation generally takes place outside the breeding grounds (Owen *et al.* 1988) and consists of numerous distinct stages requiring the co-operation of both partners in social display (Hausberger and Black 1990). The male initiates the courtship, but the female's response determines whether he proceeds from one stage to the next. With time, both birds maintain greater proximity and respond more to each other in displays (Black and Owen 1988). The geese appear to be highly selective when choosing partners, because most sample one to six potential mates in temporary 'trial partnerships' before settling with a consistent partner (Choudhury and Black 1993). Trial partnerships last from a few days to several weeks, and geese appear to use these as a means of assessing the qualities and compatibility of potential mates.

Mate choice characteristics

The majority of geese choose mates that are born in the same natal area (Owen *et al.* 1988), and in the same year (Black and Owen 1995). Experiments in captivity have shown that this is likely to occur because geese prefer to pair with familiar individuals with which they associated during early life on the breeding grounds (Choudhury and Black 1994). However, even after birds lose their first partners, which occurs at any age, they still tend to re-pair with similar-aged mates, despite the greater availability of young unpaired birds in the population (Black and Owen 1995). Most replacement mates apparently also come from the same breeding area and the new pair are likely to have previously associated with each other. No evidence of assortative mating by body size exists in Barnacle Geese (Choudhury *et al.* 1992), but large females appear to pair at an earlier age (Choudhury and Black 1993; Choudhury *et al.* 1996).

Paternity and parental care

Extra-pair fertilizations are rare in Barnacle Geese, but intraspecific nest parasitism occurs frequently in the form of egg dumping and adoptions (Choudhury *et al.* 1993; Larsson *et al.* 1995). In Barnacle Geese, parasitic females dump the eggs before and during the host female's incubation period (Forslund and Larsson 1995). The egg-dumping process lasts about 30 min during which the host female continuously attacks the parasitic female, whereas the host male usually passively stands by. This is in contrast to Lesser Snow Geese *Anser c. caerulescens*, where the parasitic pair work as a team distracting the attention of the host pair, giving the parasitic female the opportunity to dump her egg undisturbed (Lank *et al.* 1989). Parasitic Bar-headed Goose *Anser indicus* females concentrate dumping in the nests of dominant pairs, surreptitiously approaching them while host birds are absent (Weigmann and Lamprecht 1991). Bar-headed Goose females displace eggs from old nest sites prior to laying in those nests, but once they begin, foreign eggs are apparently indistinguishable from their own as no further eggs are rejected. Adoptions in geese are mainly due to brood mixing shortly after the young have hatched, but may still occur in goslings up to 12 weeks of age, when filial imprinting is complete (Choudhury *et al.* 1993; Williams 1994).

Results

The pair bond

Monogamous partnerships were by far the most common mating strategy: 99.6% of female–male pairs in 6297 pair-years. We recorded only 28 cases when a third party joined an existing pair for between 10 months and 4 years: possibly a prolonged partner-hold strategy (see Choudhury and Black 1993). In 21 of the cases an additional female accompanied the pair, and in the other seven cases two males were associated with one female. The original pair persisted in 12 of these cases. In 11 other cases the usurper displaced one member of the original pair. It was difficult to determine whether the usurper physically displaced and forced divorce upon the pair or whether an active choice was being made. Harem polygyny, which has been observed in a semi-captive goose flock with female-biased sex ratios, is likely to be a suboptimal alternative female strategy, because secondary females have a lower reproductive success than monogamous females (Lamprecht 1987).

Barnacle Geese can live up to 24 years, with the average being 8 years. Most of the geese in our population had only one mate during their lifetime (mean 1.48, maximum 6, SE 0.01). Of 2618 birds, 65% had one mate, 24.5 two, 7.5% three, and 3% four mates or more. Pair bonds persisted for 1–16 years, with a mean pair duration of 3.7 (SE 0.1) years. Mean pair duration decreased with each subsequent mate, from 4.6 years with the first mate, 3.2 years with the second, 2.4 years with the third and 1.4 years with subsequent mates (ANOVA, $F = 66.76$, $df = 3$, $n = 2452$, $P < 0.001$). No difference existed between the sexes in either number of mates or in pair duration.

Pair maintenance

Preflight and following behaviour

Waterfowl usually perform a series of head and neck movements, referred to as preflight signals, prior to taking flight. These are thought to signal intention and co-ordinate take-off between partners (Black 1988). Based on data from continuous focal samples when the observation ended in flight (n = 72), males performed significantly more preflight movements than females (mean 5.2 and 1.6, respectively; Kruskal–Wallis test, χ^2 = 33.5, df = 1, P < 0.001) and took flight first in 74% of the cases (χ^2 = 14.5, df = 1, P < 0.001); mates always follow quickly (<2 s). The distribution of this behaviour indicates that males make a substantial investment prior to flight and that females generally follow males at take-off.

Encounters with neighbours

A favoured hypothesis for explaining variation in mating systems is female–female aggression which suggests polygyny is limited by females who keep contender females from their mates (Gowaty Chapter 2). Similarly, male–male competition should limit polyandry. In the wintering flocks, which consist of about 1000 geese, nine aggressive encounters occur each minute as singles, pairs, and families compete for position and feeding space (Black and Owen 1988). Out of 426 encounters where the sex of the attacker and victim were known, females attacked both sexes with similar frequency (56 females, 64 males, χ^2 = 0.8, df = 1, NS). Males, on the other hand, did attack males significantly more than they attacked females (216 males, 90 females, χ^2 = 51.9, df = 1, P < 0.001). This is not surprising since males are by far the more active sex in agonistic activity (Black and Owen 1989*b*).

Mate familiarity effect

Barnacle Geese increase their breeding performance with increasing length of the partnership, that is, as pair members become more familiar with each other. After controlling for year and age variation, we found that reproductive success increased with the duration of the pair bond up to the seventh year and then appeared to decrease again (Fig. 5.1). There was also a significant interaction between year and pair duration, suggesting that long-term association with one partner was more beneficial in some years than in others. Since pair duration increases logically with age, including age as a factor in the model may create an unbalanced design. Therefore, we repeated the analyses for each age-class separately and obtained the same results in most cases.

In order to determine the effect of individual experience, as opposed to mate familiarity or pair duration, we included a measure of birds' cumulative experience in reproductive success (i.e. previous number of years a bird was paired during a breeding season) in the model while controlling for year, age, and pair duration. The addition of breeding experience had a slight but significant effect (ΔD = 1, Δdf = 1, P < 0.05), suggesting that the improvement of reproductive performance is also a function of breeding experience.

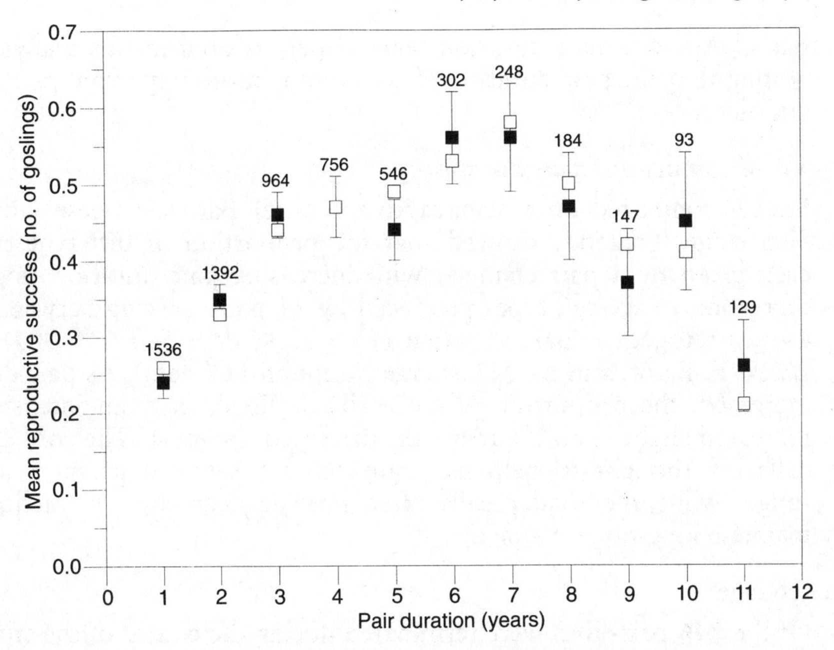

Fig. 5.1 Relationship between pair duration and reproductive success in Barnacle Geese (n = 6297 pair-years). *Filled squares* are the observed means and their SE. *Open squares* represent the fitted values of log-linear models after controlling for year and female age. Reproductive success varied significantly with pair duration ($\Delta D = 49.0$, $\Delta df = 1$, $P < 0.001$). The quadratic function of pair duration was also significant (i.e. duration^2, $\Delta D = 39$, $\Delta df = 1$, $P < 0.001$), as was the interaction between pair duration and year ($\Delta D = 43$, $\Delta df = 16$, $P < 0.001$) meaning mate familiarity may have been more useful in some years than in others.

Flock position and pair duration

Black *et al.* (1992) argue that despite the higher energetic costs due to increased vigilance and conflicts with neighbours, dominant geese value positioning themselves at the edge of the flock because they enjoy the first bite of the vegetation (see also Teunissen *et al.* 1985; Prop and Loonen 1988). One of the suspected benefits of a long-term association is enhanced team work in aggressive encounters (Black and Owen 1988) which may result in monopolization of prime feeding areas (Black and Owen 1989*b*). To test these ideas, we checked whether the length of association with a mate was effective in describing the tendency to be at the edge rather than the middle of the flock. We chose 270 pairs that were seen eight or more times in a given year. Pair duration was significantly related to flock position, with longer-term pairs spending more time in edge positions than younger pairs.[1] Age and pair duration were closely related in this analysis, suggesting that the pair duration effect became more important as the geese got older.

Vocal contact and pair duration

Studies conducted on a semi-captive flock of Barnacle Geese (at WWT-Slimbridge, England) showed that the proportion of different types of calls given by a pair changed with increasing pair duration (log-linear regression of six call types produced by 21 pairs varying between 1 and 14 years together; pair duration ($\chi^2 = 23.4$, $df = 5$, $P < 0.001$) (J. M. Black, E. Bigot, and M. Hausberger, unpublished data). As pair duration increased, the proportion of soft calls declined, medium intensity calls increased slightly, and louder calls increased the most. The soft 'contact' calls are thought to help maintain contact between partners at close range, while the louder calls often precede aggressive interactions and maintain long-range contact.

Mate change

Of the 946 pairs that were terminated during the course of the study and where both partners were ringed, 82% ended because of the death of one or both partners, 12% were divorces, and in 6% the situation for the pair break-up was not clear. The annual rate of mate change due to the death of one partner ranged from 11 to 22% in different years, with a mean of 17.1%, amounting to about 1640 'experienced' birds that became available to other unpaired birds in the population each year.

About 35% of the geese re-paired with a new mate at least once during their lifetime ($n = 2618$). If long-term monogamy is selectively advantageous in geese, we would expect individuals to incur some penalty when they change mates. Earlier work has shown that the time taken to re-pair after divorce or death of the mate may vary greatly, but is generally between 3 and 9 months (Owen *et al.* 1988). This means that birds that change partners run the risk of having to miss a breeding season, if they cannot find a replacement mate in time. The 'experienced' geese also run the risk of pairing with a young, inexperienced bird that has a lower reproductive ability (Forslund and Larsson 1991).

In spite of the superabundance of young birds in the population (e.g. 1 in every 3–9 birds), birds that lost their mate tended to re-pair with a bird their own age, often from their own natal area (Black and Owen 1995). The probability of re-pairing with a young/inexperienced versus an experienced bird was not influenced by the abundance of young birds

[1]The logistic regression model controlled for age and current number of goslings; parents with the most goslings spend most time in edge positions (Black and Owen 1989*a*). Other NS variables were dropped from the final model: bird type, year, previous breeding experience, sex, and skull size. The main effects accounted for a significant change in deviance ($\Delta D = 118.7$, $\Delta df = 6$, $P < 0.001$). The age and pair duration interaction was also significant ($\Delta D = 12.0$, $\Delta df = 1$, $P < 0.001$). From Hammond (1990).

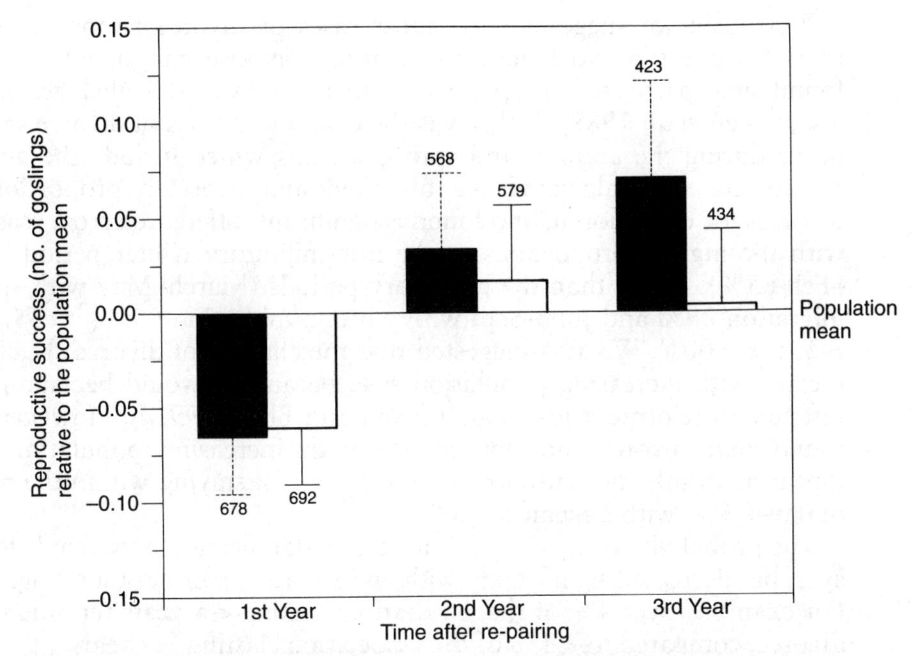

Fig. 5.2 Reproductive success of Barnacle Geese after mate change. In both males (*filled bars*) and females (*open bars*), those that changed mates returned with fewer goslings, compared to the population mean for faithful pairs, controlling for year, age, and pair duration effects in the first year with their new mates (females = 0.28, SE 0.03; males = 0.29, SE 0.03) than birds that had not changed mates (females = 0.34, SE 0.01; males = 0.36, SE 0.01) (female data set, ΔD = 13.0; Male ΔD = 14.0, both Δdf = 1, $P < 0.001$); there was no difference in the later years (ΔD = 1.5, Δdf = 1, NS).

in the population (Kruskal–Wallis test comparing the number of young 'available' birds in each year of the study, $\chi^2 = 1.63$, $df = 1$, NS).

However, those that changed mates had a lower chance of returning with a gosling in the first year with their new mates (mean 0.15, SE 0.01) than birds that had not changed mates (mean 0.36, SE 0.01; controlling for year, age, and pair duration; $\Delta D = 1$, $\Delta df = 1$, $P < 0.005$). Birds that changed mates also had fewer goslings in the first year with the new mates (Fig. 5.2). However, by the second and third year with the new mates, most appeared to achieve the same success as birds that had not changed mates.

Divorce in geese

Of the 1536 pairs where both mates survived in subsequent winter seasons (5974 pair-years), divorce was observed in 118 pairs. The incidence of divorce was not significantly different in pairs that failed to reproduce (in the year before) compared with pairs that succeeded, after controlling for year, age, and pair duration effects (male data set, ΔD = 1.0, Δdf = 1, NS; females, $\Delta D = 0.6$, $\Delta df = 1$, NS).

Previously we suggested that most cases of divorce happened when pairs became separated during migration, because pair members often found new partners before previous mates arrived (labelled *accidental loss*, Owen *et al.* 1988). If this was the case, most cases of divorce should occur during the autumn and spring seasons which include the annual migrations. We calculated the observed and expected proportion of divorces for each month, and found a significant difference across seasons, with the highest proportions in the non-migratory winter period (Oct–Feb 65%), rather than the migratory periods (March–May with spring migration 29% and June–Sept with autumn migration 6%: $\chi^2 = 65.7$, $df = 2$, $P < 0.001$). We also suggested that the chances of divorce should be greater with increasing population size, because it would become more difficult to relocate a lost mate (Owen and Black 1991*a*). However, we found that divorce did not change with increasing population size (Spearman rank correlation $r_s = 0.004$, $n = 15$), staying within the range of 0.5–5.4%, with a mean of 2.0%.

The probability of divorce did not differ significantly between different ages, but decreased significantly with an increasing pair duration (Fig. 5.3). For example, over 4% of the associations lasting < 5 years terminated in divorce, compared to < 1% of the associations lasting > 9 years.

Several other hypotheses have been proposed to explain divorce in birds (Chapters 1, 19). For example, Ens *et al.* (1993) suggest that divorce should occur when a better option is available for one partner, and that the likelihood of finding a better quality mate should increase with the availability of mates. Divorce rate, therefore, should increase with greater availability of unpaired birds in the population, but not necessarily any type of unpaired bird. In our population, the probability of divorce did not increase with mortality rate, a measure that includes a wide selection of potential mates ($r_s = 0.032$, $n = 11$, NS). It did, however, increase with the proportion of birds that lost their mates through death, that is, with the number of experienced mates becoming available in the population ($r_s = 0.538$, $n = 15$, P = 0.05). Divorce rate decreased with an increasing proportion of young (< 3 years), unpaired birds in the population ($r_s = -0.5$, $n = 15$, $P < 0.05$).

To detect whether reproductive success improves after divorce, we compared the breeding success of pairs in the years before and 2 years after 63 divorce events; the first year after was not considered, owing to the reduction in success after mate change (see above). For both sexes, mean brood success did not increase with the new mate after divorce events.[1]

[1]Reproductive success data (the number of goslings) were adjusted for annual and age variations by calculating the deviation from the annual mean for each age-class. Before divorce: males 1.20 (SE 0.10), females 1.09 (SE 0.12); with new mate: males 1.21 (SE 0.15), females 1.12 (SE 0.13); males $\Delta D = 0.06$, $\Delta df = 1$, NS; females $\Delta D = 0.88$, $\Delta df = 1$, NS.

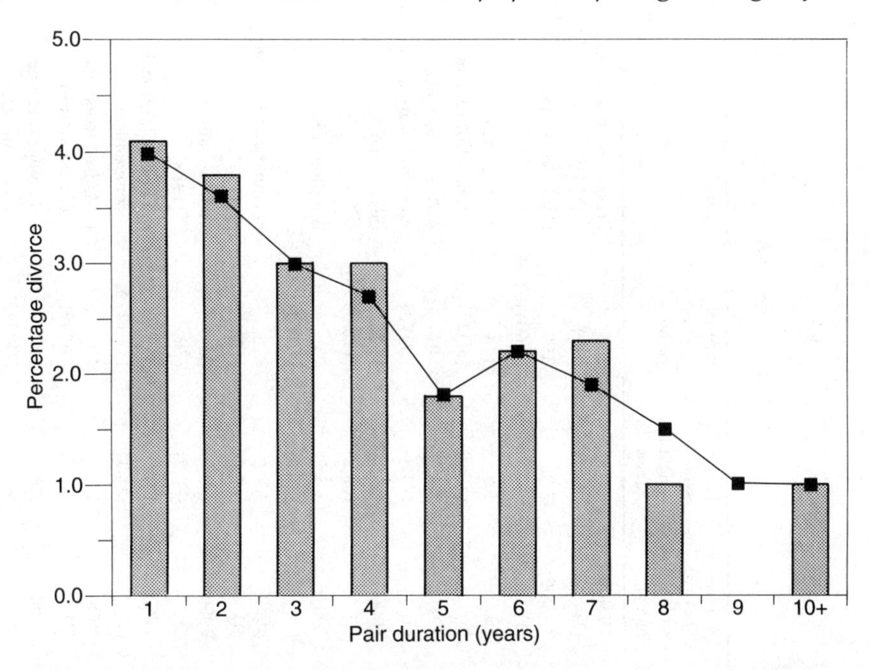

Fig. 5.3 Influence of age and pair duration on divorce. Controlling for year and age, the probability of divorce was significantly related to pair duration (n = 5637 female data set, $\Delta D = 19.4$, $\Delta df = 1$, $P < 0.001$; n = 5616 males, $\Delta D = 24.8$, $\Delta df = 1$, $P < 0.001$). However, when year and pair duration were controlled, age and divorce were not related (females, $\Delta D = 0.9$, $\Delta df = 1$, NS; males, $\Delta D = 1.6$, $\Delta df = 1$, NS).

Comparison between goose species

With data from 13 other goose studies, we attempted to explain the variation in divorce in relation to various environmental and phenotypic characteristics (Table 5.1). These included latitude (accounting for differences in nesting synchrony and time for completing a breeding attempt), mortality rates (differences in availability of new mates), migration distance, winter flock size, and degree of coloniality (i.e. risk of becoming separated, or alternatively, opportunities for extra-pair behaviours), body mass, and size (causing birds to abandon pair behaviour in favour of intensive feeding, see Johnson and Raveling 1988).

No significant correlations were achieved with these traits using two measures of divorce events (Table 5.1). The direction of the correlations with body size (tarsus) revealed that divorce was more likely in larger geese, although not significantly. This finding does not support Johnson and Raveling's (1988) idea that mate fidelity should be less well developed in smaller geese, owing to their higher degree of gregariousness and constraints in foraging. We would expect the opposite; that smaller geese would be most faithful owing to the risk of predation during incubation. It may take two small geese to thwart a raid by a fox, whereas one large

Table 5.1 Variation in divorce events in several long-term goose studies

Species—study site	Probability of divorce[1]	Percentage of pairs that divorced[2]	Latitude	Mortality (%)	Migration distance (km)	Winter flock size (coloniality index)	Female[3] winter weight (kg) (tarsus size)	*n* pairs (events/pair years)	Source[4]
Lesser Snow Goose—McConnell River, N.W.T.	0	0	61°	25	3800	10,000 (1)	2.0 (81.2)	238 (0/?)	Prevett and MacInnes (1980)
Lesser Snow Goose—La Perouse Bay	0.8	1.7	60°	25	3200	10,000 (1)	2.0 (81.2)	355 (5/604)	Cooke *et al.* (1981)[5]
Hawaiian Goose—Hawaii	0.9	2.6	20°	7	50	10 (3)	1.8 (78.0)	114 (3/327)	Black *et al.* (unpublished data)
Greenland White-fronted Goose—Wexford, Ireland	1.2	4.0	68°	14	4000	375 (3)	2.4 (74.4)	50 (2/173)	A.D. Fox (personal communication)
Barnacle Goose—Scotland	2.0	7.7	78°	10	3000	1000 (1)	1.9 (68.2)	1536 (118/5974)	this chapter
Giant Canada Goose—Marshy Point, Manitoba	2.2	5.5	50°	30	875	100 (2)	3.5 (87.0)	73 (4/183)	Raveling (1988)
Barnacle Goose—Gotland, Sweden	2.6	2.4	57°	10	750	1000 (1)	1.9 (67.8)	? (19/729)	Forslund and Larsson (1991)
Small Canada Goose—McConnell River, NWT	—	8.5	61°	25	3800	3000 (2)	2.0 (70.3)	82 (7/?)	McInnes and Dunn (1988)
Dark-bellied Brent Goose—The Netherlands	4.5	16.2	75°	16	6250	1750 (2)	1.3 (58.1)	352 (57/1274)	B. Ebbinge (personal communication)
Pinkfooted Goose—Norway	4.6	12.5	78°	8	3000	2500 (2)	2.2 (70.3)	40 (5/109)	J. Madsen (personal communication)
Barnacle Goose—The Netherlands	4.6	18.0	70°	10	3700	5000 (1)	1.6 (67.8)	100 (18/393)	B. Ebbinge (personal communication)
Greylag Goose—Southern Sweden	5.1	11.4	56°	17	2700	— (2)	3.1 (78.8)	132 (15/292)	L. Nilsson (personal communication)

Greylag Goose—Austria	8.1	27.5	47°	11	8	140 (2)	3.1 (78.8)	171 (47/566)	K. Kotrschal (personal communication)
Spearman rank correlations									
Probability of divorce	—	0.869**	0.002	−0.230	−0.304	−0.156 (0.144)	0.197 (−0.326)	−0.221	
Percentage of pairs that divorce	0.869**	—	0.22	−0.244	−0.033	−0.151 (0.209)	0.017 (−0.403)	−0.264	

** $P < 0.01$. None of the life history features was significantly correlated with divorce events.

Lesser Snow Goose *Anser c.caerulescens*. Hawaiian Goose *Branta sandvicensis*, Greenland White-fronted Goose *Anser albifrons flavirostris*, Barnacle Goose *Branta leucopsis*, Small Canada Goose *Branta canadensis hutchinsii*, Pinkfooted Goose *Anser brahcyrhynchus*, Dark-bellied Brent Goose *Branta b. bernicla*, Greylag Goose *Anser anser*.

[1] Probability of divorce = divorce events/total pair-years * 100.

[2] Percentage of pairs that divorced = number of divorces/total number of pairs * 100.

[3] Measurements from Owen (1980).

[4] Further anecdotal data not included in the comparisons due to small sample sizes: Sherwood (1967) reported no divorces in 11 Giant Canada Goose pairs (25 pair-years) over 3 years, Johnson and Raveling (1988) reported one divorce in six Cackling Canada Goose *Branta canadensis minima* pairs (10 pair-years) over 2 years, and A. Follestad (personal communication) reported one divorce in six Greylag Goose pairs—Norway (18 pair-years) in 5years.

[5] Minimum estimate due to bias towards older, known breeders, or due to lack of male philopatry.

goose may be sufficient. In studies with medium-sized Lesser Snow Geese, Bar-headed Geese, and large Canada Geese *Branta canadensis*, where males were removed or missing during incubation, their mates did succeed in hatching goslings, although subsequent gosling fitness was reduced in families attended by a single parent (Martin *et al.* 1985; Schneider and Lamprecht 1990; Paine 1992). Presumably, the lack of male help in smaller geese, like Barnacle Geese, would be noticed earlier, perhaps resulting in failed nesting attempts.

Pair characteristics and reproductive success

Age

Reproductive success of both male and female Barnacle Geese increases linearly until the age of 6 years, then levels off until the 11th year, and subsequently declines in later years (Black and Owen 1995). A similar quadratic pattern has been described for Lesser Snow Geese (Rockwell *et al.* 1993). This means that most individuals can improve their breeding performance throughout their lives, because few will experience the old-age senescence.

To investigate whether one sex was largely responsible for the reproductive success of the pair, we divided birds into three age-classes (young = 2–6 years, middle-aged = 7–11 years, and old = 12+ years), and plotted brood success (producing at least one gosling) for each male+female age combination (Fig. 5.4). The highest breeding performance was achieved by middle-aged males paired with middle-aged females. Young females and old males had a consistently low brood success, irrespective of their mate's age, suggesting that they limited the success of the pair. On the other hand, young males and old females could improve their reproductive success by pairing with middle-aged mates. This suggests that many birds should be attempting to gain the prime, middle-aged mates, the preferred option, to improve their own reproductive performance.

Size

In both sexes, larger Barnacle Geese produce more offspring, more often, than smaller geese (Choudhury *et al.* 1996). At first glance this would suggest all birds should attempt to gain the largest mates possible. Indeed, the most successful pair-types overall were large male with large female pairs and large male with medium female pairs (Fig. 5.5). However, the interaction of male+female size combinations in the model also affect brood success, with certain size combinations doing better in some years and other pair-types being more successful in other years (see also Choudhury *et al.* 1996). Interestingly, the geese maximized their reproductive performance by pairing with similar-sized mates (the preferred option). Thus, small males and females reproduced progressively worse with larger mates, and large males and females reproduced progressively

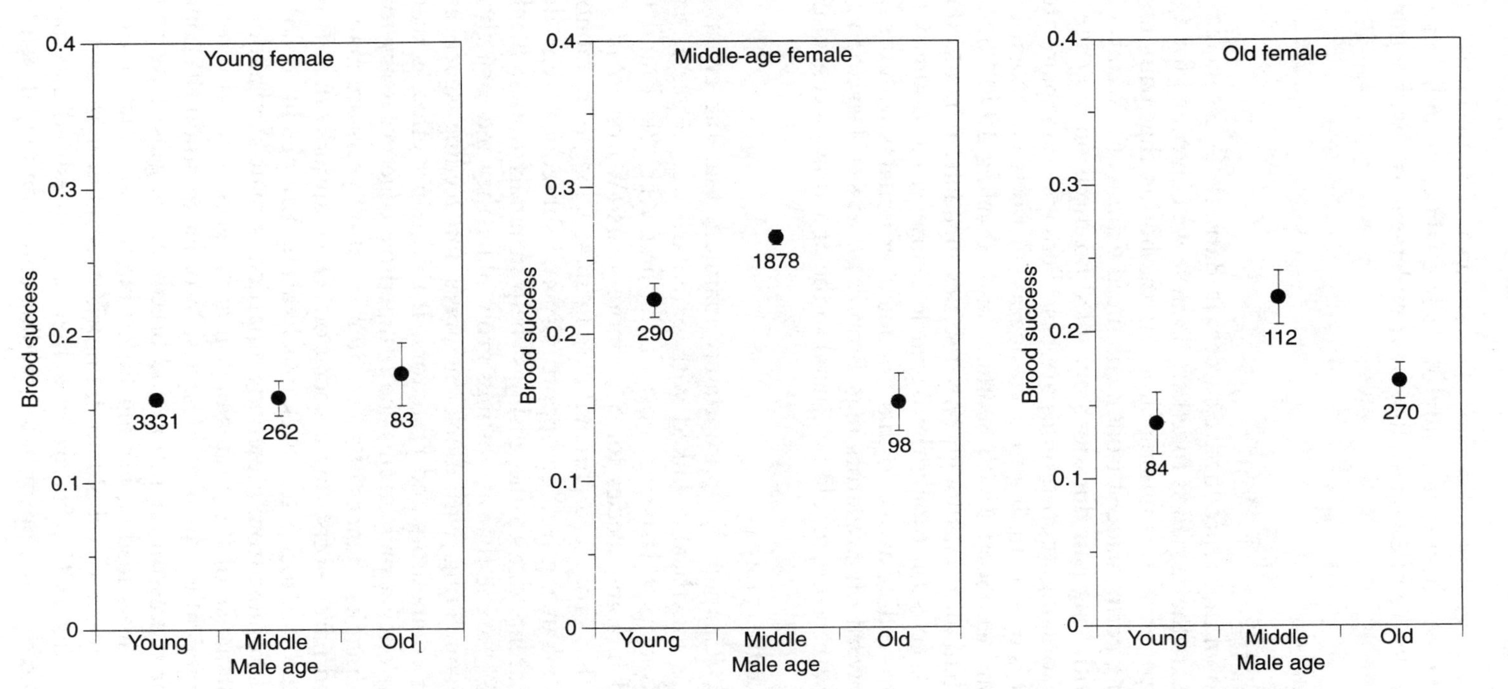

Fig. 5.4 Probability of returning to the wintering grounds with at least one gosling (i.e. brood success, and SE bars) in relation to partners' ages. The overall effect of combined ages was significant, controlling for year (probability of breeding $F = 4.81$, $df = 4$, $P < 0.001$), where middle-aged birds reproduce best. Age was partitioned into three classes: (1) the improving years (ages 2–6), (2) the prime, middle years (ages 7–11), and (3) the declining years (ages 12+). Derived from Black and Owen (1995).

worse with smaller mates (Fig. 5.5). This shows that reproductive success of a pair may be affected not only by their qualities as individuals, but also by their degree of compatibility or complementarity with a mate.

Discussion

Is long-term monogamy advantageous?

We have shown that reproductive success in Barnacle Geese increases for the first 7 years that a pair is together. Because we controlled for age and breeding experience, this finding supports the notion that pairs improve their breeding performance through enhanced team work. We also found that pairs with long pair durations occupied positions on the edge of the flock where foraging opportunities are best. Perhaps goose pairs are able to achieve this by fine-tuning their co-operative efforts in encounters and vigilance routines (Scott 1980; Sedinger and Raveling 1990). Similarly, during incubation, even though only females incubate, pair members of some goose species (including the Barnacle Goose) need to alternate foraging trips to feeding areas, so that one member remains with the nest at all times to ward off predators (e.g. Prop *et al.* 1984). Developing such partnerships, therefore, could be crucial to the breeding success of geese.

The evolution of monogamy

Mock *et al.* (Chapter 3) appropriately maintain that the evolution of monogamy is inextricably linked with parental care. In geese, parent–offspring association lasts for most of the first year (Black and Owen 1989*a*), and in some species for 5 or more years (Warren *et al.* 1993). We suggest that prolonged parental care facilitates long-term monogamy in geese by enhancing feeding opportunity and subsequent reproductive success of parents due to the goslings' help in family duties. Black and Owen (1989*a*) argue that as goslings grow older they increasingly assist parents in competing with flock members for feeding space and in scanning for competitors and predators. If this helper effect is real, it is possible that once a pair attains the competitive edge necessary for successful breeding, its chances of succeeding in future years are maintained by prolonged parent–offspring association. Once an individual is in this loop (labelled the social feedback mechanism, Lamprecht 1986), it is likely to benefit most from preserving a particular mate's collaboration.

The maintenance of the monogamy in arctic geese may also be linked to their energetically expensive lifestyles. In order to attain the condition necessary for successful migration and breeding, the geese have to spend much of their lives feeding leaving little time for other activities. Geese strive to achieve a threshold in fat/nutrient condition in each season (Drent and Prins 1987). Failure to achieve one threshold may preclude the next, such that an appropriate breeding condition may depend on an

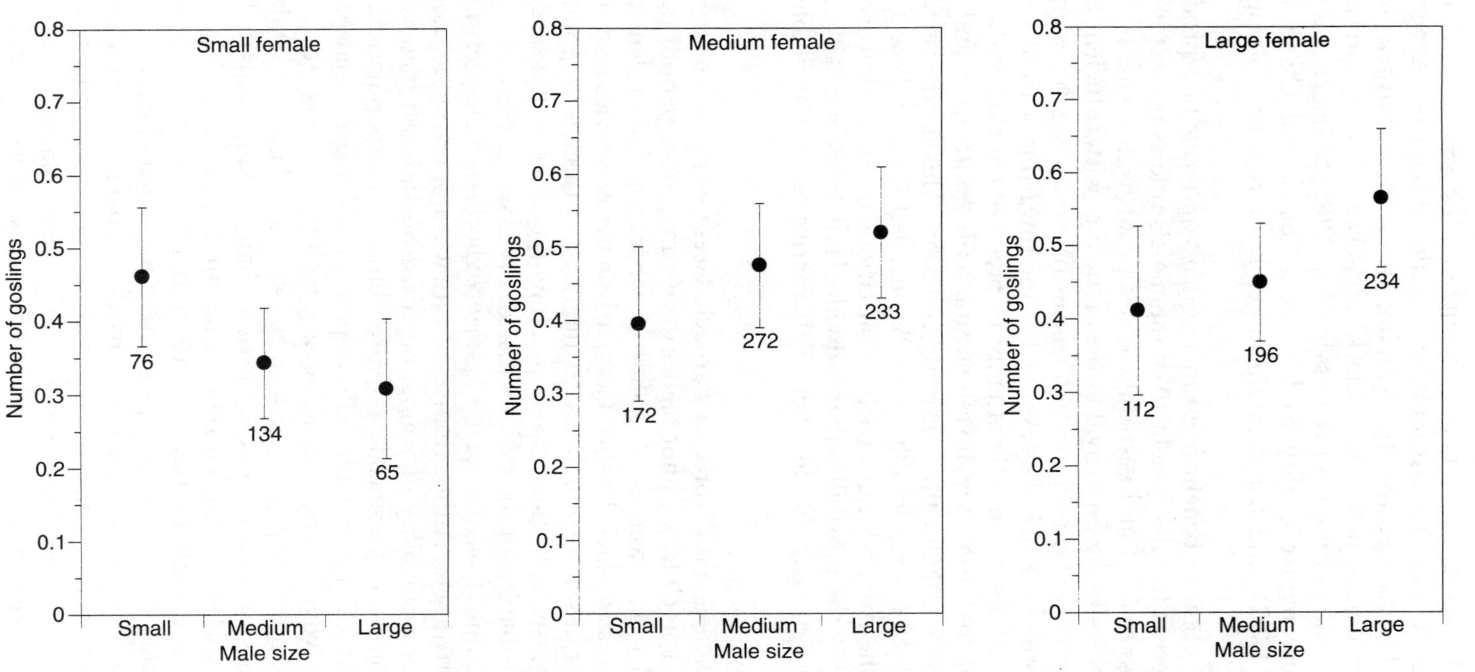

Fig. 5.5 Reproductive success (number of goslings, and SE bars) of different pair-size combinations. Barnacle Geese maximized the production of goslings when paired to a mate of similar relative body size; pair-size significantly influenced overall reproductive success, after controlling for year and age effects ($\Delta D = 21.5$, $\Delta df = 8$, $P < 0.01$). Size categories were based on the first principal component (PC1) derived from skull and tarsus measures. Medium-sized birds were within half a SD of the mean, small birds below and large birds above this. Derived from Choudhury *et al.* (1996).

individual's foraging performance throughout the year (e.g. Prop and Vulink 1992). Under this assumption, females should encourage year-round pair bonds, because by teaming up with a partner, females increase their dominance status (Black and Owen 1989*b*), are able to devote more time to foraging (Teunissen *et al.* 1985; Sedinger and Raveling 1990), use higher quality feeding areas (Black and Owen 1989*a*: Black *et al.* 1992), and hence attain larger body reserves (McLandress and Raveling 1981).

Both sexes seem to benefit from mate fidelity in terms of reproductive payoffs. Whereas females benefit from the male's role as protector which enables access to food and nest sites, males profit from access to a mate capable of reproducing on arrival in the arctic (i.e. with sufficient fat and nutrient reserves). Based on the frequencies of encounters between flock members, both sexes may be responsible for thwarting a third party to join the pair. Males apparently attempt to preserve the partnership more than females, because they invest substantially more time and effort into co-ordinating flight, thus ensuring cohesion. This is consistent with Lamprecht's (1989) report that male Bar-headed Geese follow females throughout the day and in a variety of situations on land. However, this does not nullify the possibility that individuals of either sex can improve on their choice of mate should a better alternative become available.

The role of divorce

Our previous ideas on divorce in Barnacle Geese were not supported in this study. Divorce rate was not higher during migratory periods (i.e. due to accidental loss, Owen *et al.* 1988) or when population density was greater (Owen and Black 1991*a*). Comparison across species also showed that divorce rates were unrelated to migratory distance or population density, as measured by flock size and colony type. Accidental separation of mates is probably not the major cause of divorce in geese.

The predictions from the *better option hypothesis* (Chapter 1) were partly supported here. Although divorce rate was not related to mortality rate or greater availability of young, unpaired birds, it did increase with the proportion of 'experienced', unpaired birds in the population. Since birds that lose a mate preferentially re-pair with similarly aged partners (Black and Owen 1995), increasing availability of unpaired, 'experienced' birds apparently gives pair members the chance to switch to another, perhaps better, mate. On the other hand, young, unpaired alternatives are apparently avoided by pair members. Similarly, divorce rate did not increase with mortality rate, perhaps because, in geese, the annual mortality includes a large proportion of younger birds.

If geese actively choose to divorce mates in favour of a new partner, we would expect the change to reflect an improvement in mate quality and subsequent reproductive success. We were unable to detect an improvement when comparing breeding success prior to and in the second year after divorce. Neither were incompatible partners, measured in terms of prior reproduction, more likely to divorce.

Preferred options and conflict between the sexes

Mating strategies

Differences in reproductive pay-offs for males and females may give rise to intense sexual conflict, with both males and females attempting to gain the option that increases their own reproductive success. Males should usually attempt to gain as many females as possible to increase the number of offspring carrying their genes, whilst females should favour co-operative polyandry to increase male help and genetic variation in the brood (Davies 1989). In geese, the constraints on non-monogamous mating systems include balanced sex ratios (Lamprecht 1987), the benefit of prolonged biparental care to goslings (Black and Owen 1989*b*; Schneider and Lamprecht 1990), and intrasexual aggression (at least in males, see above). Another reason why larger mating groups do not occur in geese may be related to the costs and risks involved in maintaining associations over the annual cycle, e.g. vocal and visual signals. Staying close to a partner in large flocks and over long migrations is likely to be costly. These costs would increase with additional partners.

Mating patterns

Conflict between the sexes may also apply to mate choice preferences. Whether individuals are able to realize their preferred options with regard to mate choice is likely to depend on the frequency and availability of preferred mates at the time of pairing. In Barnacle Geese we found that conflict between the sexes should arise over age and body size criteria. Since low breeding performance was attributable to young females and old males, the benefits of divorce and re-pairing could differ for the sexes at different stages of life. In general, middle-aged partners should be preferred by all birds. However, middle-aged individuals should avoid pairing with younger and older birds, because this would reduce their own reproductive success. We suspect that most geese do achieve their preferred mating options at least some time in their life by employing the strategy of age assortive pairing and lifelong monogamy (Black and Owen 1995).

Coulson (1972) suggested that the compatibility or complementarity of partners may affect the fitness of both individuals. In order to test the compatibility idea, it is necessary to demonstrate that an individual reproduces better when paired to a similar type of mate than to a dissimilar but higher quality individual. We have shown that partners can maximize their reproductive performance by pairing with a similar-sized individual rather than the largest size available (Fig. 5.5). The larger the size-mismatch the lower the reproductive success. This supports the idea that compatibility of mates may be important in determining fitness of the pair. Choudhury *et al.* (1996) also found that success of different pair-size combinations differed in different years and attributed this to the pairs' ability to cope with variation in environmental conditions. The reason for the lower fitness in partners disparate in body size may be related to intra-pair aggression and lack of co-ordination. In Pinyon Jays *Gymnorhinus cyanocephalus*, for example, aggressive encounters are rarer between partners similar in size, and this was suggested to enhance survivorship and pair duration (Marzluff and Balda Chapter 7). In geese, the male is larger than the female, and also more aggressive and dominant. Increasing size disparity of mates could increase stress and risk of injury to females during social display and copulation, which can be quite aggressive (Radesäter 1974; Black and Owen 1988). During copulations, for example, which occur in water, the male climbs onto the female's back and grasps her head. Size disparity between the partners may influence how far the female is submerged under water, how physically stressful, and how accurate the copulation process is. Small females may be overly stressed by social interactions with very large mates, thus inhibiting the flow of reproductive hormones (e.g. Greenberg and Wingfield 1987). In similar sized pairings, males may harass females less, thus enhancing the potential for co-ordination of duties, like vigilance and foraging routines.

Paternity and parental care

Behavioural observations and genetic analyses of parentage have revealed relatively low rates of extra-pair copulations and moderate to high levels of egg dumping and adoption in most geese (Table 5.2). Lamprecht's (1989) study of a semi-captive flock of Bar-headed Geese showed that proximity of pair members increases in spring/summer when females are forming eggs. Yet, when their mates were removed for hours at a time during this period, females did not engage in extra-pair behaviour with the numerous other males in the flock. Lamprecht also documented a correlation between the male's rate of warding off conspecifics and the female's time spent feeding. He argues, therefore, that mate guarding in geese may have been 'shaped in evolution to serve the female's build-up of nutrient reserves', rather than paternity protection. This form of mate guarding probably also limits the possibility of providing prolonged parental care to unrelated offspring. The situation for Black Brant *Branta bernicla* may be different. During 40 h of observations Welsh and Sedinger (1990) observed the highest value of extra-pair copulations for any goose species: 7 of 26 copulations (27%). It is difficult to assess whether extra pair copulation rates actually vary between species because no two data sets were collected in the same way (Table 5.2). However, we anticipate nontrivial variation between species.

Why did geese develop behaviour that accepts foreign eggs (egg rolling/retrieval) or offspring (brood mixing/adoption)? Parents may gain

Table 5.2 Examples of the frequency of extra-pair copulations (EPCs) and extra-pair parentage (EPPs), egg dumping, and adoption in geese; estimates based on number of goslings, nests, broods, or copulations

Species	Nest situation	EPCs and EPPs	Nest parasitism	Adoption	Source
Barnacle Goose	Colonial	Low 0% of goslings	Moderate 6.3% of goslings 11% broods	Moderate 9.4% of goslings 16% of broods	Choudhury *et al.* (1993); also see Larsson et al. (1995)
Bar-headed Goose	Colonial	Low 0 of 62 copulations in 20 h	High 27% of eggs 57% of nests	Does occur	Lamprecht (1987); Weigmann and Lamprecht (1991)
Lesser Snow Goose	Colonial	Low 2.4% of goslings	Moderate 5.6% of goslings 22% of nests	High 40% of broods	Lank *et al.* (1989); Williams (1994)
Emperor Goose	Dispersed	—	Moderate 6% of nests	Does occur	Eisenhauer and Kirkpatrick (1977)[1]
Small Canada Goose	Dispersed	—	Low c. 1% of nests	Low 2% of goslings 4% of broods	MacInnes *et al.* (1974)
Giant Canada Goose	Semi-dispersed	—	—	High 46% of broods	Zicus (1981)
Greater White-fronted Goose	Dispersed	Low 1 of 10 copulations in 356 h	—	—	Ely (1989)
Black Brant	Colonial	Moderate/high 7 of 26 copulations in 60 h	—	—	Welsh and Sedinger (1990)

EPCs are rare in most geese, whether their nests are colonial or dispersed, whereas nest parasitism and adopting foreign offspring occur more often. Barnacle Goose *Branta leucopsis*, Bar-headed Goose *Anser indicus*, Lesser Snow Goose *Anser c. caerulescens*, Emperor Goose *Anser canagicus*, Small Canada Goose *Branta canadensis hutchinsii*, Giant Canada Goose *Branta canadensis maxima*, Greater White-fronted Goose *Anser albifrons frontalis*, Black Brant *Branta bernicla nigricans*.

[1] Reference from Lank *et al.* (1989).

a net benefit from such behaviours, because the costs of caring for foreign offspring seem to be low and the benefits may be high. The costs to the host may include lower hatchability of parasitized broods (Lank *et al.* 1990*a*; Weigmann and Lamprecht 1991) and increased parental duties with increased brood size (Schindler and Lamprecht 1987; Black and Owen 1989*b*). Apparently no costs occur after hatch in terms of growth rate or survival of the host's offspring (Williams 1994; Larsson *et al.* 1995). Larsson *et al.* (1995) found no long-term costs to hosts in terms of their body condition or probability of hatching young in the following year. The benefits to retrieving foreign eggs on the periphery of nests include avoiding total nest failure through predation, because dumped eggs attract predators (Lank *et al.* 1990*b*). Additional eggs and offspring may provide benefits to hosts in terms of diluting the risk of predation, and gaining dominance status and access to more profitable foraging areas (Eadie *et al.* 1988; Black and Owen 1989*a*; Choudhury *et al.* 1993). Longer term advantages are more difficult to measure, but we suspect that parents may benefit from sharing foraging or breeding areas with grown-up offspring due to reduced aggressive conflicts and shared vigilance duties by kin groups (or perceived kin groups, i.e. adopted individuals).

Once paternity is assured by the male's guarding behaviour (even if by default), any costs of additional offspring to hosts are probably offset by the advantages of a larger family unit (Lank *et al.* 1989). Such benefits would be analogous to those in grouse (Watson *et al.* 1994), co-operative fairy-wrens (Russell and Rowley Chapter 8), and jays (Marzluff *et al.* Chapter 7). Counter selection action against an increase in brood size may be that parents have to share foraging areas with additional family members, which would be particularly costly when food is limited. Lessells (1986) showed that Canada Goose females with experimentally enlarged broods initiated clutches later in the next year. Future studies should attempt to identify the precise relationship between the individuals involved in extra-pair copulations, egg dumping and brood mixing, as these behaviours might involve kin or partners in reciprocal altruism. Sherwood (1967) suspected that one brood mixing event in Canada Geese involved a daughter's brood that was surrendered or taken into the mother's brood, whereas Weigmann and Lamprecht (1991) found that kin nests were not favoured by parasitic Bar-headed Goose females. Choudhury *et al.* (1993) detected an 'extra-pair' female dumping her egg in the nest of the male that fertilized the egg.

Summary

In Barnacle Geese, where pairs were together each day, often for life, annual reproductive success increased for the first 7 years pair members were together, even when confounding year and age effects were controlled for. Most geese had one partner throughout their lives. Those that lost a mate and re-paired suffered reproductive costs in the year after the mate change. Although divorce rarely occurred (about 2% annually) there was an indication that pair members switched mates when equal or better alternatives were available; that is, when experienced birds that lost mates were more available. The probability of divorce did not differ significantly between different ages or with prior success in breeding, but decreased with increasing duration of the pair bond. The preferred mate for both sexes was a middle-aged bird, an option that was only achieved by pairing young and remaining with the chosen mate. The preferred mate option in terms of body size was determined by relative sizes of mates. Whereas small birds reproduced best with other small birds, large birds reproduced best with large birds. This pairing strategy was not regularly achieved and its success varied with environmental conditions in different years. Geese often care for unrelated goslings due to low levels of extra-pair copulations and moderate to high levels of egg dumping and adoptions. Although the benefits of lifelong monogamy in geese seem substantial, further investigations are required to reveal how caring for unrelated goslings influences the maintenance of monogamy and mate fidelity.

Acknowledgements

We thank Jürg Lamprecht, Jouke Prop, and Glen Woolfenden for reviewing an earlier draft, Evan Cooch and Richard Pettifor for statistical direction, Sarah Hammond for sharing her analysis, and Paul Shimmings and the Caerlaverock staff for help in the field. We are grateful to B. Ebbinge, A. Follestad, A. D. Fox, K. Kotrschal, J. Madsen, and L. Nilsson for providing unpublished data on goose partnerships.

References

Black, J. M. (1988). Preflight signalling in swans: a mechanism for group cohesion and flock formation. *Ethology*, **79**, 143–157.

Black, J. M. and Owen, M. (1988). Variations in pair bond and agonistic behaviors in Barnacle Geese on the wintering grounds. In *Wildfowl in winter* (ed. M. Weller), pp. 39–57. University of Minnesota Press, Minneapolis.

Black, J. M. and Owen, M. (1989*a*). Parent–offspring relationships in wintering barnacle geese. *Animal Behaviour*, **37**, 187–98.

Black, J. M. and Owen, M. (1989*b*). Agonistic behaviour in goose flocks: assessment, investment and reproductive success. *Animal Behaviour*, **37**, 199–209.

Black, J. M. and Owen, M. (1995). Reproductive performance and assortative pairing in relation to age in barnacle geese. *Journal of Animal Ecology*, **64**, 234–44.

Black, J. M., Carbone, C., Owen, M., and Wells, R. (1992). Foraging dynamics in goose flocks: the cost of living on the edge. *Animal Behaviour*, **44**, 41–50.

Choudhury, S. and Black, J. M. (1993). Mate selection behaviour and sampling strategies in geese. *Animal Behaviour*, **46**, 747–57.

Choudhury, S. and Black, J. M. (1994). Barnacle geese choose familiar mates from early-life. *Animal Behaviour*, **48**, 81–8.

Choudhury, S., Black, J. M., and Owen, M. (1992). Do barnacle geese pair assortatively? Lessons from a long-term study. *Animal Behaviour*, **44**, 171–3.

Choudhury, S. Jones, C. M., Black, J. M., and Prop, J. (1993). Adoption of young and intra-specific nest parasitism in Barnacle Geese. *Condor*, **95**, 860–8.

Choudhury, S. Black, J. M., and Owen, M. (1996). Body size, reproductive success and compatibility in Barnacle Geese. *Ibis*, In press.

Coulson, J. C. (1972). The significance of the pair bond in the Kittiwake Gull. In *Proceedings of International Ornithological Congress*, **15**, 424–33.

Davies, N. B. (1989). Sexual conflict and the polygyny threshold. *Animal Behaviour*, **38**, 226–34.

Drent, R. H. and Prins, H. H. T. (1987). The herbivore as prisoner of its food supply. *Disturbance in grasslands* (ed. J. van Andel *et al.*), pp. 131–47. Dr W. Junk, Dordrecht.

Eadie, J. McA., Kehoe, F. P., and Nudds, T. D. (1988). Pre-hatch and post-hatch brood amalgamation in North American Anatidae: a review of hypotheses. *Canadian Journal of Zoology*, **66**, 1709–21.

Ely, C. R. (1989). Extra-pair copulation in the Greater White-fronted Goose. *Condor*, **91**, 990–1.

Ens, B. J., Safriel, U. N., and Harris, M. P. (1993). Divorce in the long-lived and monogamous oystercatcher. *Animal Behaviour*, **45**, 1199–217.

Eisenhauer, D. I. and Kirkpatrick, C. M. (1977). Ecology of the Emperor Goose in Alaska. *Wildlife Monographs*, 57, 1–62.

Forslund, P. and Larsson, K. (1991). The effect of mate change and new partner's age on reproductive success in the barnacle goose, *Branta leucopsis*. *Behavioral Ecology*, **2**, 116–22.

Forslund, P. and Larsson, K. (1995). Intraspecific nest parasitism in the barnacle goose: behavioural tactics of parasites and hosts. *Animal Behaviour*, **50**, 509–17.

Greenberg, N. and Wingfield, J. C. (1987). Stress and reproduction: reciprocal relationships. In *Hormones and reproduction in fishes, amphibians and reptiles* (ed. D. O. Norris and R. E. Jones), pp. 461–501. Plenum Press, New York.

Hammond, S. (1990). An investigation of intra-flock behaviour of Barnacle Geese. Unpublished M.Sc. thesis. University of Reading.

Hausberger, M. and Black, J. M. (1990). Do females turn males on and off in barnacle goose social display? *Ethology*, **84**, 232–8.

Johnson, J. C. and Raveling, D. G. (1988). Weak family associations in Cackling Geese during winter: effects of body size and food resources on goose social organisation. In *Wildfowl in winter* (ed. M. Weller), pp. 71–89. University of Minnesota Press, Minneapolis.

Lamprecht, J. (1986). Social dominance and reproductive success in a goose flock (*Anser indicus*). *Behaviour*, **97**, 50–65.

Lamprecht, J. (1987). Female reproductive strategies in bar-headed geese. *Behavioral Ecology and Sociobiology*, **21**, 297–305.

Lamprecht, J. (1989). Mate guarding in geese: awaiting female receptivity, protection of paternity or support of female feeding? In *The sociobiology of sexual and reproductive strategies* (ed. A. E. Rasa, C. Vogel, and E. Voland), pp. 48–66. Chapman and Hall, London.

Lank, D. B., Mineau, P., Rockwell, R. F., and Cooke, F. (1989). Intraspecific nest parasitism and extra-pair copulation in lesser snow geese. *Animal Behaviour*, **37**, 74–89.

Lank, D. B., Rockwell, R. F., and Cooke, F. (1990*a*). Fitness consequences and frequency-dependent success of alternative reproductive tactics of female lesser snow geese. *Evolution*, **44**, 1436–53.

Lank, D. B., Bousfield, M. A., Cooke, F., and Rockwell, R. F. (1990*b*). Why do geese adopt eggs? *Behavioral Ecology*, **2**, 181–7.

Larsson, K., Tegelstöm, H. and Forslund, P. (1995). Intraspecific nest parasitism and adoption of young in the barnacle goose: effects on survival and reproductive performance. *Animal Behaviour*, **50**, 1349–60.

Lessells, C. M. (1986). Brood size in Canada Geese: a manipulation experiment. *Journal of Ecology*, **55**, 669–89.

MacInnes, C. D. and Dunn, E. H. (1988). Components of clutch size variation in arctic-nesting Canada Geese. *Condor*, **90**, 83–9.

MacInnes, C. D., Davis, R. A., Jones, R. N., Lieff, B. C., and Pakulak, A. J. (1974). Reproductive efficiency of McConnell River small Canada Geese. *Journal of Wildlife Management*, **38**, 686–707.

McLandress, M. R. and Raveling, D. G. (1981). Hyperphagia and social behavior in Canada Geese prior to spring migration. *Wilson Bulletin*, **93**, 310–24.

Martin, K., Cooch, E. G., Rockwell, R. F., and Cooke, F. (1985). Reproductive performance in lesser snow geese: are two parents essential? *Behavioral Ecology and Sociobiology*, **17**, 257–263.

NAG (1993). *GLIM 4, the statistical system for generalized linear interactive modelling (2nd edn)* (ed. B. Francis, M. Green, and C. Payne). Numerical Algorithms Group, Oxford.

Owen, M. (1980). *Wild geese of the world*. Batsford, London.

Owen, M. and Black, J. M. (1989*a*). Factors affecting the survival of barnacle geese on migration from the breeding grounds. *Journal Animal Ecology*, **58**, 603–18.

Owen, M. and Black, J. M. (1989*b*). Barnacle Goose. In *Lifetime Reproduction in Birds* (ed. I. Newton), pp. 349–62. Academic Press, London.

Owen, M. and Black, J. M. (1991*a*). Geese and their future fortune. *Ibis*, **133**, S28–S35.

Owen, M. and Black, J. M. (1991*b*). The importance of migration mortality in non-passerine birds. In *Bird population studies—relevance to conservation and management* (ed. C. M. Perrins, J. D. Lebreton, and G. J. M. Hirons), pp. 360–72). Oxford University Press, Oxford.

Owen, M., Black, J. M., and Liber, H. (1988). Pair bond duration and the timing of its formation in Barnacle Geese. In *Wildfowl in winter* (ed. M. Weller), pp. 23–38. University of Minnesota Press, Minneapolis.

Paine, C. P. (1992). Costs of parental care and the importance of biparental care in Canada Geese. Unpublished Ph.D. thesis. Southern Illinois University, Carbondale.

Prevett, J. P. and MacInnes C. D. (1980). Family and other social groups in Snow Geese. *Wildlife Monographs*, **71,** 1–46.

Prop, J. and Loonen, M. (1988). Goose flocks and food exploitation: the importance of being first. *Acta Congress International Ornithology*, **19,** 1878–87.

Prop, J. and de Vires, J. (1993). Impact of snow and food conditions on the reproductive performance of Barnacle Geese. *Ornis Scandinavica*, **24,** 110–21.

Prop, J. and Vulink, T. (1992). Digestion by Barnacle Geese in the annual cycle: the interplay between retention time and food quality. *Functional Ecology*, **6,** 180–9.

Prop, J., van Eerden, M. R., and Drent, R. (1984). Reproductive success of the Barnacle Goose in relation to food exploitation on the breeding grounds, western Spitsbergen. *Norsk Polarinstitutt Skrifter*, **181,** 87–117.

Radesäter, T. (1974). Form and sequential associations between the triumph ceremony and other behavior patterns in the Canada Goose *Branta canadensis* L. *Ornis Scandinavica*, **5,** 87–101.

Raveling, D. G. (1988). Mate retention in giant Canada geese. *Canadian Journal of Zoology*, **66,** 2766–8.

Rockwell, R. F., Cooch, E. G., Thompson, C. B., and Cooke, F. (1993). Age and reproductive success in female lesser snow geese: experience, senescence and the cost of philopatry. *Journal of Animal Ecology*, **62,** 323–33.

Schindler, M. and Lamprecht, J. (1987). Increase in parental effort with brood size in a nidifugous bird. *Auk*, **104,** 688–93.

Schneider, J. S. and Lamprecht, J. (1990). The importance of biparental care in a precocial, monogamous bird, the bar-headed goose (*Anser indicus*). *Behaviorial Ecology and Sociobiology*, **27,** 415–19.

Scott, D. K. (1980). Functional aspects of the pair bond in wintering Bewick's swans (*Cygnus columbianus bewickii*). *Behavioral Ecology and Sociobiology*, **7,** 323–7.

Sedinger, J. S. and Raveling, D. G. (1990). Parental behavior of Cackling Canada Geese during brood rearing: division of labor within pairs. *Condor*, **92,** 174–81.

Sherwood, G. A. (1967). Behavior of family groups of Canada Geese. *Thirty-second North American Wildlife Conference*, 340–55.

Teunissen, W., Spaans, B., and Drent, R. H. (1985). Breeding success in the Brent in relation to individual feeding opportunities during spring staging in the Wadden Sea. *Ardea*, **73,** 109–19.

Warren, S. M., Fox, A. D., and Walsh, A. (1993). Extended parent–offspring relationships amongst the Greenland White-fronted Goose *Anser albifrons flavirostris*. *Auk*, **110,** 145–8.

Watson, A., Moss, R., Parr, R. Moundford, M. D., and Rothery, P. (1994). Kin landownership, differential aggression between kin and non-kin, and population fluctuations in red grouse. *Journal of Animal Ecology*, **63,** 39–50.

Weigmann, C. and Lamprecht, J. (1991). Intraspecific nest parasitism in bar-headed geese, *Anser indicus*. *Animal Behaviour*, **41,** 677–88.

Welsh, D. and Sedinger, J. S. (1990). Extra-pair copulations in Black Brant. *Condor*, **92,** 242–4.

Williams, T. D. (1994). Adoption in a precocial species, the lesser snow goose: inter-generational conflict, altriusm or a mutually beneficial strategy? *Animal Behaviour*, **47**, 101–7.

Zicus, M. C. (1981). Canada Goose brood rearing behavior and survival estimates. *Wilson Bulletin*, **93**, 207–17.

6 *Mate fidelity in swans: an interspecific comparison*

EILEEN C. REES, PIA LIEVESLEY, RICHARD A. PETTIFOR, AND CHRISTOPHER PERRINS

Introduction

Of the three swan species that occur in Britain, the Bewick's and Whooper Swans are migratory and the Mute Swan resident throughout the year. The Bewick's Swan *Cygnus columbianus bewickii* (see drawing above) breeds on tundra in the Russian arctic, and birds from the western population fly some 4000 km each autumn to wintering sites located primarily in the Netherlands, Britain, and Ireland. British-wintering Whooper Swans *Cygnus cygnus*, on the other hand, are predominantly from the Icelandic breeding population, and have a comparatively short migratory journey of some 800 km between the breeding and wintering grounds. They nest in a variety of habitats ranging from lowland marshes, amidst areas of intensive farming, to upland sites, where pools and lakes in the glacial moraine support aquatic vegetation, thus providing a food supply for the birds. Differences between Bewick's and Whooper Swans in the timing of their arrival and departure in Britain each season reflect the different migratory distances undertaken; Whooper Swans generally return to their wintering sites earlier than the Bewick's Swans in autumn, and leave later in the spring. In Britain the Mute Swan *Cygnus olor* is very sedentary; only some 3% of individuals move more than 100 km from their natal area (Birkhead and Perrins 1986). It occupies a wide range of waters including ponds, rivers, and gravel pits, and often frequents urban areas.

·Long-term studies of all three species indicate that the swans are generally monogamous, and that offspring remain with their parents during their first winter, although Mute Swan cygnets may begin to leave their parents in the autumn (Scott 1980*a*; Black and Rees 1984; Birkhead and Perrins 1986). Here we examine the association between pair bond characteristics and reproductive success, and assess why some partnerships are more fruitful than others. Differences between the species in the migratory distances undertaken, and also in the time available for breeding at different latitudes, may give rise to variation in the restrictions upon the migratory and reproductive cycles. We also examine, therefore, the cost of finding a new mate (measured in terms of breeding success) and consider whether a monogamous mating system appears to be more advantageous for one species than another. Mate change may be more costly for Bewick's Swans since the close synchrony of the migratory and breeding cycles allows only limited time for courtship and pair formation before the onset of the breeding season. Mate fidelity may be less important for Whooper Swan reproductive success, and of least importance to Mute Swans, since greater flexibility in the timing of the breeding season enables the birds to find a new mate and breed in a single summer. Pair duration may be significant in determining reproductive success for all three species, however, since co-operation between partners in defending the breeding territory and rearing the brood is likely to be important at all latitudes.

Background, study sites, and procedures

Data were obtained for individual Bewick's and Whooper Swans wintering at Wildfowl and Wetlands Trust Centres, and for Mute Swans breeding on the River Thames in Oxfordshire, UK. Individual Bewick's Swans wintering at Slimbridge, Gloucestershire, UK, have been identified by the variation in their black-and-yellow bill patterns since 1963 (P. Scott 1966), and since 1967 by plastic leg-rings readable with a telescope at distances of up to 200 m (Rees *et al.* 1991). Individual Whooper and Mute Swans were identified by ring codes only. The high level of site fidelity shown by swans of all three species, with the same individuals returning to particular wintering sites and breeding territories over several years, made it possible to monitor changes in their pairing and reproductive success. The three swan populations are described further in Table 6.1.

Swans were thought to be paired if they maintained close proximity to each other, co-operated in aggressive encounters, made contact signals, and co-ordinated their movements within the study area. First pairing may occur in the second summer (1 year old) in Whooper and Mute Swans (Minton 1968; Black and Rees 1984), whereas Bewick's Swans usually do not pair until the third summer (2 years old, Evans 1979). The average age at which Bewick's Swans were first recorded as paired

Table 6.1 Description of the three swan populations

	Bewick's	Whoopers	Mutes
Main study site	Slimbridge, Gloucestershire, UK	Caerlaverock, Dumfries, UK	River Thames Oxfordshire, UK
Coordinates	51°44′N 02°25′W	54°58′N 03°26′W	51°45′N 01°15′W
First year of study	1963	1980	Early 1960s
Population	NW European	Icelandic–British	British
Migration distance (km)	4000	800	<150
Population size	17000	19000	26000
Study sample size (No. of birds with at least 1 mate)	2501	607	398
Study sample size (No. of pair bonds)	1370	242	254
Maximum longevity (years)	26	14	19
Mean life span (SD) (years)	5.4 (4.2)	3.2 (2.4)	8.1 (4.4)
Age 1st pair (SD) (years)	3.4 (1.4)	2.6 (1.4)	2.5 (1.4)
Age 1st breed (SD) (years)	6.6 (2.5)	4.9 (1.9)	3.8 (1.2)
Annual mortality (%)	16.0	14.9	18.9

Bewick's and Whooper Swans studied at Welney, Norfolk, UK, and at Martin Mere, Lancashire, UK, since 1979 and 1990 respectively, are also included in the analyses. Mean age at first pairing and first breeding are given for known-age swans followed through to pairing and breeding; mean minimum life span was estimated for paired swans known or thought to have died. The life span figures are minimum values because in some cases the birds were first recorded as adults (at least 1 year old for Whooper and Mute Swans or at least 2 years old for Bewick's Swans), so their precise ages were not known. The lower longevity estimate for Whooper Swans may be due to the more recent development of the Whooper Swan study, and hence a shorter run of data, rather than to an actual shorter mean life span for these birds. Pair bonds for Whooper Swans with unringed mates were excluded from pair duration analyses since it was not possible to confirm the identity of the mate. Age at first breeding is the age at which at least one cygnet is successfully reared to fledging.

therefore was one year later than in both Whooper and Mute Swans (Table 6.1). Courtship and pair formation have been recorded in winter for both Whooper and Mute Swans (Minton 1968; Scott 1978). Although Bewick's Swans may form temporary liaisons in winter, definite pair formation has not been recorded in the wintering flock (Rees and Bowler 1991).

Paternity and parental care

Offspring of the migratory Bewick's and Whooper Swans remain with their parents during their first winter; the prolonged parental care is important in protecting offspring from feeding competition with other individuals in the flock (Scott 1980*b*). The break-up of Mute Swan families may commence in autumn, once the young have fledged, although the precise timing varies greatly between broods (Scott 1984). Reproductive success in Bewick's and Whooper Swans therefore is measured as the number of cygnets associating with a pair upon their arrival at the wintering sites, and for Mute Swans as the number of cygnets fledged in September.

Extra-pair matings and intraspecific brood parasitism are rarely recorded in swan species, perhaps because paired birds co-operate in defending the breeding territory against other swans (Coleman and Minton 1979). Extensive DNA-fingerprinting found no evidence for extra-pair fertilization in Bewick's or Whooper Swans (Meng 1990; D. T. Parkin unpublished data), although a few adoptions have been confirmed in all three species (Rees *et al.* 1990; Meng and Parkin 1991).

Data analysis

The majority of swan pairs do not succeed in raising young each year (Evans 1979; Black and Rees 1984). Most analyses of reproductive success therefore were carried out using log-linear models with Poisson error structures in GLIM (versions 3.77 and 4; NAG 1977, 1993). We controlled for the year of observation to allow for the effects of climatic or other variables influencing the reproductive success of the whole population (Evans 1979; Black and Rees 1984). The body size of the male and female were also examined since these were thought likely to influence reproductive success for the duration of the pair bond; principal component (PC1) values were calculated from skull and tarsus measurements for Bewick's and Whooper Swans, whereas skull measurements were used for Mute Swans. Many individuals included in the analyses bred over several years. Since the breeding attempts for individual pairs therefore were not independent, the results were verified by repeating the analyses using only one year's data for each individual, which was selected at random from the original database. In all cases the results obtained for data selected at random were similar to those obtained using the complete data sets.

Results

The pair bond

All the Bewick's, Whooper, and Mute Swans considered in the study were monogamous (1370, 242, and 254 pairs, respectively). Trios and foursomes were occasionally recorded for the Bewick's Swans, but these were usually due to prolonged parent–offspring and sibling–sibling associations (Evans 1979) and were therefore excluded from the analyses. Other authors have recorded rare incidents of polygamy in Mute Swans (Hilprecht 1970; P. Scott and The Wildfowl Trust 1972), which usually occur whilst the female is incubating and the male has the opportunity to establish a second association.

Most swans had only one mate during their known lifetime, although not surprisingly birds with longer life histories were more likely to have several mates (Tables 6.1 and 6.2). Individual Whooper and Mute Swans were seen with up to four mates, and one Bewick's Swan had at least

Table 6.2 Percentage of birds recorded with 1 to 3+ mates during their known lifetime (data for both sexes combined)

No. years data	Species	n	Percentage of birds with:			χ^2 values	
			1 mate	2 mates	3+ mates	a	b
1–5	Bewick's	2049	96.5	3.4	0.1		
	Whoopers	515	93.0	6.8	0.2	78.58	17.16
	Mutes	291	84.2	14.4	1.4	***	***
6–10	Bewick's	317	71.9	23.3	4.7		
	Whoopers	77	44.2	45.5	10.4	21.95	7.01
	Mutes	70	70.0	18.6	11.4	***	*
11+	Bewick's	135	46.7	32.6	20.7		
	Whoopers	15	33.3	40.0	26.6	1.04	0.32
	Mutes	35	42.9	34.3	22.8	NS	NS
All birds	Bewick's	2501	90.7	7.5	1.8		
	Whoopers	607	85.3	12.5	2.1	59.46	16.61
	Mutes	396	78.0	16.9	5.1	***	***

Individual Bewick's Swans were recorded with up to seven mates; Whooper and Mute Swans with up to four mates. The χ^2 for interspecific differences: (a) in the number of birds with one mate compared with those with two or more mates; (b) in the number of birds seen with one or two mates

seven mates. Mute Swans changed mates more frequently than the other two species; of swans seen for up to 5 years, 15.8% of Mute Swans, 7.0% of Whooper Swans, and only 3.5% of Bewick's Swans in this category were recorded with two or more mates (χ^2 = 78.58, df = 2, P < 0.001 for swans seen for 1–5 years, Table 6.2). Bewick's Swans with longer life histories also had fewer mates than Whooper or Mute Swans seen for similar periods, although there was no statistical difference between species in the number of mates recorded for swans seen for at least 11 years (Table 6.2). A high proportion of Whooper Swans seen for at least 6 years were recorded with two mates.

Pair duration recorded for swans seen for at least 5 years varied from 1 to 19 years in Bewick's Swans, 1 to 9 years in Whooper Swans and 1 to 16 years in Mute Swans. There were interspecific differences in the percentage of these birds that spent more than 2 years with the same mate; 80.2% of Bewick's Swan pairings, 44.8% of Mute Swan pairings, and only 42.9% of Whooper Swan pairings lasted for more than 2 years. Mean pair duration was higher in Bewick's Swans (5.1, SD 3.6 years, n = 724) than in Whooper Swans (2.7, SD 2.1, n = 237) or Mute Swans (3.3, SD 3.1, n = 354), although the comparatively short duration of the Whooper Swan study may have influenced the results. Mean pair duration decreased with each subsequent mate in the migratory species with the exception of three Bewick's Swans having long pair bonds (>5 years) with their fourth mates; in Mute Swans there was no decrease in pair duration between the second and third mates. Pair duration with both the first and second mates was significantly longer in Bewick's Swans (first mate 5.9, SD 3.6, second mate 3.3, SD 2.8) than in Whooper Swans (2.94, SD 2.22; 2.28, SD 1.85) and Mute Swans (3.5, SD 3.3; 2.6, SD 2.2) ($F = 72.44$, $df = 2{,}1003$, $P < 0.001$ for the first mate; $F = 4.61$, $df = 2{,}320$, $P = 0.011$ for the second mate; one-way ANOVAs).

Mate familiarity effect

Reproductive success increased with pair duration for up to 11 years of age in Bewick's Swan, 6 years in Whooper Swans, and 5 years in Mute Swans before levelling thereafter. Log-linear models confirmed that pair duration, and the square of pair duration, were associated with reproductive success in all three species after controlling for the effects of year and age (Fig. 6.1). Male and female age were not associated with reproductive success in Whooper and Mute Swans once the effects of pair duration were included in the model.[1] Age had a significant effect on Bewick's Swan reproductive success, however, in addition to the effects of pair duration, which may reflect the importance of greater breeding experience in older birds. The interactions of year with age and year with pair duration were significantly associated with reproductive success in both Bewick's and Whooper Swans, suggesting that age and long-term monogamy were more beneficial in some years (for example poor breeding seasons for the population) than in others for the migratory species. Since, on average, pair duration tends to increase with age, which potentially creates an unbalanced design, we repeated the analyses for each of the more frequent age-classes separately, and obtained similar results.

In a further analysis to differentiate between a mate familiarity effect and a measure of experience, we added the variable 'number of mates' to the previous model. After controlling for the other variables (including pair duration), reproductive success increased significantly with each subsequent mate in Whooper Swans only, although the same trend was apparent in the other species (Table 6.3). The interaction of year with mate number was significant in both Whooper and Bewick's Swans, but not in Mute Swans, indicating that mate number was more important in some years than in others for the migratory species. It is not certain, however, whether this was due simply to the greater breeding experience

[1]The GLIM model on number of fledglings included the following variables: year, age (and age^2) of both partners, associated interaction terms, and finally pair duration and the square of pair duration. The year effect was significant for all three species (ΔD ranging from 39 to 345; Δdf = 13 to 39; $P < 0.05$ to 0.001), as was pair duration (ΔD = 5 to 45, $\Delta df = 1$, $P < 0.05$ to 0.001) and pair duration squared (ΔD = 6 to 20, $\Delta df = 1$, $P < 0.05$ to 0.001). The effects of both partner's age and age^2 was significant only for Bewick's Swans (female age $\Delta D = 15$, $\Delta df = 1$, $P < 0.001$; female age^2 $\Delta D = 5$, $\Delta df = 1$, $P < 0.05$; male age $\Delta D = 20$, $\Delta df = 1$, $P < 0.001$; male age^2 $\Delta D = 12$, $\Delta df = 1$, $P < 0.001$). There was also a significant interaction between year and age (e.g. Whooper Swan females $\Delta D = 26$, $\Delta df = 14$, $P < 0.05$; Bewick's Swan females $\Delta D = 93$, $\Delta df = 39$, $P < 0.001$) and year and pair duration (e.g. Whooper Swan females $\Delta D = 55$, $\Delta df = 11$, $P < 0.001$; Bewick's Swan females $\Delta D = 102$, $\Delta df = 27$, $P < 0.001$).

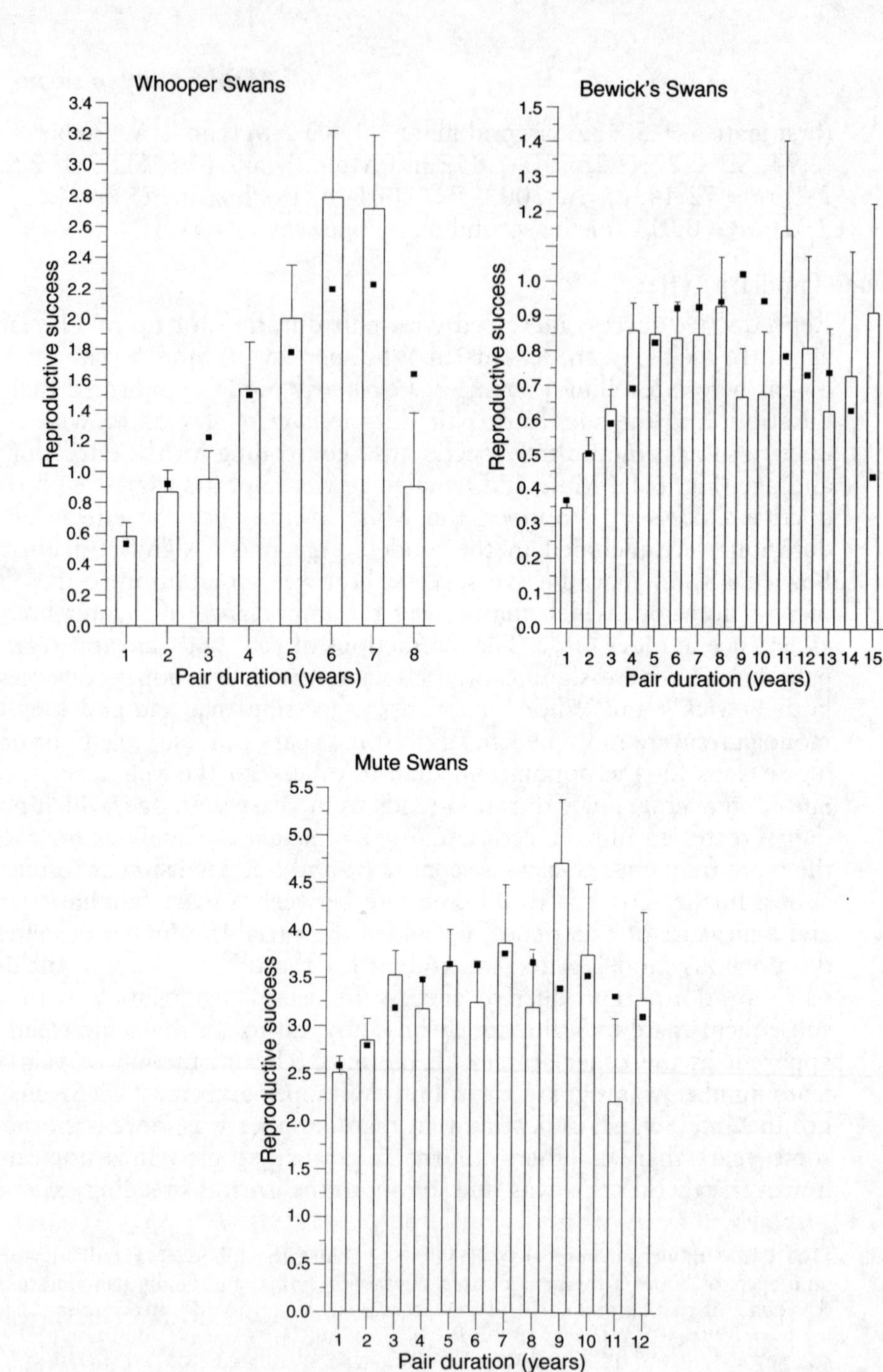

Fig. 6.1 Association between pair duration and reproductive success in Whooper, Bewick's, and Mute Swans. Reproductive success was the number of cygnets fledged for Mute Swans, and the number of cygnets associating with the pair on arrival at the winter site for Bewick's and Whooper Swans. Histograms and SE bars show mean reproductive success and black squares denote fitted values after controlling for year and age effects.

Table 6.3 Mean reproductive success of swans with their first and subsequent mates

Species	Sex	Mate no.	Mean no. Cygnets	SD	*n*	Δ*D*	Δ*df*	*P*
Bewick's	Female	1	0.56	1.09	2868			
		2	0.62	1.07	206	0.21	2	NS
		3+	0.64	1.09	97			
Bewick's	Male	1	0.56	1.09	2725			
		2	0.59	1.06	361	2.02	2	NS
		3+	0.71	1.10	85			
Whooper	Female	1	0.94	1.46	347	3.90	1	<0.05
		2+	1.10	1.64	79			
Whooper	Male	1	0.94	1.48	355	5.26	1	<0.05
		2+	1.14	1.52	71			
Mute	Female	1	3.16	2.38	365	0.26	1	NS
		2+	3.26	2.44	85			
Mute	Male	1	3.18	2.45	367	0.33	1	NS
		2+	3.18	2.14	83			

The log-linear analyses compare the reproductive success of swans with their first and subsequent mates, having controlled for the effects of year, age, and pair duration, and also for the interaction of year with these variables. For Year*MateNumber interactions, ΔD = 44 and 26, Δdf = 11 and 8, $P < 0.001$ and $P < 0.01$ for female and male Whooper Swans; ΔD = 109 and 78, Δdf = 52 and 46, $P < 0.001$ in each case for female and male Bewick's Swans; ΔD = 17 and 15, Δdf = 16, NS for female and male Mute Swans.

of swans seen with their second or third mates, or to an improvement in the swans' ability to select a better quality or more compatible partner (see below).

Dominance rank and pair duration

In a search for the mechanism behind the reproductive improvement, we assessed the dominance rank of 23 Bewick's Swan pairs whose pair durations ranged from 1 to 16 years; pair bonds exceeding 10 years were grouped into a 10+ year category. Rank score was calculated from a matrix of wins and losses for each diad during encounters in foraging flocks (after Scott 1978). Initial results found that dominance rank increased linearly with pair duration ($F = 1.96$, $df = 9,65$, $P = 0.06$), suggesting that co-operation between the pair during aggressive encounters may improve over several years. Further analysis to control for variation in the breeding success of the pair, however, found that there was no association between pair duration and dominance rank for years when the birds accompanied by young were treated separately ($F = 1.08$, $df = 8,23$, $P = 0.41$ when recorded as families; $F = 1.27$, $df = 9,33$, $P = 0.29$ when recorded as paired but without young). The increase in dominance rank associated with pair duration therefore may be attributable to the improved breeding success of the pair rather than to improved fighting ability, as occurs in geese (Black and Owen 1989).

Consequences of mate change

Short-term effects on reproduction

The mean time taken to re-pair following the loss of a mate was 2.6 years (SD 1.7, range 1–9, n = 275) for Bewick's Swans and 1.9 years (SD 1.2, range 1–7, n = 105) for Whooper Swans (test between species: W = 56 136, P = 0.001, Mann–Whitney U-test). There was no significant difference between the sexes for either species. Data on time to replacing a mate have not been collected for Mute Swans, but it is likely to occur within weeks. The mean time taken to resume breeding after replacing a mate was 2.3 years for Bewick's Swans (SD 1.5, range 1–8, n = 133), 1.5 years for Whooper Swans (SD 0.8, range 1–4, n = 39) and 1.5 years (SD 0.8, range 1–4, n = 109) for Mute Swans. Bewick's Swans took significantly longer to breed with a new partner than both Whooper Swans and Mute Swans (W = 18 737, P = 0.0014 and W = 12 379, $P < 0.0001$, respectively); Whooper and Mute Swans did not differ (W = 2958, P = 0.82). There were no differences between the sexes in the time taken to replace mates or breed. Reproductive success of Bewick's Swans was lower in the first year after pairing with a new mate compared to the average value for those that were paired with the same partner, after controlling for year and age effects (adjusted means = 0.4, SD 0.2 cygnets with a new mate and 0.7, SD 0.4 cygnets with the same mate; log-linear model, quasi-likelihood ratios: F = 7.33, df = 1,1886 $P < 0.01$). Whooper Swan breeding success was not significantly reduced in the year following a change of mate (adjusted means = 0.6, SD 0.3 and 1.3, SD 0.66 cygnets for birds with a new mate and the same mate, respectively; F = 1.00, df = 1,298 NS). Mute Swans produced significantly fewer fledglings after replacing a mate when the data set included birds that had no previous breeding experience (adjusted means = 2.5 cygnets and 3.3 cygnets for birds with new mates and the same mates, respectively; log-linear models $\chi^2 = 6.59$, df = 1, $P < 0.01$), but not when the data set was restricted to birds that had been previously paired ($\chi^2 = 2.74$, df = 1, NS).

Long-term effects on reproduction

Mate change did not appear to incur a long-term cost to an individual's lifetime reproductive success, since its productivity (measured as the total number of cygnets raised in its lifetime divided by the number of years seen) was not affected by the number of mates recorded during its known history.[1]

Analysis of variance in the productivity of Bewick's Swans seen for at least 10 years, and of Whooper and Mute Swans seen for at least 5 years,

[1]GLIM analysis of number of cygnets/lifetime in years. Bewick's Swan females ΔD = 0.66, Δdf = 3, NS; males ΔD = 0.64, Δdf = 3, NS; Whooper Swan females ΔD = 1.97, Δdf = 2, NS; males ΔD = 2.86, Δdf = 1, NS).

Table 6.4 Reproductive success of swans with different numbers of mates (individuals seen for a minimum of 10 years for Bewick's Swans, minimum of 5 years for Whooper and Mute Swans)

Mean no. cygnets raised per year	Number of mates in lifetime				ANOVA		
	1	2	3	4	*F*	*df*	*P*
Bewick's Males							
Mean	0.82	0.57	0.71	0.70	1.18	3, 84	0.32
SD	0.61	0.52	0.56	0.61			
n	33	40	8	7			
Bewick's Females							
Mean	0.57	0.74	0.59	0.52	0.56	3, 73	0.64
SD	0.56	0.45	0.29	0.31			
n	45	18	8	7			
Whooper Males							
Mean	0.89	0.83	1.30	—	0.54	2, 68	0.59
SD	1.05	0.99	0.21	—			
n	43	22	6	—			
Whooper Females							
Mean	1.16	0.73	0.96	1.04	0.89	3, 66	0.45
SD	1.20	0.91	1.02	0.24			
n	33	30	4	3			
Mute Males							
Mean	2.91	3.12	3.00	—	0.08	2, 36	0.93
SD	1.86	1.14	1.36	—			
n	20	14	5	—			
Mute Females							
Mean	2.85	3.11	2.54	—	0.29	2, 40	0.75
SD	1.82	1.80	1.02	—			
n	22	12	9	—			

One-way analysis of variance is used to test for differences in the mean number of cygnets raised during the known lifetime for birds with 1, 2, 3, or 4 mates.

to control for any bias in the results due to the inclusion of birds with short life spans, similarly showed that although mean reproductive success tended to diminish with the number of mates, this was not statistically significant (Table 6.4).

Divorce

Most pair bonds ended by the death or disappearance of one or both members of the pair: 94.2% of 276 Whooper Swan pairs, 96.3% of 163 Mute Swan pairs, and 100% of 919 Bewick's Swan pairings known to have ended. Although swan species appear to improve their reproductive success by remaining with the same partner, 32 (5.8%) Whooper Swans (16 males and 16 females) and four (3.7%) Mute Swans paired with a new mate whilst their original mate was alive. Two Whooper Swans divorced twice, bringing the total number of Whooper Swan divorces to

17. The probability of divorce was 5.6% in 305 pair-years for Whooper Swans, 0.7% in 603 pair-years for Mute Swans, and 0% in 2220 pair-years for Bewick's Swans.

Of 16 Whooper Swan pairs that divorced, and whose reproductive success was known, four (25.0%) had bred in the year prior to divorce. By comparison, of 360 Whooper Swan pairs that did not divorce between one year and the next, 47.5% had bred. Similarly, 2.3% of 175 Whooper Swans that bred in one year had divorced by the next, compared with 6.0% of 210 pairs that had not bred. However, the frequency of divorce was not related to the breeding success of the pair in the previous year ($\chi^2 = 3.09$, $df = 1$, NS). Annual productivity recorded prior to the divorce was 0.6 cygnets (SD 1.2, $n = 16$), compared with 1.3 cygnets (SD 1.32 $n = 8$) for females with their new mates and no cygnets for males with their subsequent mate ($n = 8$); the results did not reach statistical significance ($F = 1.73$, $df = 23$, $P = 0.20$, for females and $F = 1.81$, $df = 21,1$, $P = 0.19$ for males, one-way ANOVAs). There were no significant differences in size between the divorced and subsequent mates of either sex. There was also no evidence to suggest that one sex was more likely than the other to re-pair after separation from the first mate; nine males and nine females that divorced were subsequently seen with new partners, as opposed to remaining unpaired.

Dominance rank and mate change

Changes in dominance rank following a change of mate were investigated for nine Bewick's Swans (six males and three females) where dominance rank with both partners was known over several years. After removing the effects of brood size we found a significant increase in the dominance rank following a change of mate ($F = 6.01$, $df = 1,20$, $P < 0.05$), with mean dominance ranks of 0.4 (SD 0.3, $n = 10$) with the first mate and 0.7 (SD 0.3, $n = 13$) with the second mate. Unfortunately, we could not control for other confounding variables. Perhaps the higher dominance rank recorded with the second mate was due to an improvement with age or in the birds' ability to select a better quality or more compatible partner.

Pair characteristics and reproduction

The reproductive success of a pair may be influenced not only by the characteristics of the individuals but also by a combination of their qualities. The combination of male and female characteristics may therefore describe the 'quality' of the pair, where pair bond quality is measured in terms of their reproductive success. Here we consider for paired birds whether their individual and combined ages and body sizes affect their reproductive success, and also whether birds appear to select mates that are likely to improve their reproductive performance.

Age

Reproductive success increased rapidly with age during the first 6–7 years for Bewick's and Whooper Swans, and during the first 4 years for Mute Swans, before reaching a plateau thereafter. To determine whether the combination of male and female age influenced the reproductive success of a pair, the birds were grouped into three age categories; young, middle-aged, and old. Categories were based on reproductive performance: the improving years (up to 5 years for Bewick's and Whooper Swans; up to 4 years for Mute Swans), the middle years (6–15 years in Bewick's Swans, 6–9 years in Whooper Swans, and 5–11 years in Mute Swans), and the declining years (16+ years in Bewick's Swans, 10+ years in Whooper Swans, and 12+ years in Mute Swans).

The combination of male and female age groupings influenced reproductive success in Bewick's Swans and Whooper Swans, but not in Mute Swans, after controlling for year and pair duration.[1] In Bewick's Swans, young males paired with young females had the lowest breeding success, whereas middle-aged and old birds of both sexes were most successful when paired with swans from the same age categories (Fig. 6.2). Although young Bewick's Swans of either sex were more successful when paired with middle-aged or old mates, most young birds were not paired with older mates (Fig. 6.2); there was some evidence for assortative pairing by age in Bewick's and Mute Swans ($P = 0.058$ and $P = 0.002$, respectively), but not Whooper Swans ($P = 1.00$; Fisher exact tests). Old Bewick's Swan females had a high level of reproductive success, irrespective of the age category of their mates, indicating that their breeding performance was not reduced when paired with younger males.

Body size

In determining the effect of body size on reproductive success each swan was classified as being small, medium, or large, according to its size relative to the mean for each sex. The medium group included birds whose body sizes were within half a SD of the mean value; large birds were above and small birds below this level. We adjusted for the effects of year, female and male age, pair duration, and the squares of these variables, before testing the effect of body size on reproductive success. Male size was associated with reproductive success in all three species; in Bewick's Swans large males tended to produce more cygnets, whereas in Whooper and Mute Swans medium-sized males had the highest reproductive success ($\Delta D = 11$, 6, and 7, respectively; $\Delta df = 2$, $P < 0.05$ in each case). Female size was also associated with reproductive success when considered in combination with the characteristics of the male for

[1]GLIM analysis of the effect of female and male age-class combinations, after controlling for year and pair duration: Bewick's Swans $\Delta D = 15.92$, $\Delta df = 4$, $P < 0.01$; Whooper Swans $\Delta D = 37.49$, $\Delta df = 4$, $P < 0.001$ and Mute Swans $\Delta D = 8.29$, $\Delta df = 4$, NS.

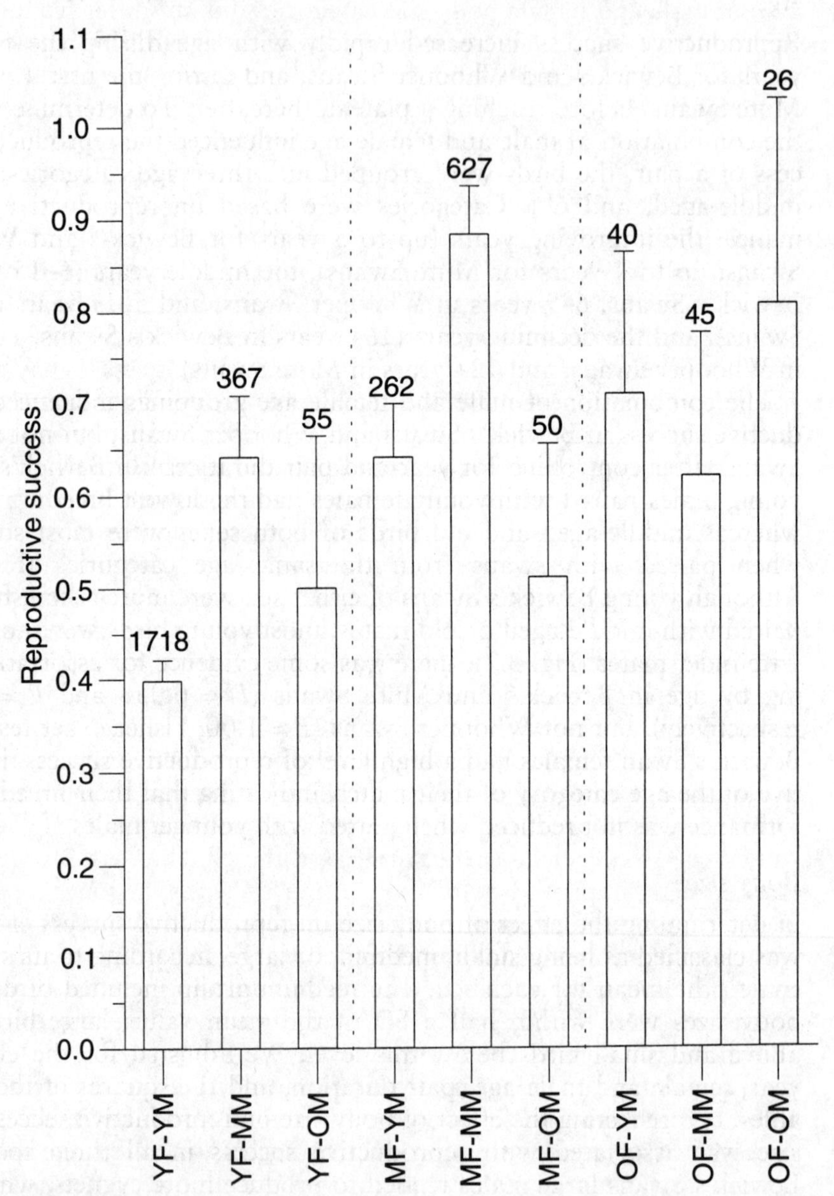

Fig. 6.2 Reproductive success (number of cygnets associating in winter) of Bewick's Swan pairs in relation to the combination of the male and female age categories; young (Y), middle-aged (M), and old (O) for females (F) and males (M). Sample sizes are shown above the histograms.

the migratory species, but not in Mute Swans.[1] The observed combination of male and female body size categories did not differ from the combination that would have been expected had the birds paired at random in all three species, indicating that size alone does not influence selection of a mate.[2]

Discussion

Advantages of monogamy

Long-term monogamy is advantageous in terms of improved reproductive success for all three swans species. Previous studies have found an increase in Bewick's Swan reproductive success over the early years of the pair bond (Scott 1988), but we found that the increase may continue for up to 11 years, even after controlling for year and age factors. Several authors have argued that maintaining the pair bond from one season to the next reduces the costs of courtship and establishing new partnerships, facilitates a prompt start to the breeding season, and improves the birds' access to feeding sites even in the winter months (Coulson 1966; Scott 1980*a*). Since the swans' reproductive success continues over several years following pair formation, there may also be a long-term development in the adults' ability to co-ordinate their activities, to the extent of producing more offspring and ensuring the survival of the young. Detailed behavioural studies would be necessary, however, to describe such fine-grained changes in interactions between members of a pair. The level of co-operation between partners may be more important for reproductive success in Bewick's Swans since both sexes incubate, or at least sit on the eggs, and the co-ordination of incubation duties may be crucial for the survival of the clutch. Certainly reproductive success improved with pair duration over a longer period in Bewick's Swans than in the other two species. In Whooper and Mute Swans incubation is usually undertaken by the female only, although the male has an active role in defending the breeding territory from other swans, and the brood from predators, throughout the breeding season (Birkhead and Perrins 1986).

Annual variation in an individual's reproductive success, evident in all three species, may be associated with annual variation in the breeding success of the total population, which in turn is usually due to climatic conditions (Minton 1968; Evans 1979; Black and Rees 1984; Bacon and Andersen-Harild 1989). The interaction of year with pair duration influenced Bewick's and Whooper Swan reproductive success, indicating

[1]Interaction between female and male size in the GLIM model: Bewick's Swans $\Delta D = 15.85$, $\Delta df = 4$, $P < 0.01$; Whooper Swans $\Delta D = 11.94$, $\Delta df = 4$, $P < 0.05$; Mute Swans $\Delta D = 5.42$, $\Delta df = 4$, NS.

[2]Log-likelihood test for assortative pairing by body size: Bewick's Swans $G = 4.9$, $df = 4$; Whooper Swans $G = 5.8$, $df = 4$; Mute Swans $G = 6.5$, $df = 4$. All values were not significant.

that long-term monogamy is more advantageous in some years than in others for swans breeding at high latitudes. Similarly, the interaction of year with age (and mate number), where age (and mate number) is thought to reflect an individual's breeding experience, also affected reproductive success for the migratory species. Perhaps pair duration and breeding experience are less important in determining individual success in years when conditions are favourable, and a high proportion of the nesting birds raise cygnets. In poor breeding seasons, however, prior knowledge of the site, and the level of co-operation between the pair in defending their territory and offspring, may become more important for the survival of the young (see also Marzluff *et al.* Chapter 7). The interaction of year with other variables did not have a significant influence on the number of Mute Swan cygnets fledged, perhaps because some of the factors causing annual variation in Mute Swan breeding success, such as the level of human interference (Minton 1968), could not be mediated by pair bond quality. Also, in poor breeding years some Bewick's and Whooper Swan pairs may not attempt to breed, whereas all the Mute Swans included in the analyses had at least built a nest.

Age influenced reproductive success in Bewick's Swans, even after controlling for pair duration; age was not significant in the other two species once pair duration was taken into account; Perrins *et al.* (1994) found that clutch size and hatching success increased with age in colonial-breeding Mute Swans, but this analysis did not control for pair duration. Since Bewick's Swans show a high level of site fidelity in both winter and summer (Evans 1980; E. C. Rees unpublished data) the increase in reproductive success with age may reflect prior experience of variation in conditions at the breeding grounds over several years, such as the areas that are the first to thaw in late springs. This would be particularly important for Bewick's Swans since their arctic breeding sites are vulnerable to adverse weather conditions and the shortness of the arctic summer does not allow much flexibility in the timing of the breeding programme. Reproductive success in the 4 years after pair formation was also influenced by age in Bewick's and Whooper Swans, but not in Mute Swans; older birds were more successful initially and achieved maximal reproductive success more rapidly than pairings between younger birds. The ability conferred by age and experience in one partner to compensate for any inexperience in the other therefore appears to be particularly important in the migratory species (see below).

Constraints on mate fidelity imposed by migration

The migratory and reproductive cycles of both Bewick's and Whooper Swans are constrained to varying extents by the distance between the breeding and wintering sites, by food availability at their staging areas, and by the shortness of the arctic and subarctic breeding seasons. Bewick's Swans are wholly migratory; the birds must breed, moult, and start their 4000 km return migration in just 4 months while the tundra is habitable. Although most of the Icelandic-breeding Whooper Swan population migrates some 800 km to Britain and Ireland, several hundred birds remain in Iceland throughout the winter including a high proportion of family parties (Gardarsson and Skarphedinsson 1985). Long-term pair bonds may be especially advantageous for the Bewick's Swans since they have little time for courtship or pair formation upon arrival in the breeding range; most courtship activity appears to occur in the non-breeding flocks (E. C. Rees personal observations). The evolution and maintenance of mate fidelity may also be attributed in part to the high energetic cost of long distance migration (see also Black *et al.* Chapter 5), which may result in Bewick's Swans, and to a lesser extent Whooper Swans, needing to spend a high proportion of their time in feeding. The presence of a mate appears to improve the feeding rate of an individual at the staging areas (Rees and Bowler 1991) and also in winter (Scott 1980*a*).

If mate fidelity in swans is indeed associated with constraints imposed by the migratory and reproductive cycle it might be predicted that maintaining long-term pair bonds would be most advantageous to Bewick's Swans, of intermediate value to the Icelandic-breeding Whooper Swan population, and less important for Mute Swans. Bewick's Swans incurred the highest short-term costs after mate change, taking some 2.6 years to re-pair and 2.3 years to breed with a new mate, compared with 1.9 and 1.5 years, respectively in Whooper Swans. Whooper and Mute Swans perform courtship displays and develop pair bonds in winter (Minton 1968; Scott 1978; Black and Rees 1984), whereas Bewick's Swans rarely initiate pair formation in the wintering range. These interspecific differences may be attributable to the fact that Whooper and Mute Swans spend less time in feeding, as they do not have the energetically expensive migratory cycle of the Bewick's Swan, and also to greater flexibility in the timing of their migratory and breeding programmes.

Although reproductive success increased with pair duration in all three species, there were no obvious long-term costs of mate change to an individual. The lack of any long-term costs may perhaps be due to the overriding effect of climatic conditions and predator levels on breeding success, which makes the annual success of individual pairs somewhat erratic (Ebbinge 1989). Reduced breeding success due to the time taken to find a new mate would therefore be obscured by the effects of poor breeding seasons when considering an individual's lifetime productivity. Moreover, if the new mate is of higher quality (or the newly formed pair are more compatible) then the initial shortfall in reproductive success may be made up subsequently. There was some tentative evidence to suggest that mate selection improved with experience. In all three species, individuals were more successful with their second or subsequent mates, although the results were significant only for Whooper Swans. The interaction of year with mate number was associated with reproductive success for both the migratory species. Bewick's Swans also achieved a higher dominance rank with the second partner, suggesting that experienced birds may be better able to select a compatible mate.

Monogamy and parental care

Several authors suggest that the evolution of monogamy may be closely associated with the development of parental care (see Mock *et al.* Chapter 3). For the migratory swans, parental care is prolonged, with parents and offspring associating throughout the first winter or longer (Evans 1979; Scott 1980*b*). The Mute Swan's parent–offspring association is usually much shorter (Scott 1984). The benefits to family members include higher dominance rank and better access to food than pairs or single birds in the wintering flock (Scott 1978; Black 1988). Bewick's Swan cygnets and females from dominant families achieve better winter condition than those from less successful groups, whereas, the males from dominant families are generally thinner due to their reduced feeding time and greater investment in encounters than subordinate males (Bowler 1995). Perhaps cygnets assist in the competition for feeding areas, thereby enhancing feeding opportunity and potential breeding success for their parents in the manner described for geese (Black and Owen 1989).

Reasons for divorce

The idea that divorce is due to mates losing contact during migration (i.e. due to *accidental loss*. Owen *et al.* 1988; Black *et al.* Chapter 5) was not supported in this study. Divorce was highest in Whooper Swans that migrate short distances and nonexistent in Bewick's Swans that migrate long distances.

The observation that mate replacement is more costly, albeit in the short-term, to Bewick's Swans than Whoopers or Mute Swans suggests that Bewick's Swan divorce rates should be correspondingly lower. No cases of divorce have previously been recorded for Bewick's Swan pairs known to have bred (Evans 1979; Scott 1988). Our results confirmed that divorce is extremely rare in Bewick's Swans; none of the birds included in the analyses changed mates whilst the previous partner was known to be still alive. Indeed, there is only one anecdotal record of divorce in Bewick's Swans, involving a pair observed at Welney, Norfolk, that separated after 7 years together, during which time they failed to rear any young. The female was recorded with a new mate and bred in 3 of the 4 years seen with her second partner; the male was seen for a further 2 years but is not known to have paired with a new mate (D. K. Scott personal communication). Mute Swans changed their mates more frequently than both Bewick's and Whooper Swans, but the rate of mate change attributable to divorce was highest for Whooper Swans. It seems, therefore, that the frequency of mate change for Mute Swans included in this study may be associated with higher mortality levels, which in turn may be due to the high incidence of lead poisoning (Sears and Hunt 1991), rather than to forming temporary pair bonds. The probability of divorce of 0.7% of Mute Swan pairs may be an underestimate, owing to the combined effects of high mortality levels reducing the probability of both partners surviving from one season to the next, and the inclusion only of pairs known to have bred in the Mute Swan analyses. Other studies have estimated Mute Swan divorce rates at 3% per annum among established breeding pairs, and at around 10% among pairs that failed to breed (Minton 1968).

Advantages of divorce were considered in further detail for the Whooper Swans. The previous breeding experience of the pair, in both the short and longer term, had no apparent effect upon the probability of divorce in this species. Divorce appeared to improve reproductive success for females and to reduce reproductive success for males, but the results did not reach statistical significance.

Mating patterns

The interaction of male and female size, and also of male and female age groups, was associated with reproductive success in Bewick's and Whooper Swans, but not in Mute Swans, indicating that the combination of male and female characteristics is important for breeding success in the migratory species. However, swans did not appear to pair assortatively on the basis of size. Similar results obtained for the northern geese have been attributed to the over-riding effects of other variables influencing the migratory and breeding cycles (Ebbinge 1989). In Bewick's Swans both males and females were more successful with mates of a similar age, except that young birds appeared to improve their reproductive success when paired with older mates, suggesting that the greater breeding experience of older swans may compensate for an inexperienced partner. Young birds did not usually achieve their preferred option of increasing their reproductive success by pairing with older mates, despite old females proving successful irrespective of the age of the mate.

Summary

Reproductive success improved with pair duration in three swan species after controlling for the effects of other variables. The effect of pair duration was more pronounced in some years than in others for the migratory Bewick's and Whooper Swans, which breed at high latitudes. Fighting ability of pairs (i.e. dominance rank) did not appear to be the mechanism behind the reproductive improvement. The combination of male and female age, and of male and female size, also influenced reproductive success in Bewick's and Whooper Swans, indicating that the compatibility of individual characteristics may determine the quality of the pair. Mate replacement resulted in higher short-term costs for Bewick's Swans than for Whooper or Mute Swans. Interspecific differences in the benefits of long-term monogamy, measured in terms of increased reproductive success, were attributed to variation in limitations imposed on the swans' breeding programmes by the migratory distances undertaken and shortness of the arctic summer. Mate change did not appear to affect an individual's lifetime reproductive success.

Acknowledgements

We are grateful to John Bowler, Mary Evans, Philippa Scott, Linda Butler, Dafila Scott, Chris Tomlinson, Jenny Roberts, Sue Carmen, Olafur Einarsson, and Richard and Carol Hesketh for information over many years on the pairing and breeding success of individual Bewick's and Whooper Swans. Mike Birkhead, Philip Bacon, and Jane Sears each made substantial contributions to the Mute Swan study and Sverrir Thorstensen to the Whooper Swan study. John Bowler kindly made available his data on the dominance ranks of Bewick's Swans wintering at Slimbridge. Ute Zillich edited the Bewick's Swan database and prepared the figures. Jeff Black, Myrfyn Owen, and Dafila Scott made constructive comments on a draft of the text.

References

Bacon, P. J. and Andersen-Harild, P. (1989). Mute Swan. In *Lifetime reproductive success in birds* (ed. I. Newton), pp. 363–386. Academic Press, London.

Birkhead, M. E. and Perrins, C. M. (1986). *The mute swan.* Croom Helm, London.

Black, J. M. (1988). Preflight signalling in swans: a mechanism for group cohesion and flock formation. *Ethology*, **79**, 143–57.

Black, J. M. and Owen, M. (1989). Agonistic behaviour in goose flocks: assessment, investment and reproductive success. *Animal Behaviour*, **37**, 199–209.

Black, J. M. and Rees, E. C. (1984). The structure and behaviour of the Whooper Swan population wintering at Caerlaverock, Dumfries and Galloway, Scotland: an introductory study. *Wildfowl*, **35**, 21–36.

Bowler, J. M. (1995). The condition of Bewick's Swans *Cygnus columbianus bewickii* in winter as assessed by their abdominal profiles. *Ardea*, **82**, 241–8.

Coleman, A. E. and Minton C. D. T. (1979). Pairing and breeding of Mute Swans in relation to natal area. *Wildfowl*, **30**, 27–30.

Coulson, J. C. (1966). The influence of the pair bond and age on the breeding biology of the kittiwake gull *Rissa tridactyla. Journal of Animal Ecology*, **35**, 269–79.

Ebbinge, B. (1989). A multifactor explanation for variation in breeding performance of Brent Geese *Branta bernicla. Ibis*, **131**, 196–204.

Evans, M. E. (1979). Aspects of the life cycle of the Bewick's Swan based on recognition of individuals at a wintering site. *Bird Study*, **26**, 149–62.

Evans, M. E. (1980). The effects of experience and breeding status on the use of a wintering site by Bewick's Swans *Cygnus columbianus bewickii. Ibis*, **122**, 287–97.

Gardarsson, A. and Skarphedinsson, K. H. (1985). The wintering of Whooper Swans *Cygnus cygnus* in Iceland. *Bliki*, **4**, 45–56.

Hilprecht, A. (1970). *Hockerschwan, Singschwan, Zwergschwan*. Neue Brehm-Bucherei, Wittenberg.

Meng, A. (1990). DNA finger-printing and minisatellite variation in swans. Unpublished Ph.D. thesis. University of Nottingham.

Meng, A. and Parkin, D. T. (1991). Alloparental behaviour in Mute Swans *Cygnus olor* detected by DNA finger-printing. In *Proceedings of the Third IWRB International Swan Symposium, Oxford* (ed. J. Sears and P. J. Bacon), *Wildfowl*, Supplement No. 1, 310–318.

Minton, C. D. T. (1968). Pairing and breeding of Mute Swans. *Wildfowl*, **19**, 41–60.

NAG (1977). *GLIM 3.77 Reference Manual*, (2nd edn), (ed. R. J. Baker) Numerical Algorithms Group, Oxford.

NAG (1993). *GLIM 4 The Statistical System for Generalized Linear Interactive Modelling*, (2nd edn) (ed. B. Francis, M. Green and C. Payne) Numerical Algorithms Group, Oxford.

Owen, M., Black, J. M., and Liber, H. (1988). Pair bond duration and the timing of its formation in Barnacle Geese. In *Wildfowl in winter* (ed. M. Weller), pp. 23–38. University of Minnesota Press, Minneapolis.

Perrins, C. M., McCleery, R. H., and Ogilvie, M. A. (1994). A study of the breeding Mute Swans *Cygnus olor* at Abbotsbury. *Wildfowl*, **45**, 1–14.

Rees, E. C. and Bowler, J. M. (1991). Feeding activities of Bewick's Swans *Cygnus columbianus bewickii* at a migratory site in the Estonian SSR. In *Proceedings of the Third IWRB International Swan Symposium, Oxford* (ed. J. Sears and P. J. Bacon), *Wildfowl*, Supplement No. 1, 249–55.

Rees, E. C., Bowler, J. M., and Butler, L. (1990). Bewick's and Whooper Swans: the 1989–90 season. *Wildfowl*, **41**, 176–81.

Rees, E. C., Gitay, H., and Owen, M. (1991). The fate of plastic leg-rings used on geese and swans. *Wildfowl*, **41**, 43–52.

Scott, D. K. (1978). Social behaviour of wintering Bewick's Swans. Unpublished Ph.D. thesis. University of Cambridge.

Scott, D. K. (1980*a*). Functional aspects of the pair bond in Bewick's swans (*Cygnus columbianus bewickii*). *Behavioral Ecology and Sociobiology*, **7**, 323–7.

Scott, D. K. (1980*b*). Functional aspects of prolonged parental care in Bewick's swans. *Animal Behaviour*, **28**, 938–952.

Scott, D. K. (1984). Parent–offspring association in mute swans. *Zeitschrift für Tierpsychologie*, **64**, 74–86.

Scott, D. K. (1988). Reproductive success in Bewick's swans. In *Reproductive success* (ed. T. H. Clutton-Brock), pp. 220–36. University of Chicago Press, Chicago.

Scott, P. (1966). The Bewick's swans at Slimbridge. *Wildfowl Trust Annual Report*, **17**, 20–6.

Scott, P. and The Wildfowl Trust (1972). *The swans*. Michael Joseph, London.

Sears, J. and Hunt, A. E. (1991). Lead poisoning in Mute Swans *Cygnus olor* in England. In *Proceedings of the Third IWRB International Swan Symposium, Oxford* (ed. J. Sears and P. J. Bacon), *Wildfowl*, Supplement No. 1, 383–8.

7 *Breeding partnerships of two New World jays*

JOHN M. MARZLUFF, GLEN E. WOOLFENDEN, JOHN W. FITZPATRICK, AND RUSSELL P. BALDA

Introduction

Two New World jays with different lifestyles, the Pinyon Jay (PJ) *Gymnorhinus cyanocephalus* and the Florida Scrub Jay (FSJ) *Aphelocoma coerulescens* (see drawing above), form long-term, monogamous breeding partnerships. PJ pairs live in large flocks of breeders and prebreeders that roam widely over home ranges encompassing thousands of hectares, lack territories, and breed colonially (Marzluff and Balda 1992). FSJ pairs live alone, or with a few prebreeding offspring that act as helpers, in permanent territories, and individuals rarely move more than 1 or 2 km from their natal site during their entire lifetimes (Woolfenden and Fitzpatrick 1984, 1986). We discuss the establishment of long-term, socially monogamous pair bonds, the influence of mate fidelity on reproduction, and the factors that promote mate fidelity in these phylogenetically related, but ecologically dissimilar corvids. Although both species form long-term pair bonds, clear differences exist between many features of these partnerships. We conclude that the disparities result from marked differences in sociality, use of space and use of resources.

Background, study sites, and procedures

The PJ study began in 1968 and continues. Most of the results presented here derive from intensive observations of one colour-ringed flock inhabiting the environs of Flagstaff, Arizona, USA. During 1972–1987 nearly every member of this flock was uniquely marked (over 1000 individuals), and mating partners, productivity, and survivorship were monitored. Here we report on 141 pairs and a total of 279 pair-years. The study area, observational methods, and statistical procedures are detailed in Marzluff and Balda (1992).

The FSJ study began in 1969 at Archbold Biological Station in south-central Florida, USA, and also continues. All young produced in a study tract of about 500 ha and all immigrants to the tract are colour-ringed. We establish genealogies by finding all nests, and gather data on individual fitness by following marked individuals until they die. All jays within the study tract are censused at least monthly, and extensive searches for dispersers are conducted periodically. A total of 477 pairs and 960 pair-years was used in analyses where breeding history was not needed (breeding history was known for both members of 248 pairs over 522 pair-years).

PJs are found in the mountainous western United States and northern Mexico in close association with the Pinyon Pine *Pinus edulis*, (Fig. 1 in Marzluff and Balda 1992). The tree and its jay have close coevolutionary histories; the pine provides the jay's principal food and the jay acts as the pine's principal seed disperser. In years of poor pine seed production PJs may roam widely through many different habitat types from desert scrub and grassland to mixed conifer forest.

FSJs are found only in peninsular Florida, where they are restricted to oak scrub. Several species of oaks *Quercus* are among the shrubs important to FSJs, because they produce both cover and acorns. The habitat is maintained by frequent fire. After a decade or two of no fire, the shrubs become tall and dense, bare ground becomes scarce, trees, especially pines, become more abundant, and the resident FSJ population dwindles (Fig. 26.1 in Woolfenden and Fitzpatrick 1991).

A difference between the habitats of the PJ and FSJ, which may have affected profoundly their social systems, is the predictability of their primary plant foods. Seed crops of pinyon pines vary tremendously from one year to the next. Bumper crops and years with virtually no seeds follow each other in an unpredictable pattern (Fig. 4 in Marzluff and Balda 1992). In contrast the scrub oaks of Florida produce acorns in considerable numbers annually, without crop failures (DeGange *et al.* 1989).

Social milieu

PJs reside in one of the largest permanent social groups known among birds. Flocks range in size from 50 to 500 individuals and consist of numerous extended families. Most PJs in a flock are not related, but relatives from successful families within a flock commonly span three generations. The pair bond is the strongest bond among flock members and acts as a social cement, linking members of different families into a cohesive group. As the breeding season approaches, a PJ flock begins to visit one of several traditional breeding grounds and spends increasingly longer portions of the mornings on these grounds before selecting one on which to initiate a breeding colony. At the breeding site, pairs isolate themselves from the foraging flock and initiate subtle courtship displays. Colonies of one flock ranged in size from 2 to 32 nests, spaced an average of 33–430 m apart (Marzluff and Balda 1992). Nonbreeding PJs form a subgroup of the flock that becomes more autonomous as the breeding season approaches. As breeders become tied to the nesting colony, nonbreeders roam the home range, usually reuniting with breeders sporadically throughout the day and nightly at a communal roost that contains all nonincubating flock members.

In great contrast, FSJs reside as pairs, sometimes accompanied by a few prebreeders that act as helpers (Woolfenden and Fitzpatrick 1984; Mumme 1992), in permanent territories varying in size from 5 to 20 ha (mean 9 ha). As members of a pair, FSJ interactions with other breeders are virtually confined to aggressive interactions at territory borders. Territories are defended all year. Courtship displays between the members of the pair are inconspicuous and tend to be performed away from all other jays. Prior to dispersal (normally beginning in the 14th month postfledging), young FSJs remain with their parents or stepparents. As they gain mobility and experience, they join their parents and any helpers present, and together conduct the daily activities of foraging, territory defence, predator surveillance, and body maintenance. Although family members usually stay near each other and work together, individuals of different status assume different proportions of the various activities (Francis *et al.* 1989; McGowan and Woolfenden 1989).

Pair formation

In PJs, competition within a sex and choice of partners by the opposite sex are likely to occur in the flock. Laboratory experiments indicate that both male and female PJs exercise choice of prospective partners (Johnson 1988*a, b*). In FSJs, choosing a partner almost always occurs within the territory of a bird that lost its mate or the budding territory of a novice. The exceptions (inherited natal territories or *de novo* territories) are rare (Woolfenden and Fitzpatrick 1984). Prebreeders of like sex often compete to become the mate of opposite-sex birds that have lost their mates, in the recently unmated bird's territory.

Commonly PJs form subsequent pair bonds with individuals with whom they have spent their entire adult lives. Only rarely are subsequent mates obtained from outside the flock. We suspect that every PJ can recognize each member of its flock and can associate specific attributes such as age, previous breeding history, and social bonds, with these future potential mates. As with initial bonds, FSJs form subsequent pair bonds either with dispersers or with nearby breeders who have lost their mates. Long-distance dispersers may have had few or no prior interactions with their new mate. Although FSJs almost certainly recognize many individuals, they rarely interact with more than the 10–20 adult jays that comprise their extended family and neighbouring territory-holders. Pair formation sometimes appears to take many months. Especially during the breeding season, however, pairing may be completed in a few days.

Pair maintenance

Pair bonds in both species are maintained throughout the year. Year-round association of partners in flocks (PJ) or in proximity on territories (FSJ) probably helps maintain persistent pair bonds. Pair bonds are not obvious within the flocks of PJs outside of the breeding season, but laboratory experiments suggest that mates are vocally recognized throughout the year and this may help them maintain their pair bonds within large flocks (Marzluff and Balda 1992). No doubt the caching behaviour of FSJs also promotes their staying in one place and maintaining pair bonds. Annually, in the autumn, each FSJ caches thousands of acorns in its territory for use in winter and early spring (DeGange *et al.* 1989). The sentinel system FSJs use as defence against hawks and conspecific intruders (McGowan and Woolfenden 1989) probably also promotes staying with the same individuals.

Breeding

In both PJs and FSJs, breeders are monogamous and function as a coordinated team. Exceptions to social monogamy are unknown in 489 PJ nesting attempts and limited to 1 out of 960 FSJ pair-years. The exception was one male with two females (Woolfenden 1976). Division of labour is conspicuous in both species, especially early in the nesting cycle; females incubate and brood, males feed and guard (Woolfenden and Fitzpatrick 1986; Marzluff and Balda 1992). Observations of the polygynous pairings of FSJs (e.g. Woolfenden 1976), and at nests of both PJs and FSJs in which breeders have lost their mates, indicate that individuals of neither species are able to raise nutritionally independent young alone (Marzluff and Balda 1992). Extra-pair copulations may occur in PJs. Although observations of copulations among wild PJs are rare, Nancy Stotz (personal communication) once observed a male displace another male that had just copulated with a female. To date DNA studies to determine parentage have not been conducted in PJs. In FSJs, DNA-fingerprinting (Quinn *et al.* 1990) has detected no exceptions to behavioural mates being the parents of the offspring they tend.

For both PJs and FSJs, modal age of first breeding is 2 years. Breeding at age 1 year is frequent in PJs but very rare in FSJs, and first breeding past age 3 years is unknown for PJs but documented up to age 7 years for FSJs. Certainly for PJs, possibly for FSJs, more females than males first breed at age 1 year. Both species breed annually and rarely produce more than one brood of nutritionally independent young per year. PJs are among the earliest breeding passerines in North America, mainly nesting from February to July. The FSJ breeds only from March to June (Fig. 8.1 in Woolfenden and Fitzpatrick 1984). Average clutch sizes (3.7 and 3.4 eggs, respectively), fledging brood sizes (1.2 and 2.0), and production of nutritionally independent young (0.9 and 1.2), are similar in PJs and FSJs. Slightly lower average productivity of fledglings and independent young by PJs reflects greater rates of predation on nests of PJs than of FSJs. Both species renest after failure early in the season, but PJs try more times (4–5 times) than FSJs (four clutches by a pair has only been recorded twice).

In PJs productivity and survival of all age cohorts is extremely variable from year to year and is correlated with climatic variability that typifies the mountain regions of western USA. This variability produces annual variation in the age and sex structure of a flock (Figs 29–32 in Marzluff and Balda 1992). In FSJs annual nesting success and survival of fledglings also vary considerably. As a result, the number of unpaired jays (i.e. prebreeding helpers) varies. Because large territories are necessary for breeding, and the total amount of available breeding habitat is stable from year to year, breeding densities of FSJs are fairly stable. As a result, unpaired jays accumulate in years following high reproduction, and become scarce during successive years of low reproduction (Fig. 26.4 in Woolfenden and Fitzpatrick 1991). Survival of breeders also fluctuates annually. However, on average more prebreeders are produced than space exists for establishing territories (Fitzpatrick and Woolfenden 1986). PJ sex ratios for both adults and yearlings prior to dispersal typically are male biased to varying degrees (Marzluff and Balda 1989). The sex ratio of FSJs as breeders, and also (apparently) at age 1 year, is equal. However, the sex ratio of older prebreeders is biased towards males because dispersal forays of females begin earlier in their lives and take them farther from home than is true for males (Woolfenden and Fitzpatrick 1984), thereby increasing the mortality of prebreeding females (Fitzpatrick and Woolfenden 1986).

Young PJs fledge when approximately 3 weeks old and remain in a crèche for approximately 4 more weeks gaining nutritional independence and perfecting their social skills. Here, parent–young recognition is well developed (McArthur 1982). Young FSJs fledge between 2 and 3 weeks after hatching and reach nutritional independence about 12 weeks after hatching (McGowan and Woolfenden 1990). Typically FSJs remain in the family unit for more than a year, where they increase their knowledge of the environment and improve their social skills (McGowan 1987). In this chapter we use the number of nutritionally independent young as the measure of reproductive success of jay partnerships.

Results

The pair bond

Pair duration of both species was similar, averaging just over 2 years (Table 7.1). However, because a few individuals of both species lived as long as 15 or 16 years, occasionally pair bonds persisted for up to 10 years. Death accounted for the dissolution of almost all pair bonds in both PJs and FSJs. In both species the endurance of pair bonds was approximated by multiplying the survivorship of breeders of one sex by that of the other (Fig. 7.1). Male PJs had higher survival than females (Marzluff and Balda 1992), but survival of males and females was equal in FSJs (216 deaths in 876 male breeder-years, 217 deaths in 893 female breeder-years). Sexual asymmetry in mortality has an important consequence in the PJ: males often outlived their mates (46.2% of pair bonds were broken by death of the female, but only 30.8% were broken by death of male). Males therefore had more opportunity to select subsequent mates than did females and averaged slightly more mates in their lifetimes than females (Table 7.1). The average number of mates obtained by FSJs did not differ between the sexes (Table 7.1). Pair duration may be influenced by the properties of the pair. We suggest that compatible PJ partners may enhance each other's survival, whereas incompatible ones reduce their survival. PJs died the same autumn or winter as their mates died 20.8% of the time. This was much more often than expected by chance (6.8%; Marzluff and Balda 1992). This situation did not occur in FSJs. From 1988 through 1993, mortality of FSJ females whose mate died (n = 59 female-years; 0.020 deaths per month after death of the mate) was nearly identical to that of paired females (n = 323; 0.025 deaths per month).

Table 7.1 Social demography of two New World Jays

Demographic parameter	Pinyon Jay	Florida Scrub Jay
Average adult survivorship (%)	73.9	77.5
Expectation of further life[1]		
Breeding females	4.1 years	4.5 years
Breeding males	5.2 years	4.5 years
Maximum lifespan		
Males	16+ years	15 years
Females	14 years	15 years
Average pair bond duration (SD)	2.5 years (1.8)	2.1 years (1.4)
Maximum pair bond duration	10 years	10 years
Average mates per lifetime		
Males (SD)	1.63 (0.93)	1.55 (0.97)
Females (SD)	1.43 (0.65)	1.56 (0.84)

[1] Assumes no senescence.

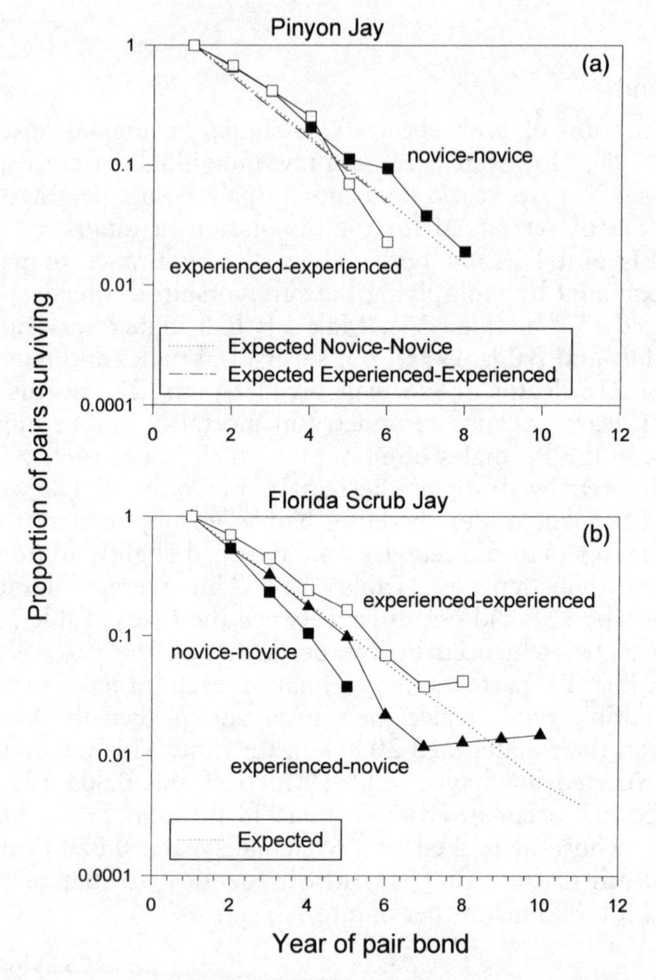

Fig. 7.1 Observed survivorship of pair bonds compared to those expected from individual survival rates of breeders. Survival of Pinyon Jay partnerships (a) formed between two first-time breeders (novice–novice) and two previously experienced breeders (experienced–experienced) are plotted separately (see Marzluff and Balda 1988*b* for calculation of expected frequencies). Survival of Florida Scrub Jay partnerships (b) between two novices (*solid squares*), two experienced jays (*open squares*), and mixed pairs (*solid triangles*) are plotted separately.

Mate fidelity and reproduction

Increasing pair duration did not confer a reproductive advantage to PJs (Fig. 7.2). Increased pair duration among experienced pairs was associated with higher average productivity, but variation around the means in Fig. 7.2 was substantial (Table 16 in Marzluff and Balda 1992). Pair duration was not correlated with the number of independent young produced by a pair ($r = 0.08$, $n = 271$, $P = 0.26$). This correlation changed

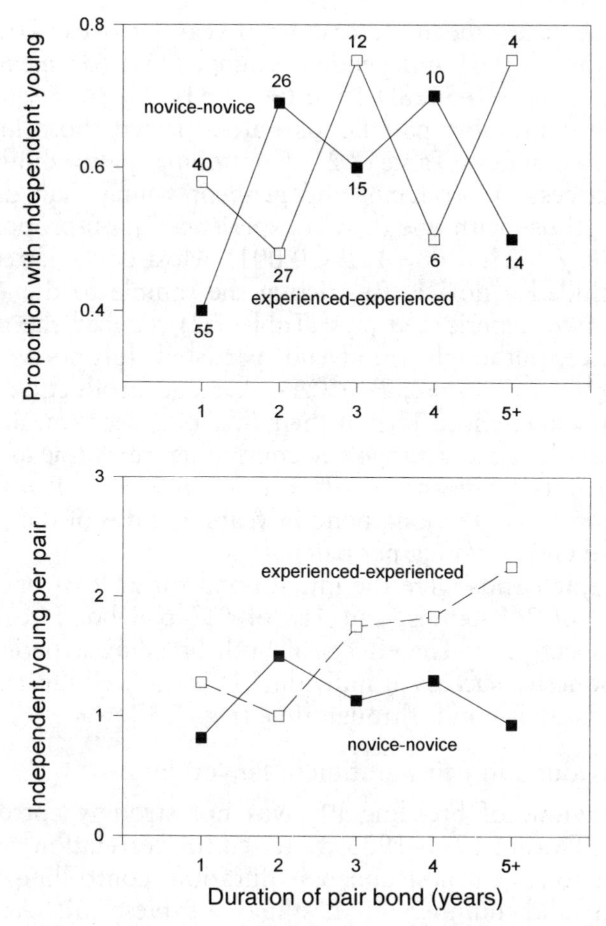

Fig. 7.2 Reproductive success of Pinyon Jays during successive years of a pair bond. Average success is plotted as a function of pair duration separately for bonds initiated between two novices (*solid squares*) and two experienced breeders (*open squares*). Sample sizes are shown on upper graph.

very little if parental age and experience were controlled (partial r = –0.05, P = 0.49, controlling for male age, female age, male experience, and female experience). It should be noted that data on the same pairs in multiple years were included in these correlations. Therefore the data may not be entirely independent and probability levels may be artificially inflated. This conservative bias makes the lack of a significant relationship between pair duration and productivity even more robust. The tendency for novice pairs to increase their productivity from their first to their second year appeared to result from accumulation of experience and not from some advantage of remaining with the same mate. We have data on 18 novices that remained paired and 18 that changed mates between their first and second breeding attempts. Both samples showed similar increases (mean increase from year 1 to year 2 for pairs remaining together = 0.61 independent young, SD 1.61; mean for individual changing mates = 0.56, SD 1.95; $t = -0.93$, $df = 34$, $P = 0.93$).

FSJs in continuing pair bonds outperformed those in their first year together as mates (Table 7.2). Continuing pairs showed significantly higher success at producing independent young than did first-pairings, including those with one or two experienced jays (proportion 0.657 versus 0.441, $\chi^2 = 28.3$, $df = 1$, $P < 0.001$). Most of the latter pairings, however, included a novice. Restricting the sample to the 47 new pairings between two experienced jays (Table 7.2) reduced the difference below significance, although the trend persisted (proportion 0.657 versus 0.562, $\chi^2 = 2.6$, $df = 1$, $P = 0.1$). Average production of independent young by experienced jays in their first pair-year (mean 1.02, SD 1.31) was lower than the average for continuing pairs (mean 1.31, SD 1.21), although not significantly so ($t = 1.54$, $df = 313$, $P = 0.13$). Offspring production peaked among pairs in years 3 and 4 of the pair bond (mean 1.47 independent young per pair).

Most pair bonds were the initial bond for at least one member of the pair (207 of 248 bonds, and 416 of 522 pair-bond years, or 84% and 80%, respectively). The effects of both breeding experience and helpers on reproductive success of individual FSJs are well illustrated by tracking these initial pair bonds through time (Fig. 7.3).

Parental behaviour and pair duration in Pinyon Jays

The behaviour of breeding PJs was not strongly correlated with pair duration. During 1981–1983 we tested for correlation between the average behaviour at a nest and pair duration, controlling for date of nest initiation and number of nestlings (Table 7.3). As pair duration increased, males spent more time cleaning their nests; however, this did not reduce the amount of time their mates spent cleaning the nest. Thus, males may contribute more to raising their young after they have been mated to the same female for several years, but females do not garner energetic savings from such an adjustment. Parental behaviour was correlated with age and breeding experience to a greater extent than with pair duration (Marzluff and Balda 1992).

Causes and consequences of mate change

Divorce was rare in both PJs and FSJs (Marzluff and Balda 1992; Woolfenden and Fitzpatrick 1984). In PJs divorce occurred in only 3 of 104 pairs (2.9%; 3 of 279 pair-years), and in FSJs it occurred in only 25 of 477 pairs (5.2%; 25 of 960 pair-years). One instance of divorce in PJs suggests that partner compatibility may be important; the pair that divorced was morphologically more dissimilar than any other pair (Marzluff and Balda 1992). Another instance of divorce in PJs and three in FSJs occurred when one member of the pair was sick or injured. Three

Table 7.2 Mate change, pair duration, and production of independent young by Florida Scrub Jays

	New pair bonds						Continuing pair bonds							
	Novice × novice (0.19)		Novice × experienced (0.40)		Experienced × experienced (0.60)		2nd year pair bonds (0.51)		3rd year pair bonds (0.55)		4th year pair bonds (0.58)		>4th year pair bonds (0.70)	
	Without helpers	With helpers	Without helpers	With helpers	Without helpers	With helpers	Without helpers	With helpers	Without helpers	With helpers	Without helpers	With helpers	Without helpers	With helpers
Pairings (n)	56	13	83	55	19	28	66	68	32	39	15	21	8	19
Independent young (total)	42	19	47	50	19	29	67	94	42	57	19	39	6	28
Production per pair	0.75	1.46	0.57	0.91	1.00	1.04	1.02	1.38	1.31	1.46	1.27	1.86	0.75	1.47
	Pooled		Pooled		Pooled		Pooled		Pooled		Pooled		Pooled	
Pairings (n)	69		138		47		134		71		36		27	
Independent young (total)	61		97		48		161		99		58		34	
Production per pair	0.88		0.70		1.02		1.20		1.39		1.61		1.26	

Proportion of pairs with helpers are indicated in parentheses.

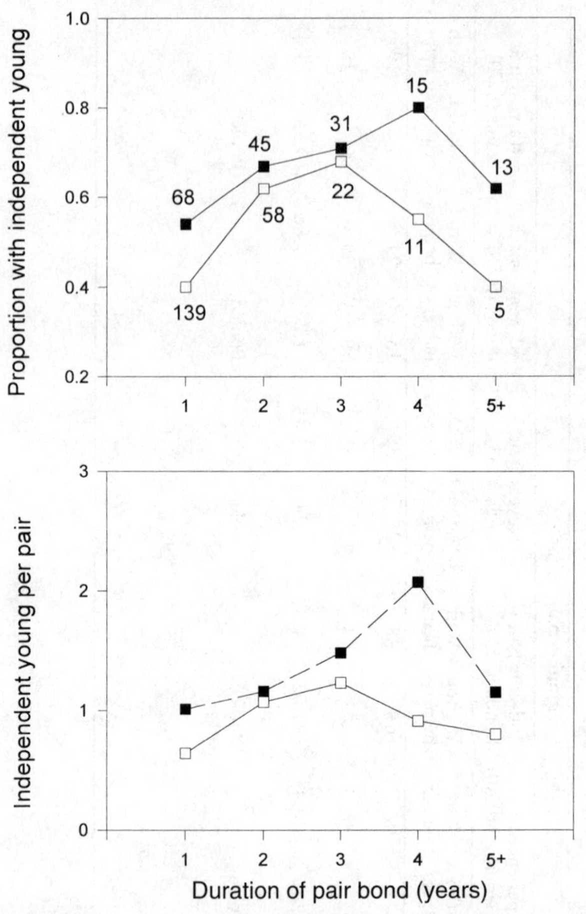

Fig. 7.3 Reproductive success of Florida Scrub Jay pairs during successive years of their initial pair bonds. Samples for each year are divided into pairs with helpers (*solid squares*) and without helpers (*open squares*). Sample sizes are shown on upper graph.

FSJ pair bonds broke after persistent efforts at territorial defence failed to establish a viable territory. In general, we suspect that poor performance at certain duties by a member of a pair accounted for most divorces (Woolfenden and Fitzpatrick 1986; Marzluff and Balda 1992).

PJs did not show an increased tendency to divorce after failure to produce independent young, even after several successive failures (Marzluff and Balda 1992). Nearly every pair we have studied failed in at least one year but remained paired during the subsequent year. Ten pairs failed in each of their first two seasons together without divorcing and five pairs failed for four or more consecutive years without divorcing. FSJs, however, were slightly more prone to divorce after failure early in the pair bond. Of 25 divorces, 15 occurred after the pair's first breeding season together, and only one of these had produced independent young.

Table 7.3 Correlations between pair duration and parental behaviour of Pinyon Jays

Behaviour	*n*	*r*	*P*
Time male feeds nestling/h	25	−0.22	0.32
Time female feeds nestling/h	25	−0.16	0.46
Time male cleans nest/h	25	+0.61	0.002
Time female cleans nest/h	25	−0.22	0.32
Percentage of feeding done by male	27	+0.15	0.47
Percentage of cleaning done by female	25	−0.37	0.09
Number of feeding trips by pair	27	−0.26	0.21
Time pair feeds nestlings/h	25	−0.41	0.06
Percentage of time parents at nest	27	−0.22	0.30
Percentage of nest visits by 1 parent	26	−0.07	0.75
Time male in nest tree before reaching nest	27	−0.10	0.63
Time female in nest tree before reaching nest	26	+0.27	0.20

Partial correlations holding the number of nestlings and date of nest initiation constant are reported.

Divorce rate among successful first pairs was 0.005 (1 of 186), while that among failed first pairs was 0.061 (14 of 228; $\chi^2 = 9.2$, $df = 1$, $P < 0.005$). In contrast, of the 10 multi-year FSJ pairs that later divorced, eight did so within a year after producing independent young. This suggests that factors other than past reproductive success cause divorce among established pairs.

We have identified three consequences of mate change in FSJs that help explain why continuing pairs outperform first-time pairs. First, continuing pairs initiated reproduction significantly earlier than first-time pairs. The mean date of first incubation for continuing pairs was 28 March (Julian date 88.4, SD 11.9, $n = 255$), that for first-time pairs was 6 April (Julian date 96.8, SD 14.7, $n = 223$). This difference was significant ($t = 6.8$, $df = 476$, $P < 0.001$) and held even when first-time pairs were both experienced (mean Julian date 95.3, SD 15.4, $n = 44$; $t = 3.4$, $df = 297$, $P < 0.002$). This difference of 9 days translated into a drop in expected nest success from approximately 65% (for nests initiated in the last week of March) to approximately 55% (for nests initiated between the first and second weeks of April; Woolfenden and Fitzpatrick 1984, p. 214). Second, newly established pairs always involve at least one member occupying unfamiliar ground, where new sites for roosting, foraging, and acorn-caching must be learned. Third, the frequency of having helpers (which significantly improves breeding performance) is greater for continuing pairs (57%) than for pairs immediately following mate replacement (48%). Helpers occasionally are driven off the territory by replacement mates, and sometimes do not follow replacement mates to their new territories.

Pair characteristics and reproduction

PJs use a variety of morphological and social cues in their assessment of potential mates and choice of a mate. Laboratory experiments and field observations indicate that age, physical size, plumage brightness, previous breeding history, and dominance are used by jays to assess mates (Johnson 1988*a, b*; Marzluff and Balda 1992). Males and females do not always assess mates equally with respect to these criteria, suggesting that this is an area of conflict between monogamous partners. In the laboratory, males prefer relatively large, dominant females, but large females do not prefer large males (Johnson 1988*a, b*). This difference is also likely to occur in nature because the fitness benefits accrued by PJs depended on the size of their mate. Males tended to live longer and produced more offspring if they were mated to large (heavy mass) females, but females did not benefit from mating with large males (Table 7.4). Above average-sized females were significantly more fit than smaller females if they mated with below average-sized males (Fig. 7.4). Thus, pairs of large females and small males appeared to be the most productive and they were also the most similar in absolute body size owing to slight sexual dimorphism in this species. Neither sex was a clear winner in the quest for an optimal-sized mate. Large females were as likely to pair with large males as with small ones and the resulting distributions of size differences between partners were statistically random (Marzluff and Balda 1988*b*).

Partner age and previous breeding experience may influence choice to a greater extent than physical characteristics. PJs exhibit strong assortative pairing for these characters and the age difference between partners influenced several fitness components (Marzluff and Balda 1988*a, b*). Partners that differed in age by 1 year or less tended to produce more yearlings (which also live longer and produce more offspring) than partners differing in age by a greater extent (Marzluff and Balda 1988*b*). A closer look at the influence of partner age on reproduction suggested that pairs including males that were more than 1 year older than females were especially unproductive. We used an ANCOVA (controlling for

Table 7.4 The influence of mate size on fitness in Pinyon Jays

Type of pair	Life span (years)	Yearlings while paired	Yearlings produced by sons	Life span of sons (years)
Males mated to heavy females	7.62	1.50	1.43	5.2
	3.48	2.28	1.51	2.9
	13	18	7	11
Males mated to light females	5.91	1.11	0.86	4.2
	2.12	1.27	0.77	2.1
	11	9	14	21
Females mated to heavy males	6.38	1.69	0.67	4.1
	3.40	2.56	0.82	2.1
	13	13	6	8
Females mated to light males	7.08	1.07	1.20	4.7
	2.75	1.27	1.15	2.5
	13	14	15	24

We report the average longevity, fecundity, and fitness of sons produced by birds mated with heavier than average and lighter than average mates. Table entries, from top to bottom, are the average, SD, and sample size. From Marzluff and Balda (1992).

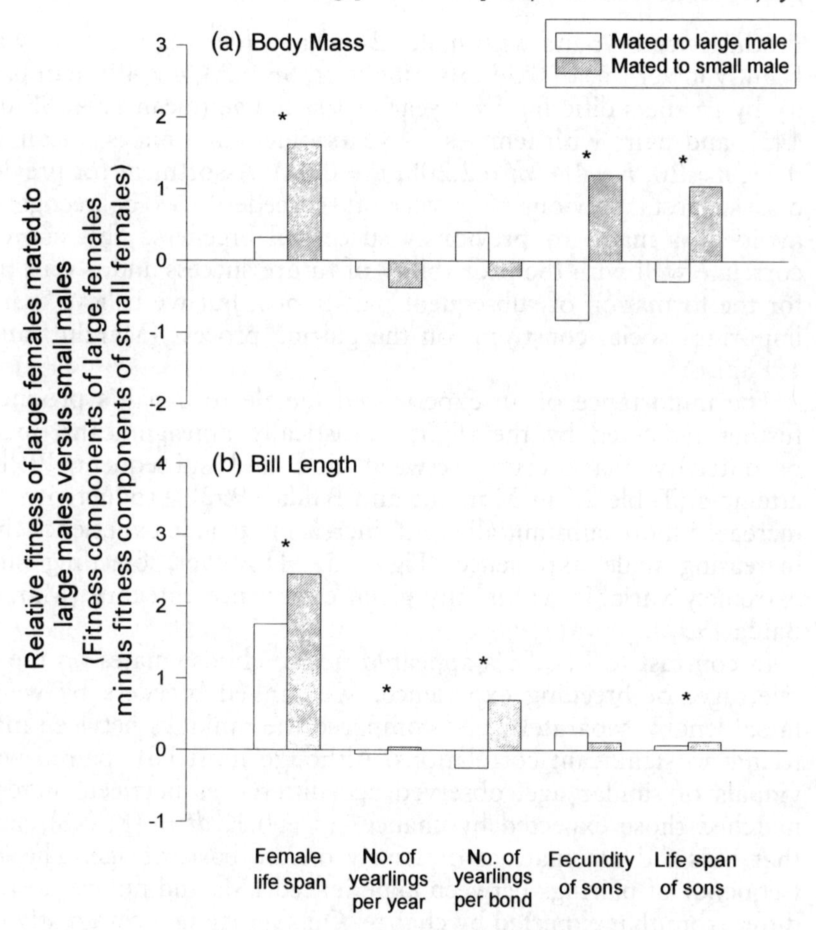

Fig. 7.4 Differences in fitness components of large female Pinyon Jays and small females mated to large and small males. For example, in our first comparison, life span of heavy females minus life span of light females equals 0.3 years for females mated to heavy males and 1.6 years for females mated to light males. Thus, relative to small females, large females live five times longer if they mate with small males (those more similar to them in size) than if they mate with large males. Positive differences indicate large females are more fit than small females. Negative differences indicate small females are more fit. Five components of fitness are compared for two measures of female size: body mass (a) and bill length (b). *Asterisks* indicate advantages of large females mated to small males. From Marzluff and Balda (1992).

pair duration, male and female breeding experience) to compare the number of offspring at the crèche stage produced per year by pairs where:

(1) males and females differed in age by 1 or fewer years;
(2) males were 2 or more years older than females;
(3) females were 2 or more years older than males.

Productivity by pairs with males 2+ years older than females was significantly lower (mean 0.34 offspring/year, SE 0.23, $n = 49$) than productivity by partners differing by a year or less in age (mean 1.04, SE 0.11, $n = 145$), and pairs with females 2+ years older than males (mean 1.51, SE 0.31, $n = 20$; $F = 4.4$, $df = 2{,}208$, $P = 0.01$). Assortment for previous success suggests previously unsuccessful breeders can be recognized and avoided as mates by previously successful breeders. This cue does not correlate well with the probability of future success and is only available for the formation of subsequent pair bonds, but we believe that it is an important social constraint on the pairing process (Marzluff and Balda 1988a).

The importance of an experienced female to a pair's productivity is further indicated by the slight, statistically nonsignificant, increase in productivity that occurs between first and subsequent PJ breeding attempts (Table 25 in Marzluff and Balda 1992). Productivity of a pair increased more substantially with increasing female experience than with increasing male experience (Fig. 7.5). However, breeding success is extremely variable within any given experience category (Marzluff and Balda 1992).

In contrast to PJs, FSJs appeared not to choose mates on the basis of size, age, or breeding experience. We ranked breeders by weight and tarsal length, separately, and compared the rankings between mates and found no significant correlations. Although most FSJs paired with individuals of similar age, observed age differences between mates closely matched those expected by chance ($\chi^2 = 0.1$, $df = 18$, NS), suggesting that FSJs did not mate assortatively on the basis of age. The observed frequency of pairings between experienced FSJs and novices also did not differ from that expected by chance. On average novices greatly outnumbered unmated experienced jays as potential mates in the population. In the absence of assortative mating based on experience, 69% of birds that lost mates were expected to pair with novices and 31% with other birds that lost mates. Observed proportions did not differ significantly from the expected values: of 185 re-pairings by experienced birds, 138, or 74.6%, were with novices ($\chi^2 = 0.7$, $df = 1$, NS).

For FSJs, average annual reproductive success increased with breeding experience for several years, then levelled off and later declined slightly (Table 7.5). Pairs in which at least one member had 3–7 years of breeding experience consistently showed the highest probabilities of producing independent offspring (Fig. 7.6). Only about half of all FSJs in their first or second year of breeding produced independent young (Table 7.5 $n = 238$ males, 233 females; proportions successful = 0.496 and 0.506, respectively), while two-thirds of those in breeding years 3–7 succeeded ($n = 238$ males, 235 females; proportions = 0.651 and 0.651; $\chi^2 = 11.8$ and 10.0 for males and females, respectively, $df = 1$, both sexes $P < 0.001$). Because the proportion of pairs having helpers increases with

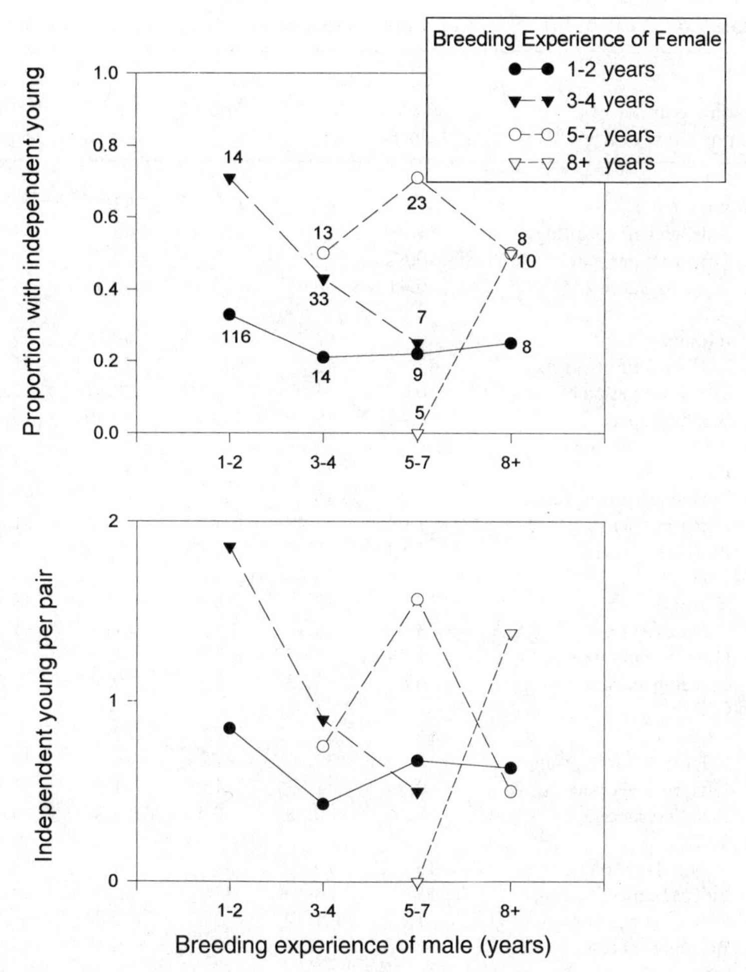

Fig. 7.5 Reproductive success of Pinyon Jays in relation to partner breeding experience. The influence of male breeding experience is tracked by comparing points connected by a line of equal breeding experience. The influence of female breeding experience is tracked by comparing points between lines at a single level of male breeding experience. Sample sizes are shown on upper graph.

breeding experience and helpers improve the performance of FSJ breeders (Woolfenden and Fitzpatrick 1984; Fitzpatrick and Woolfenden 1989), testing for effects of experience alone requires that we control for presence of helpers. Without helpers, breeders during their first 2 years (n = 310, sexes pooled) were indeed less successful than during breeding years 3–7 (n = 200; proportions successful: 0.445 versus 0.635; $\chi^2 = 17.6$, df = 1, $P < 0.001$). With helpers, breeders during years 1–2 (n = 161) were only slightly less successful than during years 3–7 (n = 273; proportions successful = 0.609 versus 0.663; $\chi^2 = 1.3$, df = 1, NS).

Table 7.5 Year of breeding and reproductive success for Florida Scrub Jays

Breeding year (proportion with helpers)	Males: Without helpers	Males: With helpers	Males: Pooled	Females: Without helpers	Females: With helpers	Females: Pooled
1 (0.30)						
n (pairs)	97	43	140	93	37	130
Independent offspring	65	57	122	66	34	100
Offspring per pair	0.67	1.33	0.87	0.71	0.92	0.77
Breeding success	0.41	0.65	0.48	0.44	0.51	0.46
2 (0.40)						
n (pairs)	56	42	98	64	39	103
Independent offspring	42	45	87	52	41	93
Offspring per pair	0.75	1.07	0.89	0.81	1.05	0.90
Breeding success	0.45	0.59	0.51	0.50	0.67	0.56
3 (0.47)						
n (pairs)	51	31	82	31	42	73
Independent offspring	54	46	100	42	68	110
Offspring per pair	1.06	1.48	1.22	1.35	1.62	1.51
Breeding success	0.59	0.74	0.65	0.81	0.71	0.75
4 (0.63)						
n (pairs)	18	40	58	26	36	62
Independent offspring	23	66	89	27	45	72
Offspring per pair	1.28	1.65	1.53	1.04	1.25	1.16
Breeding success	0.67	0.73	0.71	0.58	0.61	0.60
5 (0.71)						
n (pairs)	16	31	47	10	33	43
Independent offspring	15	32	47	16	46	62
Offspring per pair	0.94	1.03	1.00	1.60	1.39	1.44
Breeding success	0.56	0.58	0.57	0.70	0.67	0.68
6 (0.57)						
n (pairs)	13	19	32	14	17	31
Independent offspring	15	26	41	12	27	39
Offspring per pair	1.15	1.37	1.28	0.86	1.59	1.26
Breeding success	0.77	0.58	0.66	0.50	0.59	0.55
7 (0.53)						
n (pairs)	8	11	19	13	13	26
Independent offspring	14	17	31	11	20	31
Offspring per pair	1.75	1.55	1.63	0.85	1.54	1.19
Breeding success	0.75	0.64	0.69	0.46	0.69	0.58
>7 (0.61)						
n (pairs)	17	25	42	15	25	40
Independent offspring	14	27	41	16	35	51
Offspring per pair	0.82	1.08	0.98	1.07	1.40	1.28
Breeding success	0.47	0.52	0.50	0.47	0.64	0.58

FSJ 'WIMPs'

About half of FSJ breeders first pair at age 2 years. Males that remain nonbreeders for 2 or more years, and then pair and breed, perform less well than the early-breeding sample in several respects (Fitzpatrick and Woolfenden 1989; McDonald *et al.* MS). These later-breeding males had

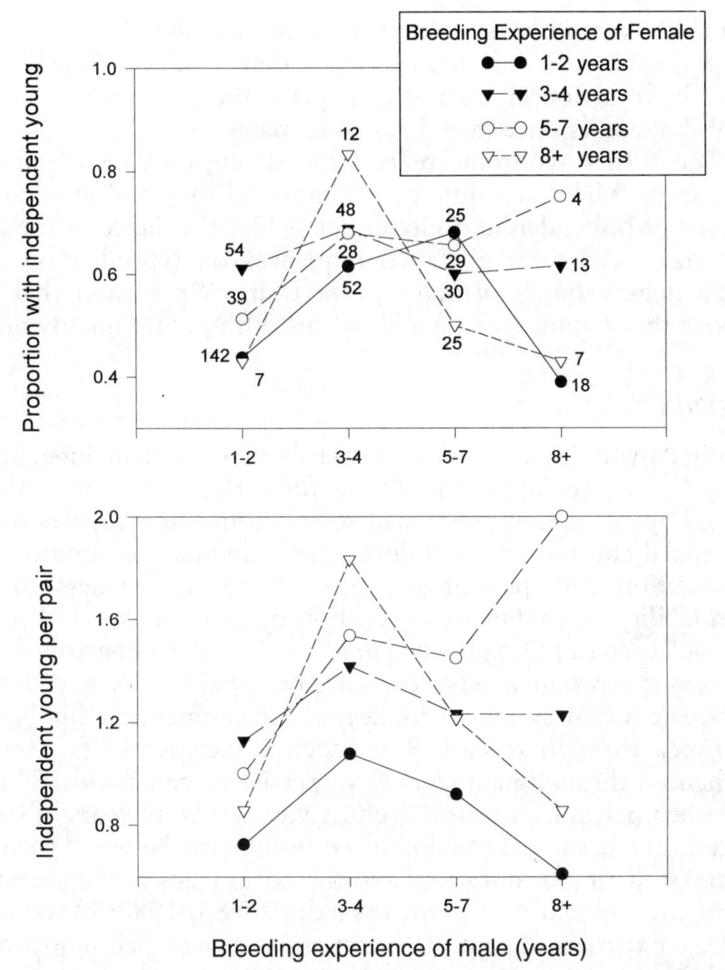

Fig. 7.6 Reproductive success of Florida Scrub Jay pairs composed of males and females with different years of breeding experience. Lines connect samples in which females have similar years of breeding experience. Sample sizes are shown on upper graph.

significantly lower survival and lower reproductive success. Furthermore, juvenile offspring of later-breeding males exhibited lower survival than did those of early-breeders. Early- and late-breeding females did not exhibit these differences in survival and reproduction.

These differences suggest variation in overall quality among males. Indeed, a behaviour pattern exists among certain male prebreeders that strengthens this interpretation. When a male breeder dies the neighbourhood bachelors assemble and vie for the vacancy. Losers in these competitions often include males that have wandered from one vacancy to another, always losing. These wandering, inept, male prebreeders

(WIMPs) often take several years to become breeders, or they die trying. The combination of characteristics that produces WIMPs is probably mostly behavioural, and may explain the poor performance of later-breeding males once they do acquire mates.

The WIMP-syndrome reflects the strong division of labour between the sexes. Males are more active in establishing and maintaining the territory (Woolfenden and Fitzpatrick 1977). Variance in the ability to do so may have more direct consequences on reproduction and survival than does variance in many other traits. We suspect that females can detect this variance, and avoid pairing with poorer quality males.

Discussion

Both Pinyon Jays and Florida Scrub Jays form lifelong, monogamous pair bonds, yet appear to do so for different reasons. Almost all PJs breed by age 2, and permanent social monogamy appears to be imposed by social constraints. Well-developed individual recognition and lifelong association with present and future breeding partners constrain mate availability and influence perception of mate quality in PJs. Many FSJs do not breed until age 3 or later, and social monogamy is favoured by increased reproductive performance once paired. These differences reflect differing social systems and degrees of competition for breeding space between the two species. Reproductive success of PJs is not obviously enhanced through maintenance of persistent pair bonds. The correlation between pair duration and productivity across all years of study was not significant because behavioural co-ordination between partners did not increase with pair duration, and the advantages of long-term pair bonds vary with annual conditions (Marzluff 1983, 1988; Marzluff and Balda 1992). Experienced pairs are better at concealing their nests from predators and nest earlier in the spring than less experienced pairs. We suspect, therefore, that pair duration is positively correlated with productivity in years when predation on eggs and nestlings is high and spring snowfall is minimal, and the opposite is true when predation rates are low and snowfall is extreme, because early nesting, experienced pairs have nests destroyed by snow (see Marzluff and Balda 1992). This annual variation in the relationship between pair duration and productivity probably confounds the ability of natural selection to favour the evolution of mate fidelity because the lifetime fitness of individuals that retain mates during the several, variable years of their life is equal to the fitness of individuals that change mates.

Criteria used in mate choice among PJs appear to allow compatible partners to pair. Partners are similar in age, previous reproductive success, and size. We suggest that similar partners are more compatible than dissimilar ones because they have similar demographic schedules, previous experiences, and social status. Similarity influences productivity, but does not change with pair duration. Therefore pair duration is not correlated with productivity. When mate choice is closely tied to an individual's physical or social attributes that influence fitness, then the initial choice of a partner establishes a pair's reproductive potential, and subsequent association with the individual has little chance of changing this potential. However, in systems such as that of the FSJ, competition for resources appears to drive variance in fitness. In this case, continued association with the same mate may promote more effective defence of the resources, and static attributes of the individual may be less important.

Choosing a mate based on similarities appears to be more important for PJs than for FSJs. PJs use many criteria in selecting appropriate mates (Johnson 1988*a, b*; Marzluff and Balda 1988*a b*), which can increase the chances of selecting a high quality mate (Burley 1981). Perhaps for this reason, PJs show little increase in reproductive success with pair duration. Although FSJ females do avoid poor quality males, mate choice focuses more on acquisition and defence of space than on physical characteristics. As the pair bond endures, increased efficiency and frequency of helpers improve reproductive success.

FSJs that remain with the same mate virtually always have higher reproductive output than FSJs that switch to a new mate. The only possible exception is a second-year breeder that lacks helpers (Table 7.2): its options include remaining with the same mate (expected reproductive success: 1.0 independent young) or changing to an experienced jay with helpers (expected reproductive success: 1.0 independent young). In this instance, mate fidelity and mate change on average produce equal results. In all other combinations of experiences and helpers, however, a paired jay can expect higher reproductive success by remaining with its current mate. For FSJs, therefore, mate fidelity confers a clear reproductive advantage over mate change.

The benefits of long-term pair bonds to FSJs are probably related to their reliance on all-purpose territories, in a demographic environment where breeding space is limited. Territory ownership is a requisite for breeding, and territories are difficult to obtain. Most jays that acquire breeding space have succeeded in competition for a jay that lost its mate and the unmated jay's territory. Jays that first pair with novices usually must also establish a new territory. The slightly lower survival of these novice–novice pair bonds (Fig. 7.1 (b)) perhaps reflects the difficulty of maintaining these new territories.

In FSJs, an individual's reproductive success is tied to its ability to use and defend resources on the territory. We suspect that this ability increases with pair duration. For example, as tenure on the land increases, each member of the pair becomes more familiar with the distribution of resources in the territory and with neighbouring, potentially competing families.

Although long-term breeding partnerships enhance FSJ reproduction, they do not enhance FSJ survival (Fig. 7.1(b)). FSJ breeders live year round in one all-purpose territory with the same mate, until one dies. After a breeder dies, its mate actively works to remain on its territory, because dispersal greatly increases the probability of death (Fitzpatrick and Woolfenden 1986). A disperser leaves behind its known resources, including acorn stores, and faces competition with territory-holders and other dispersers across the landscape. Even at home, neighbouring jays, and sometimes even resident helpers, try to usurp some or all of the unmated jay's territory. However, unmated jays usually retain their territory and obtain a new partner. We suspect that natural selection favours remaining in the same territory and that retaining the same mate may be a residual effect of this selection.

Because FSJ territories are held and defended year round, and close incest is avoided, establishing a new pair bond means that at least one member is new to the territory. Perhaps for this reason, new pairs often lose ground to neighbouring groups. Increased problems with territorial defence and the process of pairing itself cause new pairs to begin breeding later than continuing pairs. Later nests are less successful than early nests (Woolfenden and Fitzpatrick 1984). Finally, replacement breeders sometimes drive existing helpers away from the territory in the process of establishing their dominance (Woolfenden and Fitzpatrick 1977). Other times, jays that lose mates lose their own helpers by moving to a neighbouring territory while the helpers stay on their own natal ground. Establishing a new pair bond decreases the probability of breeding with helpers, which further decreases expected reproductive success. All these factors contribute to the importance of FSJ breeders remaining on and defending the same piece of ground year after year. Rewards from this strategy are symmetrical between the sexes. Moreover, both sexes are required for the successful rearing of young. The result is a system of mate fidelity in which FSJs stay mated both to one another and to their space, until they are parted by death.

Summary

The evolution of mate fidelity and social monogamy in two New World jays differs because of differences in social organization and use of space and resources. Reproductive success in the nonterritorial PJ appears to be determined more by the qualities of an individual and compatibility with its mate than by resources defended by a pair. As a result, mate choice is accurate and based on many physical and experiential traits. The reproductive potential of a pair appears to be set at the time of pair bond formation and increases little with increasing pair duration. Mate fidelity is high in PJs because divorce is constrained by a social system that features extensive individual recognition and memory of individual past performance. Poor reproducers, who seemingly would benefit most by divorce, are not selected as mates by previously successful, high quality individuals. In contrast, reproductive success in the territorial FSJ is tied to an individual's ability to use and defend resources on the territory. An individual's physical and experiential attributes are not essential to successful exploitation of a territory and therefore mate choice does not appear to be based on these criteria. Rather, successful exploitation is enhanced with each successive year a pair spends on the territory. These leads to a positive relationship between pair duration and reproductive success for FSJs, which might explain why mate fidelity has evolved in this species. However, mate fidelity may simply result from the equal importance to males and females of remaining on and defending the same piece of ground year after year and the fact that both sexes are required to rear young successfully.

Acknowledgements

JMM and RPB thank the many students and volunteers that have aided the PJ study, especially Gene Foster, Katharine Bartlett, Jane Balda, Elaine Morrall, and Bill and Judy Burding. Financial support for the PJ study was provided by Northern Arizona University, Sigma Xi, The Chapman Memorial Fund, The Wilson Ornithological Society, and Achievement Rewards for College Scientists. JWF and GEW thank Archbold Biological Station for its continued support of their long-term study. Two grants from the US National Science Foundation BSR 896276 and BSR 90210902) also supported the FSJ study. We greatly appreciate the assistance of numerous volunteers and students during hundreds of monthly censuses.

References

Burley, N. (1981). Mate choice by multiple criteria in a monogamous species. *American Naturalist*, **117**, 515–28.

DeGange, A. R., Fitzpatrick, J. W., Layne, J. N., and Woolfenden, G. E. (1989). Acorn harvesting by Florida scrub jays. *Ecology*, **70**, 348–56.

Fitzpatrick, J. W. and Woolfenden, G. E. (1986). Demographic routes to cooperative breeding in some New World jays. In *Evolution of behavior* (ed. M. Nitecki and J. Kitchell), pp. 137–60. University of Chicago Press, Chicago.

Fitzpatrick, J. W. and Woolfenden, G. E. (1989). Florida Scrub Jay. In *Lifetime reproduction in birds*. (ed. I. Newton), pp. 201–18. Academic Press, London.

Francis, A. M., Hailman, J. P., and Woolfenden, G. E. (1989). Mobbing by Florida scrub jays: behaviour, sexual asymmetry, role of helpers and ontogeny. *Animal Behaviour*, **38**, 795–816.

Johnson, K. (1988*a*). Sexual selection in pinyon jays I: female choice and male–male competition. *Animal Behaviour*, **36**, 1038–47.

Johnson, K. (1988*b*). Sexual selection in pinyon jays II: male choice and female–female competition. *Animal Behaviour*, **36**, 1048–53.

McArthur, P. D. (1982). Mechanisms and development of parent–young vocal recognition in the pinion jay (*Gymnorhinus cyanocephalus*). *Animal Behaviour*, **30**, 62–74.

McDonald, D. B., Fitzpatrick, J. W., and Woolfenden, G. E. (MS). Actuarial senescence and demographic heterogeneity in the Florida Scrub Jay.

McGowan, K. J. (1987). Social development in young Florida Scrub Jays (*Aphelocoma c. coerulescens*). Unpublished Ph.D. thesis, University of South Florida, Tampa.

McGowan, K. J. and Woolfenden, G. E. (1989). A sentinel system in the Florida scrub jay. *Animal Behaviour*, **37**, 1000–6.

McGowan, K. J. and Woolfenden, G. E. (1990). Contributions to fledgling feeding in the Florida scrub jay. *Journal of Animal Ecology*, **59**, 691–707.

Marzluff, J. M. (1983). Factors influencing reproductive success and behavior at the nest in pinyon jays (*Gymnorhinus cyanocephalus*). Unpublished M.Sc. thesis. Northern Arizona University, Flagstaff.

Marzluff, J. M. (1988). Do pinyon jays use prior experience in their choice of a nest site? *Animal Behaviour*, **36**, 1–10.

Marzluff, J. M. and Balda, R. P., (1988*a*). The advantages of, and constraints forcing, mate fidelity in Pinyon Jays. *Auk*, **105**, 286–95.

Marzluff, J. M. and Balda, R. P., (1988*b*). Pairing patterns and fitness in a free-ranging population of Pinyon Jays: what do they reveal about mate choice? *Condor*, **90**, 201–13.

Marzluff, J. M. and Balda, R. P., (1989). Causes and consequences of female-biased dispersal in a flock-living bird, the pinyon jay. *Ecology*, **70**, 316–28.

Marzluff, J. M. and Balda, R. P., (1992). *The pinyon jay: behavioral ecology of a colonial and cooperative corvid*. T. and A. D. Poyser, London.

Mumme, R. L. (1992). Do helpers improve reproductive success? An experimental analysis in the Florida scrub jay. *Behavioural Ecology and Sociobiology*, **31**, 319–28.

Quinn, J. S., Woolfenden, G. E., Fitzpatrick, J. W., and White, B. N. (1990). DNA fingerprinting analysis of Florida Scrub Jay parentage. *Abstract no. 128, American Ornithologists' Union*, annual meeting, Los Angeles.

Woolfenden, G. E. (1976). A case of bigamy in the Florida Scrub Jay. *Auk*, **93**, 443–50.

Woolfenden, G. E. and Fitzpatrick, J. W., (1977). Dominance in the Florida Scrub Jay. *Condor*, **79**, 1–12.

Woolfenden, G. E. and Fitzpatrick, J. W. (1984). *The Florida scrub jay: demography of a cooperative-breeding bird*. Monographs in Population Biology No. 20, Princeton University Press, Princeton.

Woolfenden, G. E. and Fitzpatrick, J. W. (1986). Sexual asymmetries in the life history of the Florida Scrub Jay. In *Ecological aspects of social evolution:*

birds and mammals. (ed. D. Rubenstein and R. W. Wrangham), pp. 87–107. Princeton University Press, Princeton.

Woolfenden, G. E. and Fitzpatrick, J. W. (1991). Florida Scrub Jay ecology and conservation. In *Bird population studies: relevance to conservation and management*. (ed. C. M. Perrins, J,-D. Lebreton, and G. J. M. Hirons), pp. 542–65. Oxford University Press, Oxford.

8 *Partnerships in promiscuous Splendid Fairy-wrens*

ELEANOR RUSSELL AND IAN ROWLEY

Introduction

Splendid Fairy-wrens *Malurus splendens* are small insectivores living in the dry shrublands of Australia, where they forage under and through the undergrowth, generally within a metre of the ground. Groups of two to eight birds remain together all year round in contiguous territories (1–9 ha) persisting in much the same locations from year to year. The basic social unit is the pair that remains together for as long as both survive. However, mating is promiscuous, with more fertilizations occurring outside the pair than within it. Many offspring remain in the family group after they reach sexual maturity, helping in territory defence and care of young, in a co-operative breeding system. The size and composition of a group depends on breeding success in the previous season and whether vacancies became available to which helpers could have dispersed. Many males and some females fill breeding vacancies in their natal territory, so that about 20% of pairs are closely related.

We studied a population of Splendid Fairy-wrens in eucalypt woodland for 17 years, and traced the apparent breeding success of 206 pairs from 1304 colour-banded individuals. All the suitable habitat in the study area was occupied by group territories. In this chapter, we ask why pairs, often close relatives, stay together as a social unit while mating promiscuously outside the pair. We investigate whether Splendid Fairy-wrens are more productive with the same partner, and whether some partners are better than others, and show that because vacancies are limited, the choice of a social partner is dictated by the availability of a breeding vacancy in a territory, while extra-pair copulation provides a choice of mates.

Background, study site, and techniques

Splendid Fairy-wrens are small (10 g), sexually dichromatic passerines. Adult males are entirely brilliant blue and black and females are plain grey-brown; males become blue in the first spring after they hatch, but enter a plain female-like eclipse plumage in winter. The pair or group forage and defend the territory as a unit, resting and allopreening together during the day, and roosting together at night. Both members of a pair feed the nestlings, although the female generally makes more feeding visits. When helpers are present, feeding rates of nestlings are not greatly increased, but the work load of the senior female is reduced (Rowley 1981; Rowley *et al.* 1989).

Our study site covered 120 ha on the western edge of the Darling Range east of Perth (31°56′S, 116°03′E). South-western Australia has a Mediterranean climate, with mild, wet winters and hot, dry summers. The habitat is fire prone; a period free from fire, 1978–1984, was followed by a fire in January 1985 that burnt more than 90% of the study area (Rowley and Brooker 1987; Brooker and Rowley 1991). As the vegetation became denser during the fire-free period, survival, productivity, and population density increased from 40 birds/100 ha in 1978 to a peak of 101 birds/100 ha in 1983. Some new territories were established, but the increase in population density led chiefly to an increase in group size and in the proportion of the population that were helpers (Russell and Rowley 1993*b*).

A group studied for one year was counted as one group-year, and from 1973 to 1990 the study encompassed 346 group-years. All birds were individually colour-banded (n = 1304), the majority as nestlings 6–7 days old. From August to January, daily observations were made, to locate all nests and to record the birds attending them; 830 nesting attempts were recorded. We estimated survival and dispersal from sightings of marked individuals, and censused all groups each autumn (March–April) and winter (June–July). Pair duration was based on all pairs known from 1973 to 1989. A pair was a senior male and breeding female persistently sighted in a territory. Changes of partner, including divorce, were obvious at census.

Pair duration was estimated by the number of nests and breeding seasons during which a male and female were associated. Reproductive success of females (per nest and per breeding season) was assessed by the numbers of fledglings and yearlings produced; the latter is a minimum figure because some young females disperse. Breeding life span was estimated as the number of breeding seasons in which a male or female was associated with one or more nests.

We consider all pairs that built one or more nests, a total of 360 pair-years, with information on reproductive success for 351 pair-years. Annual mate change was the number of pairs in which a partner was replaced during the year from 1 September. In considering the characteristics of replacement males and females we have included the formation of 16 new groups in which one or more of the birds was of known history.

We used generalized linear models (GLIM 4) to examine relationships between reproductive success (annual production of fledglings), pair duration, and partner characteristics. Since replacement of females could occur during the breeding season, we classified reproductive success scores according to whether the female was present for the whole or part of the breeding season (factor PartBS). The assumption of normally distributed errors was justified using Q–Q plots (Baker and Nelder 1978).

Demographics

Splendid Fairy-wrens breed in the austral spring and summer, with eggs laid from September to January. Clutch size is almost always three; frequently two broods are raised, resulting in 2.9 fledglings, 2.0 independent young and 0.93 yearlings (Russell and Rowley 1993*b*). Nestlings remain in the nest for 11–12 days, and fledged juveniles are fed and escorted by the group for approximately a month. After independence, juveniles remain with the group, and do not disperse until at least 6 months old, if at all (Rowley *et al.* 1991; Russell and Rowley 1993*a, b*). Predators accounted for 24% of nest failures and 22% (0–52%) of nests were parasitized by Horsfield's Bronze-cuckoo *Chrysococcyx basalis*.

Annual survival of senior and helper males was 70% and that of breeding females 59% (n = 17 years). Survival of juveniles from fledging to 1 year was 31% (11–59%). In most years, the sex ratio was male biased (mean 1.3). Although winters were not severe, a combination of rain, cold, and food shortage sometimes led to increased mortality; survival was better in years of low rainfall (Rowley *et al.* 1991; Brooker and Rowley, in press). On average, males lived longer than females (4.2 years, SE 0.2, range 1–13: 3.0 years, SE 0.1, range 1–9).

Extra-pair behaviour

Older males (including helpers) make frequent excursions into neighbouring territories both before and during the breeding season, which we interpret as advertising for or seeking extra-pair copulations. Philandering males perform striking displays to females, emphasizing their brilliant plumage. In one display, 'petal carrying', males pick a flower petal of a colour that contrasts with their plumage and carry it in approaches to a female. These displays are directed by a male to a neighbouring female, rarely to his social partner (Rowley and Russell 1990; Rowley 1991). A female may accept or reject a displaying male.

During the winter and early in the breeding season (August–September), there are considerable age-related differences in the plumage of males. Most males enter an eclipse female-like plumage after the autumn moult (February–April). Older males resume their blue nuptial plumage earlier (May–July), while immature males do not do so until October, and may be imperfect even then. Females may use this variation in timing to assess the relative quality of prospective extra-group mates (Mulder and Magrath 1994).

An allozyme study showed that between 65% and 100% of nestlings could not have been fathered by any male within the social group to which the female belonged, but all nestlings were compatible with her (Brooker *et al.* 1990). In some broods, all young were incompatible with the male or males in the group, in others, only one or two were fathered by an outside male; in 2 out of 30 broods, the young were fathered by more than one outside male.

Because of this high degree of promiscuity, we refer to the oldest, territory-holding male in the group as the senior male rather than the 'breeding male'. This male consorts closely with his female throughout the year, and during the breeding season he accompanies his group female during nest building and when she is foraging between bouts of incubation; helpers remain apart until nestlings require feeding. We have little evidence of mate guarding and obviously cuckoldry is common; helper males may also father young (Brooker *et al.* 1990), and all males in the group care for nestlings. Although the senior male may philander in nearby territories, he contributes significantly to the feeding of nestlings and fledglings in the group and to their defence.

Results

The pair bond

Because mating was promiscuous and incompletely understood, pair bonds in Splendid Fairy-wrens refer only to a long-lasting social partnership between the senior male and breeding female. In 346 group-years, 66% of the groups consisted of more than a simple pair; the extra birds were mature adults, but they rarely achieved a mating. In years of high population density, 1981–1988, 8.7% of helper females (usually the daughter of the breeding female) nested as secondary females in their natal territory (Rowley *et al.* 1989). Until long-term DNA studies of paternity are completed (at present in progress for Superb Fairy-wrens *M. cyaneus*, Mulder *et al.* 1994), it is prudent not to call various groups polygynous, polyandrous, or polygynandrous.

Senior males (n = 118) held their position for 3.1 (SE 0.2) breeding seasons with 1.9 (SE 0.1, range 1–7) partners. Females (n = 146) remained as breeders for 2.3 (SE 0.1) seasons with 1.5 (SE 0.1, range 1–4) partners. Pairs were associated for 1.7 (SE 0.1, n = 206) breeding seasons. Fifty-seven of the 206 pairs were associated for only one nest; the high mortality of these, mostly unproductive, novice females biases the mean and median pair duration and detracts from the significance of the fewer, longer pair bonds. Pairs had a mean of 3.7 nests (median 2, mode 1). In this multi-brooded species, a pair could have three nests during one breeding season, or three nests over 2 years, with a much longer pair duration.

Relatedness between partners

Partners in a social pair were likely to be related, with 40% of replacements coming from within the group (Table 8.1). Because of female dispersal and frequent extra-pair copulations, the probability of close inbreeding was lower for females than for males. In 18.5% of pair bonds, the partners were apparently brother–sister, father–daughter, or mother–son pairs (Table 8.2). Apparent brother–sister pairs had the same mother, but may have been only half-siblings, with different fathers. Some females paired with their 'social' father, but there is a high probability that he was unrelated. Males paired with their mother were definitely closely related (r = 0.5), but her offspring were more likely to be fathered by extra-group males than by her son. Eighteen of 53 males (34%) that 'inherited' a place in their natal group paired with a female that had replaced their mother. Experienced females paired with their sons produced 1.8 fledglings (SE 0.2) from 51 nests, no less productive than experienced females with unrelated males; 1.7 fledglings (SE 0.1) from 282 nests (Rowley *et al.* 1993). Since at least 66% of nestlings are the result of extra-pair copulations (Brooker *et al.* 1990), there is clearly no inbreeding cost from pairing socially with a close relative.

Pair formation, mate choice, and partner characteristics

In general, if one member of a pair disappeared, the survivor remained on the territory and acquired a new partner from within the group or from outside (Table 8.1). Groups and territories persisted for many years despite the turnover of individuals. Suitable habitat was fully occupied by territories all year round in most years; territories became vacant only occasionally after fire and were filled by new pairs as the vegetation recovered. A vacancy occurred at any time of year and was filled rapidly. We saw no obvious competition. In Superb Fairy-wrens, vacancies created by experimental removals were filled within hours (Pruett-Jones and Lewis 1990; Ligon *et al.* 1991).

The annual rate of mate change (31% males and 40% females) translates to one or two vacancies per month, the occurrence of which could not be predicted. A surplus of helpers was available to fill any vacancy; depending

Table 8.1 The origin and age of 206 replacement mates in existing pairs and new pairs

Age of replacement mate (years)	Male				Female			
	From outside (n=32)	Helper (n=53)	New pair (n=16)	Total (n=101)	From outside (n=97)	Helper (n=32)	New pair (n=16)	Total (n=145)
<1	3.0	12.9	4.0	19.9	13.1	9.7	2.8	25.6
1+	20.8	20.8	9.9	57.5	44.8	8.3	6.2	59.3
2+	6.0(2)	6.9	1.0	13.9	4.8(4)	3.4(2)	0.7	8.9
3+	1.0(1)	7.9	—	8.9	3.5(2)	0.7(1)	—	4.2
4+	1.0	4.0	1.0	6.0	0.7(1)	—	1.4(1)	2.1
Total	31.8	52.5	15.9		66.9	22.1	11.0	

Replacement mates were identified as coming from within the group (=helper) or from outside the group. Ages shown are minimum ages. Values are percentages of males or females. Values in parentheses indicate birds with previous breeding experience, as a senior male in another territory, as a secondary female in her natal territory, or as a breeding female in another territory (3 males, 13 females).

Table 8.2 Pair bonds formed between related individuals

	Mother/ son	Father/ daughter	Brother/ sister	Half-sibling	Other[1]	Not related	Unknown	Percentage with r >0.25
Pair-years (n=351)	8.8	6.3	3.7	1.1	3.1	58.1	18.8	12.5
Pair bonds (n=206)	9.7	4.9	3.9	1.0	2.9	60.7	17.0	13.6

Many pairs were less closely related than their apparent relationship indicates, because of high levels of extra-pair copulations. Values are percentages.

[1] 'Other' includes putative cousins and aunt/nephew relationships. Some were related (r = 0.125) through the female line. Only mother/son and brother/sister were definitely related with r at least 0.25.

on the time of year, from 6 to 8 (range 1–37) male and from 2 to 3 (range 0–14) female helpers were present (see Fig. 1 in Russell and Rowley 1993*a*).

A mate replacement came from within the group if a helper of the right sex was available. Otherwise, a vacancy was filled by a bird from an adjoining territory. Single male helpers rarely left their natal groups to fill such vacancies elsewhere; when more than one male helper was present, younger rather than older helpers dispersed. On the other hand, single female helpers moved to a vacancy whenever one became available (Russell and Rowley 1993*a*). Few birds of either sex bred more than two territories distant from their natal territory. More than half of all replacement males were previous helpers within the group (Table 8.1), so that established females generally acquired familiar partners. Twenty-two per cent of replacement females had been helpers within their group. Most replacement males and females, and birds forming a new pair where none had been before, were breeding for the first time (Table 8.1).

In 32% of newly formed pairs both sexes were young birds (1 or 2 years old); in the remainder, an older male or female acquired a younger partner. Since most females had become established as breeders by their third year, 71% of replacement females were 1 or 2 years old. Evidence of assortative mating by age was slight; the correlation between the ages of males and females at pair formation was low (r = 0.24, n = 197, $P < 0.05$). Novice males and females made up 23% of new pairings; in 11%, both male and female were experienced. The remaining 66% involved experienced females with novice males (26%) and experienced males with novice females (40%).

The pattern of staggered replacement means that, in general, the duration of individual pair bonds was less than the total breeding life of either male or female partner. Competition for a few available vacancies and the mechanism of pair formation suggest that there was little opportunity in the process to exercise a choice of social partner. Mate choice was probably of most significance in female choice of a male partner in extra-pair copulations.

Mate change

Over the 17-year study period, 56% of 360 pairs that were together for at least part of the breeding season were not together at the beginning of the next. Annual mate change ranged from 25 to 82% (mean 54.7%). Each year, a mean of 31% of males and 40% of females were replaced. In 203 cases of mate change, 97% were due to disappearance (death) of one or both partners and 3% were due to divorce. Of the seven cases of divorce, six occurred after successful breeding seasons. Divorced pairs had therefore succeeded in 85.7% of cases, whereas the long-term population figure was 84.5% (Rowley *et al.* 1991). This and another unusual pair history where a pair remained together for 4 years although the female never laid an egg (he later re-paired and raised young) suggests that reproductive performance did not influence mate fidelity.

Breeding males and females that disappeared were replaced so quickly that even if the replacement took place during the breeding season, the surviving partner never failed to breed in that year (Russell and Rowley 1993*a*). Of 185 new pairs, half (n = 91) involved novice males pairing for the first time and the other half involved replacing previous partners. Of these experienced males, 78% of their replacement partners were novice females. Novice females were relatively unsuccessful breeders (Rowley *et al.* 1991) so that a novice replacing an experienced breeder represented a cost to the male in few fledglings produced in that year, and in the following year there could be increased costs of territory defence and vigilance, if he had fewer helpers. For females, there was no drop in productivity associated with a new inexperienced partner (see below).

Reproductive success

Previous analyses revealed that several factors influenced annual reproductive success in Splendid Fairy-wrens, including environmental factors affecting the length of the breeding season, the incidence of nest predation, group size (some of which are helpers), and the age and experience of both sexes (Rowley *et al.* 1991; Russell and Rowley 1993*b*). In the following analyses we are concerned with the effect of mate familiarity and partner characteristics on reproductive success. Therefore, we controlled for year, age of partners, number of helpers, and whether the female was present for all or part of the breeding season (PartBS).[1]

Mate familiarity effect

After controlling for year, PartBS, female age and number of helpers, there was no effect of pair duration on reproductive success ($\Delta D = 0.35$, $\Delta df = 1$, NS).

Partner characteristics

To investigate which sex had most influence on the reproductive success of the pair, we divided birds into three age-classes (male, 1–2, 3–4, and >4 years; female, 1, 2–3, and >4 years). The highest reproductive success was for older females with males 3–4 years old (Fig. 8.1). Young females breeding for the first time at 1 year had consistently low reproductive success regardless of male age, whereas male age had little effect on the success of older females. This suggests that there was little incentive, on the grounds of age alone, for a female to choose between male social partners. For males, older females were the preferred option, and this was achieved by males more than 3 years old; younger males tended to mate with younger females. Perhaps the surest way for a male to improve his reproductive success was to remain with a female until she was a year older.

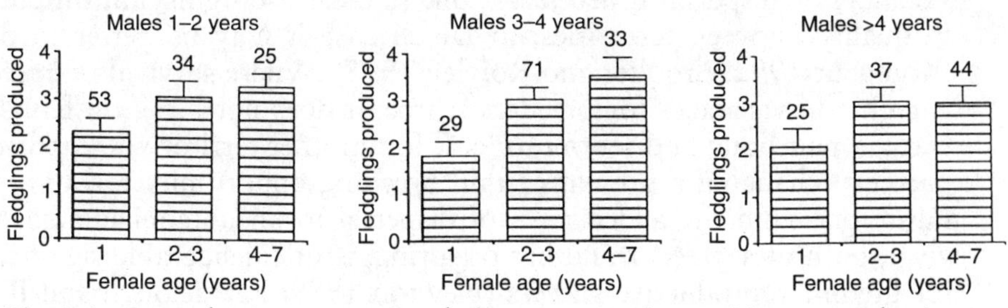

Fig. 8.1 Reproductive success in relation to partner's age. Mean reproductive success (fledglings per year) for three age-classes; values above columns indicate the number of females. Older females should be preferred by males of any age.

[1]Controlling for confounding variables (year, PartBS), female age significantly affected reproductive success ($\Delta D = 66.5$, $\Delta df = 1$, $P < 0.001$). Male age had no influence on reproductive success, after controlling for year and PartBS ($\Delta D = 0.82$, $\Delta df = 1$, NS).

Discussion

In a resident, territorial species such as Splendid Fairy-wrens, relatively long-lived for such a small bird, the pair bond is part of everyday life for several years, unlike the short-term partnerships that are renewed each year during the breeding season, as happens in many northern hemisphere passerines, where choosing a mate is usually coupled with choosing a territory (e.g. Payne and Payne Chapter 17). In situations such as our study, where all available habitat is occupied all year round by established pairs or groups, an important limiting resource for males and females is a breeding vacancy in an established territory. The social structure of permanent territories occupied by resident pairs or groups is coupled with a promiscuous mating system; choice of a social partner and choice of a male to fertilize eggs are separate. What we know about the filling of vacancies in social pairs suggests little opportunity for choice. In other words, rather than choice of pair bond members, Splendid Fairy-wrens may exercise choice for copulation partners (Rowley and Russell 1990).

The availability of territories and competition for them are important factors in the choice of partner and territory (Rowley and Russell 1990; Russell and Rowley 1993*a*). Choice is rarely between several territory vacancies; rather, birds have the option of remaining in their natal territory, moving to a known vacancy nearby, or wandering in search of a chance opening. Few young birds took the last course; the majority accepted a vacancy in an established territory with a bird of the opposite sex already present.

We suggest that it is the importance of the permanent territory that contributes to the persistence of the pair bond. For females, a territory is a prerequisite for breeding. For a long-lived species in a fully occupied habitat, it is important for a female to take the first available territory vacancy, by dispersal if necessary, unless there are significant differences in quality between territories, in which case it may be better to delay (Komdeur 1992; Brooker and Rowley 1995). Where survival of breeders is high and vacancies for breeders scarce, it does not benefit a female to break a pair bond and set out to look for another territory (and breeding partner), changing a present certain breeding opportunity for a possible future one with the added risks of dispersal to an unfamiliar area. Survival, both of a breeder and her offspring, is of considerable significance for lifetime reproductive success (Newton 1989; Barraclough and Rockwell 1993); the presence of helpers in some circumstances can increase both the productivity of a female and her survival. The territory also provides a focal point where she can be found by neighbouring males (Russell and Rowley 1988; Brooker and Rowley 1995).

The costs of a persistent pair bond depend on the quality of partner or territory. Mate quality can be addressed by appropriate choice of extra-pair copulations. In a strictly monogamous mating system, a male would be limited to the production of offspring from only one female who might be of poor quality. By inseminating more than one female, the risks of failure are spread and reproductive success may be higher for a few preferred males. Since males survive better than females, a female could be paired with a male of inferior quality for the whole of her breeding life span; promiscuous mating provides a choice of males, and the chance to mate with high quality males. Any loss of help from the senior male because of his extra-pair copulation activities is compensated for by the helpers in the co-operatively breeding group. Thus the promiscuous mating system compensates the female for the lack of choice of territory.

For a male living in a population with a male–biased sex ratio, the possession of a territory is a prerequisite to gaining a partner. Keeping a current female partner is more certain than competing for another one when females are scarce. A new partner would most likely be young and inexperienced and therefore less productive. A persistent pair bond ensures that he retains the higher productivity of an older female even though he does not father all her progeny. The territory provides a familiar area in which to forage, and from which to make excursions to surrounding territories, seeking extra-pair copulations with familiar females. The production of offspring that remain as helpers has several benefits to a male, regardless of whether they are his progeny. They help to defend the territory and they assist the female, releasing him to pursue extra-pair copulations. The presence of female helpers increases the reproductive success of the breeding female, and the size of a territory increases with the number of male helpers, increasing the number of adjacent territories and thus the senior male's access to extra-pair copulations (Brooker and Rowley 1995). While the advantages of this system to the senior male may not be as great if he fathered all the young in his group, they seem to be substantial because of the number he does father.

Splendid Fairy-wren pairs are together in their territory not only for the 4–5 months of the breeding season, but for the whole of one or more years with the same partner. Lifetime reproductive success depends on survival and length of life as much as on annual reproductive success. The life of the pair outside the breeding season, their survival, and that of other group members in the territory, assume an importance not seen in birds that live together for a short breeding season and live independent lives for the rest of the year. It is entirely possible that familiarity with a territory and a partner may enhance survival. Indeed, as Marzluff and Balda (1992) suggest for the Pinyon Jay *Gymnorhinus cyanocephalus*, there may be an added benefit in compatibility, so that as well as 'good birds' (Coulson and Thomas 1985), there are 'good pairs'.

Summary

In the co-operatively breeding Splendid Fairy-wren, socially monogamous pairs or groups live in contiguous territories throughout the year. Mating is promiscuous, with high levels of extra-pair copulations, and more offspring are sired by extra-group males than by the female's social partner. Divorce was rare, and not related to reproductive failure. Apparent reproductive success did not improve with an increasing number of years with the same mate. For males, apparent reproductive performance improved with increasing female age. We suggest that in Splendid Fairy-wrens breeding vacancies are a limited resource, filled as soon as they occur with little opportunity for choice of territory or partner. The opportunity for mate choice occurs each breeding season, with males seeking extra-pair copulations, and the female choosing who fertilizes her eggs. We suggest that the persistent pair bond is a consequence of the importance of permanent territory maintained all the year round in saturated habitat.

Acknowledgements

Many people helped to find nests and groups over the years. We thank Craig Bradley, Michael, Lesley and Belinda Brooker, Graeme Chapman, Joe Leone, Bob and Laura Payne. Censuses in the 1990 breeding season were done by the Brookers and Joe Leone. We are indebted to Graeme Arnold and Lesley Brooker for help with the analysis of reproductive success. We thank Jeff Black and two anonymous referees for helpful criticisms of earlier drafts of the paper.

References

Baker, R. J. and Nelder, J. A. (1978). *The GLIM system*. Royal Statistical Society, Oxford.

Barraclough, G. F. and Rockwell, R. F. (1993). Variance of lifetime reproductive success: estimation based on demographic data. *American Naturalist*, **141,** 281–95.

Brooker, M. G. and Rowley, I. (1991). The impact of wildfire on nesting of birds in heathland. *Wildlife Research*, **18,** 249–63.

Brooker, M. G. and Rowley, I. (1995). The significance of territory size and quality in the mating strategy of the splendid fairy-wren. *Journal of Animal Ecology*, **64,** 614–27.

Brooker, M. G., Rowley, I., Adams, M., and Baverstock, P. R. (1990). Promiscuity—an inbreeding avoidance mechanism in a socially monogamous species? *Behavioral Ecology and Sociobiology*, **26,** 191–200.

Coulson, J. C. and Thomas C. (1985) Differences in the breeding performance of individual Kittiwake Gulls, *Rissa tridactyla* (L.). In *Behavioural ecology:*

ecological consequences of adaptive behaviour (ed. R. M. Sibley and R. H. Smith), pp. 489–503. Blackwell, Oxford.

Komdeur, J. (1992). Importance of habitat saturation and territory quality for evolution of cooperative breeding in the Seychelles warbler. *Nature*, **358,** 493–5.

Ligon, J. D., Ligon, S. H., and Ford, H. A. (1991). An experimental study of the bases of male philopatry in the cooperatively breeding superb fairy-wren *Malurus cyaneus*. *Ethology*, **87,** 134–48.

Marzluff, J. M. and Balda, R. (1992). *The pinyon jay: behavioural ecology of a colonial and cooperative corvid.* T. & A. D. Poyser, London.

Mulder, R. A. and Magrath, M. J. L. (1994). Timing of pre-nuptial molt as a sexually selected indicator of male quality in superb fairy-wrens. *Behavioral Ecology*, **5,** 393–400.

Mulder, R. A., Dunn, P. O., Cockburn, A., Lazenby-Cohen, K. A., and Howell, M. J. (1994). Helpers liberate female fairy-wrens from constraints on extra-pair mate choice. *Proceedings of the Royal Society, Series B.*, **255,** 223–9.

Newton I. (1989). Synthesis. In *Lifetime reproduction in birds* (ed. I. Newton), pp. 441–69. Academic Press, London.

Pruett-Jones, S. G. and Lewis, M. G. (1990). Sex ratio and habitat limitations promote delayed dispersal in superb fairy-wrens. *Nature*, **348,** 541–2.

Rowley, I. (1981). The communal way of life in the splendid wren *Malurus splendens*. *Zeitschrift für Tierpsychologie*, **55,** 228–67.

Rowley, I. (1991). Petal-carrying by fairy-wrens of the genus *Malurus*. *Australian Bird Watcher*, **14,** 75–81.

Rowley, I. and Brooker, M. G. (1987). The response of a small insectivorous bird to fire in heathlands. In *Nature conservation: the role of remnants of native vegetation* (ed. D. A. Saunders, G. W. Arnold, A. Burbidge, and A. J. M. Hopkins), pp. 211–18. Surrey Beatty, Sydney.

Rowley, I., and Russell, E. M. (1990). 'Philandering'—a mixed mating strategy in the splendid fairy-wren *Malurus splendens*. *Behavioral Ecology and Sociobiology*, **27,** 431–7.

Rowley, I., Russell, E. M., Payne, R. B., and Payne, L. L. (1989). Plural breeding in the splendid fairy-wren *Malurus splendens* (Aves: Maluridae), a cooperative breeder. *Ethology*, **83,** 229–47.

Rowley, I., Brooker, M. G., and Russell, E. M. (1991). The breeding biology of the Splendid Fairy-wren *Malurus splendens*. *Emu*, **91,** 197–221.

Rowley, I., Russell, E. M., and Brooker, M. G. (1993). Inbreeding in birds. In *The natural history of inbreeding and outbreeding: theoretical and empirical perspectives* (ed. N. W. Thornhill), pp. 304–28. University of Chicago Press, Chicago.

Russell, E. M. and Rowley, I. (1988). Helper contribution to reproductive success in the splendid fairy-wren (*Malurus splendens*). *Behavioral Ecology and Sociobiology*, **22,** 131–40.

Russell, E. M. and Rowley, I. (1993*a*). Philopatry or dispersal: competition for territory vacancies in the splendid fairy-wren *Malurus splendens*. *Animal Behaviour*, **45,** 519–39.

Russell, E. M. and Rowley, I. (1993*b*). Demography of the cooperatively breeding Splendid Fairy-wren, *Malurus splendens* (Maluridae). *Australian Journal of Zoology*, **41,** 475–505.

Part-time partnerships

The term *part-time partnership* is used to describe a relationship where pair members temporarily separate during the nonbreeding season. Nine case studies, including nine species and a cross-species review of the penguins are presented in this section. The chapters are ordered according to the relative amount of time partners associate together. This ranges between 9 and 2 months of the year, and between 24 and 2 h per day (see Table 1.1 for values from each case study).

John Coulson's study of the pair bond in Kittiwake Gulls *Rissa tridactyla,* nesting on window ledges of a North Shields warehouse in northern England, is perhaps the most notable contribution on long-term monogamous birds with *part-time partnerships* (Coulson 1966, 1972; Coulson and Thomas 1980, 1983, 1985). Coulson proposed the *incompatibility hypothesis*, where it is suggested that asynchrony in shared incubation duties is the reason why divorce occurs in poor breeders. He also described how *arrival asynchrony* at the colony in spring was the probable mechanism for divorce.

Most experimental studies testing relevant hypotheses in birds with *part-time partnerships* considered within-season rather than between-season mate fidelity. One useful study, by Lifjeld and Slagsvold (1988), tested why some Pied Flycatcher *Ficedula hypoleuca* pairs reunited for their second nesting attempt within a season. Manipulations included removing first clutches (forcing a second nesting attempt), erecting extra nest boxes (offering additional mating opportunities) and 'handicapping' a sample of males and females by removing two primary feathers (altering the condition of mates). They found that females in good condition divorced previous mates by moving to alternative territories and males that reunited with their previous partners had darker, black plumage, whereas handicapped males tended to lose their mates. Hence, the decision of mate fidelity was influenced by the condition of both sexes and the attractiveness of the male. Ollason and Dunnet (1986) asked whether Fulmar *Fulmarus glacialis* chick growth was better in pairs that reunited than in new pairs. They swapped eggs between old and new pairs of various ages. Since the fledglings raised by the old pairs were about 5% larger than those raised by new pairs, it was suggested that the old pairs had become more co-ordinated in parental care duties. Fulmars, which can live 50 years, also reunite on return to the colony after a temporary separation during the nonbreeding season (Ollason and Dunnet 1978, 1988). The longest Fulmar partnership lasted for 31 years (G. M. Dunnet personal communication).

This section begins with a study of the European Blackbird *Turdus merula* in the Cambridge Botanical Gardens in England. It proceeds to the Canadian tundra and the mountains of Colorado for two species of ptarmigan, to an island west of San Francisco, California, for Cassin's Auklets *Ptychoramphus aleuticus*, down to the south-eastern tip of Australia for Short-tailed Shearwaters *Puffinus tenuirostris*, back to the woodlands of Europe for Great Tits *Parus major* and Sparrowhawks *Accipiter nisus*, down again to the Antarctic wilderness for penguins, back to the New Zealand coast for Red-billed Gulls *Larus novaehollandiae*, and finishes with Indigo Buntings *Passerina cyanea* in the rural habitats of the mid-western state of Michigan, USA.

Comparison between the nine chapters shows that mate fidelity and long-term pair bonds seem to carry little advantage in species that separate during the nonbreeding season; the *mate familiarity effect* and the *cost of mate change* are not as influential. Divorce may be constrained in these species by the high costs of searching for alternative partners, fighting rivals, and the risk of failing to find a new mate.

References

Coulson, J. C. (1966). The influence of the pair-bond and age on the breeding biology of the kittiwake gull, *Rissa tridactyla*. *Journal of Animal Ecology*, **35**, 269–79.

Coulson, J. C. (1972). The significance of the pair bond in the Kittiwake *Rissa tridactyla*. *International Ornithological Congress*, **15**, 423–33.

Coulson, J. C. and Thomas, C. S. (1980). A study of the factors influencing the duration of the pair bond in the Kittiwake Gull *Rissa tridactyla*. *International Ornithological Congress*, **27** 823–33.

Coulson, J. C. and Thomas, C. S. (1983). Mate choice in the Kittiwake Gull. In *Mate choice* (ed. P. Bateson), pp. 361–76. Cambridge University Press, Cambridge.

Coulson, J. C. and Thomas, C. S. (1985). Differences in the breeding performance of individual kittiwake gulls *Rissa tridactyla*. In *Behavioural ecology* (ed. R. M. Sibley and R. M. Smith), pp. 489–503. Blackwell, Oxford.

Lifjeld, J. T. and Slagsvold, T. (1988). Mate fidelity of renesting pied flycatchers *Ficedula hypoleuca* in relation to characteristics of the pair mates. *Behavioral Ecology and Sociobiology*, **22**, 117–23.

Ollason, J. C. and Dunnet, G. M. (1978). Age, experience and other factors affecting the breeding success of the fulmar *Fulmarus glacialis* in Orkney. *Journal of Animal Ecology*, **47**, 961–76.

Ollason, J. C. and Dunnet, G. M. (1986). Relative effects of parental performance and egg quality on breeding success of Fulmars *Fulmarus glacialis*. *Ibis*, **128**, 290–6.

Ollason, J. C. and Dunnet, G. M. (1988). Variation in breeding success in Fulmars. In *Reproductive success* (ed. T. H. Clutton-Brock), pp. 263–78. University of Chicago Press, Chicago.

9 *Divorce in the European Blackbird: seeking greener pastures?*

ANDRÉ DESROCHERS AND ROBERT D. MAGRATH

Introduction

The Blackbird *Turdus merula* is one of Europe's most familiar birds. Originally shy inhabitants of woodland, Blackbirds are now a common sight, searching for earthworms (*Lumbricus* sp.) on lawns in urban parks and gardens. Males are striking, their sooty black plumage contrasting with a yellow orange bill and eye ring; females are brown, and their eye ring and bill are usually darker. Blackbird populations can be migratory or, like the one we studied, sedentary. Despite occasional records of polygyny, the species is behaviourally monogamous, with the pair breeding on an all purpose territory. In his classic study of the Blackbird, Snow (1988) found that pairs usually formed in the 2 months before egg laying started, but that, once formed, pairs usually stayed together from year to year unless one individual died. Our study is based on 6 years of data on 297 individually marked pairs. We focus on the effects of age and territory quality on reproductive success and the frequency of divorce. We find no advantage in staying with a familiar mate, and argue that divorce is a side effect of females leaving low quality sites in search of greener pastures, that is, higher quality territories. We also suggest that in general any advantages of breeding with a familiar or 'high quality' mate for short-lived passerines may be small compared with the demonstrably important advantages of seeking a good territory and, at least for males, an older mate.

Background, study site, and procedures

We studied Blackbirds in the Cambridge Botanic Garden, Cambridge, UK (52°12′N 0°7′E), between 1985 and 1991. Each winter, we mist netted throughout the 16-ha study area, so that most birds were colour-ringed (details in Desrochers and Magrath 1993*a*). One-year-old birds (thereafter, yearlings) can be identified from their pale brown primaries and secondaries, which contrast with the rest of the males' and females' plumage. Because of this plumage variation, coupled with the recruitment of birds ringed as nestlings, the age of a majority of individuals was known (Desrochers and Magrath 1993*b*). Pairs were determined at the beginning of each breeding season, when males conspicuously follow their mates. We defined divorce as occurring if both members of a pair were individually marked in one year and survived to the next without reuniting.

The position of each nest was recorded with an accuracy of 30 m. In 1985–1990, we made daily surveys from a few weeks before most birds began to breed to the end of the breeding season to find nests, to record dates of clutch initiation, clutch size, and the number of fledglings. Clutches were checked at least twice and each brood was checked on several occasions. All nestlings were marked individually at 8 days of age. We systematically searched for fledglings in the 2 weeks following the fledging date, to confirm fledging success and juvenile survival (details in Magrath 1991). The number of 2-week-old fledglings was used as our main and latest measure of annual reproductive success. We used other measures as circumstances and analytical methods dictated (details given with analyses). In 1991, we only recorded nest building and laying, to document the fate of the pairs of 1990.

Demographics

Of all the birds recruited into the Cambridge population, at least 31% of the males and 17% of the females were hatched in our study area. Once recruited into the breeding population Blackbirds were very site faithful, with a median breeding dispersal distance of 15 m for birds remaining with their mates (range 0–275 m; Desrochers and Magrath 1993*b*). There were few nonterritorial males and females during the breeding season, and they were usually colour-ringed. Annual survival rates of territorial males and females were 79% and 74%, respectively, leading to a large proportion of old birds in the population. Females could lay up to five clutches in a single season, which usually lasted from early March to late June. The clutch size ranged between two and five, with larger clutches being more common in the middle of the season (Desrochers and Magrath 1993*a*). Despite their high fecundity and ability to raise up to three broods successfully, Blackbirds rarely raised more than three independent offspring a year. The largest cause of nestling mortality was predation by cats, squirrels, and corvids (A. Desrochers and R. Magrath unpublished data), although at times starvation was common (Magrath 1989; R. Magrath and A. Desrochers unpublished data).

Mate choice criteria

Owing to the high individual survivorship and low divorce rates (see below), the ages of mates were highly correlated ($r_s = 0.40$, $df = 48$, $P < 0.01$). To determine whether assortative mating by age occurred, we calculated expected frequencies of parental age combinations with the method developed by Perrins and McCleery (1985) for Great Tits *Parus major*, which accounts for pair survivorship. With two age-classes (yearling and older), we found that even after accounting for pair survivorship, pairs of same aged individuals were more frequent than by chance ($\chi^2 = 18.5$, $df = 3$, $P < 0.001$). We found no evidence of assortative mating by size from data based on 230 pairs ($-0.07 < r < 0.09$ between sexes, for wing length, mass, tarsus, and culmen).

Parental care

Male and female Blackbirds differ in their parental roles. While females alone build the nest and incubate, males play a more important role in territory defence and the feeding of nestlings. Males have more time to defend nests, whereas females spend from 10 to 70% of time brooding nestlings, depending on the nestlings' age. Furthermore, when foraging, males give a greater proportion of earthworms that they find to their nestlings than females do (Desrochers 1991). After fledging at about 14 days old, the young are fed for about 3 weeks. Early in the season, the male often cares for the fledglings while the female renests, but late in the season, both parents care for the young (Edwards 1985). Males also seem to play a major role in reducing the incidence of nest predation by chasing off intruders (A. Desrochers and R. Magrath unpublished data).

Although only females brood, paternal care seems necessary for successful reproduction. In four of the five cases where the male died after the chicks hatches, all nestlings starved. In the remaining case, one nestling fledged.

Results

The pair bond

Most Blackbirds in this population chose new partners more than once in their lifetime. Over the 6 years of our study, we recorded 70.0% (SD 6.6) mate fidelity between years ($n = 6$ years). From Table 9.1, we calculate that the 'life expectancy' of pairs was less than 2 years, because of the joint occurrences of mate loss and divorce. The expected pair duration is therefore lower than an individual's 3–4 year life expectancy from

Table 9.1 Fate of breeding pairs in the Cambridge Blackbird population, 1985–1990

Year	All pairs		Both mates survived	
	n	% Mate change due to partner's death	n	% Divorcing
1985	10	20.0	8	25.0
1986	29	48.3	15	20.0
1987	49	34.7	32	34.4
1988	81	30.5	56	28.6
1989	63	42.9	36	36.1
1990	65	44.6	36	36.1
All years	297	38.4	183	31.7

Adapted from Desrochers and Magrath (1993*b*). Divorce rates did not vary significantly among years ($\chi^2 = 2.1$, $df = 5$, $P = 0.8$). Sample sizes based on colour-ringed pairs only.

the recruitment into the breeding population. Although most ringed Blackbirds were sedentary, there was little evidence that they remained mated all year. In fact, most Blackbirds probably split up each year, because former mates often chose different foraging areas during the nonbreeding season. We did not find cases where a bird was involved in two or more nests simultaneously.

Consequences of mate fidelity

Familiarity with a mate

It is widely believed that partners with past experience together are more effective parents than new pairs (Black Chapter 1), but we could find no evidence that this was true of Blackbirds. The main problem in analyses of a 'pair familiarity' effect is that age itself affects reproductive success (see below). How can one tell whether an increase in reproductive success with pair duration is due to a partnership effect *per se* or simply to both mates ageing and thus getting more proficient at rearing young? We have addressed this problem by removing yearling females from the analysis, because age effects occur mainly with females and are most pronounced between yearlings and older birds (Desrochers and Magrath 1993*a*). After removing yearling females from the analysis, we found no obvious relationship between pair duration and reproductive success defined as the number of fledglings that reach 2 weeks of age (Fig. 9.1). Whatever apparent relationship there was between pair duration and reproductive success seemed almost entirely due to the confounding effect of the yearling females' poor reproductive performance.

Familiarity with a territory

Although pair duration did not affect reproductive success, staying with a partner usually entailed keeping a territory, with the result that mate and site familiarity were closely linked (2 × 2 contingency table, χ^2 =

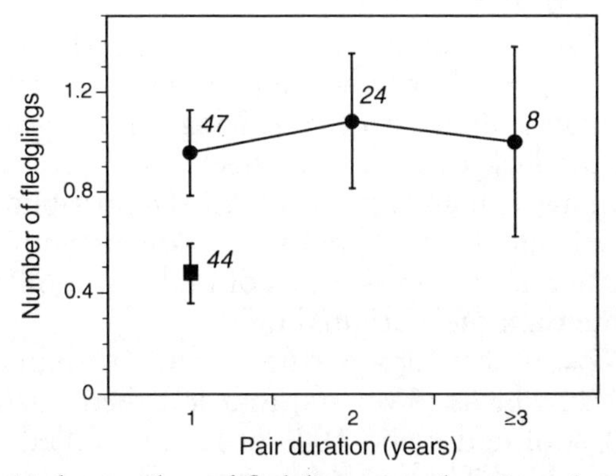

Fig. 9.1 Changes in the number of fledglings raised to 2 weeks with pair duration. *Circles* = yearling females excluded, *square* = pairs with a yearling female. Vertical bars = standard errors. Numbers of pairs in italics (including longitudinal data from some pairs). There was no statistical evidence for a pair duration effect after excluding yearling females (regression, $F = 0.2$, $df = 1{,}77$, $P = 0.7$).

102.2, $n = 207$, $P < 0.001$). We suspect that any benefits of site familiarity are likely to be greater for males than for females, because males are more territorial. We asked whether site familiarity was beneficial by looking at the effect of whether both pair members nested in a particular site for the first time. Sites were defined here as 60 m × 60 m squares on the study area (see below). Before comparing site familiar pairs with other pairs, we needed to control for a confounding effect: birds were likely to stay in good sites, thus becoming familiar (below; Newton and Marquiss 1982; Chapter 14). We did so by including site quality as a covariate in a logistic analysis. We defined the quality of a site as its mean annual number of young raised to fledging. After controlling for site quality, the probability of raising chicks to 2 weeks after fledging was not significantly affected by a pair's familiarity with its breeding site ($\chi^2 = 1.4$, $df = 1, 102$, $P = 0.2$). Accounting statistically for mate familiarity prior to testing for site familiarity led to the same result. These results contrast with the apparent benefit of familiarity using numbers of birds fledged as the measure of reproductive success (Desrochers and Magrath 1993*b*). We attribute this discrepancy to the borderline significance of the published result (based on ANOVA) and the change of statistical method. We believe a logistic analysis is more appropriate here because our measures of reproductive success were heavily biased towards zero, thus making it difficult to analyse with ANOVA. We conclude, therefore, that there is no obvious annual fitness cost associated with mating with an unfamiliar bird or in an unfamiliar site.

Consequences of divorce

If there is little to be gained by staying with a particular partner or in a particular place, then divorce may seem to carry little cost. But, given that 58 of the 60 divorces observed happened between years, they entailed leaving an old partner and running the risk of gaining a young one. For females, the cost of re-pairing with a yearling male may seem small from the lack of male age effects on reproductive success, but for males, changing a mate opens the costly possibility of mating with an inexperienced female, with poor reproductive prospects. Indeed, from the 58 inter-annual divorces, 58% of males and 68% of females gained a younger mate than their original one.

Another cost of leaving a partner is the possibility of not finding a new one. Of the 116 birds (58 pairs) that left their partner from one year to the next, 21% of males and 10% of females failed to find a new mate in the following year. The sex difference in re-pairing success was not statistically significant ($\chi^2 = 2.4$, $df = 1$, $P = 0.1$).

Finally, birds that divorce may experience an increase in productivity with a new mate in the year following a divorce. We found that change in reproductive success between one year and the next was unaffected by whether the birds divorced or not. Birds that divorced as yearlings gained 0.99 (SE 1.0) more fledglings in the following year than faithful yearlings, and older birds that divorced gained 0.43 (SE 0.78) fledglings compared with faithful older birds (divorce effect: ANOVA, $F = 1.2$, $df = 1,59$, $P = 0.3$, after accounting for age). These effects were similar for males and females. Also, none of these comparisons included birds that failed to get a mate, and were thus not breeding.

Frequency of divorce

Divorce from one year to the next was common (Table 9.1), but it rarely happened between nesting attempts within a breeding season. Of 21 birds that changed mates during a breeding season, 19 did so because of the death of their mate (as evidenced by ring recoveries, Desrochers and Magrath 1993*b*). The proportion of faithful pairs did not differ significantly between years (Table 9.1).

Correlates of divorce

There was a significant effect of prior reproductive success on the probability of divorce, after accounting for pair duration; i.e. poor breeders divorced more. By contrast, there was no significant association between divorce rate and how long partners had been together, after accounting for the effect of reproductive success (Fig. 9.2). The relationship between the probability of divorce and prior reproductive performance may have been confounded by the fact that old pairs had a higher annual reproductive success than young ones (see below). However, after accounting for the effects of reproductive success and mate age, divorce rates were independent of age (males: logistic regression $\chi^2 = 0.8$, $df = 1,93$, $P = 0.4$; females: $\chi^2 = 0.1$, $df = 1,93$, $P = 0.8$).

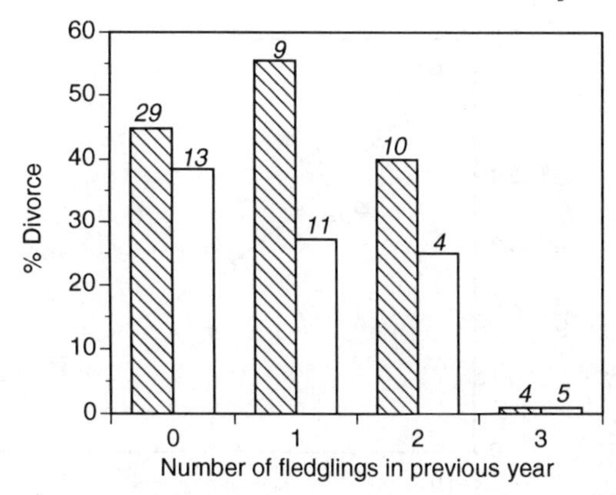

Fig. 9.2 Change in the probability of divorce with prior reproductive success for established pairs (*open bars*) and new pairs (*hatched bars*). Analyses showed that the variation of probabilities was mostly due to the number of 2-week fledglings raised in the breeding season before the divorce event (logistic regression, pair duration after prior fledglings' effect: $\chi^2 = 1.4$, $df = 1{,}82$, $P = 0.2$; prior fledglings after pair duration: $\chi^2 = 4.0$, $P = 0.04$). Numbers of pairs shown above bars.

Who initiated divorces?

We suspect that females initiated divorce events more often than males did, from indirect evidence. Females, which socially dominated males when paired, tended to move further than their former mate following divorce (medians 46 m versus 30 m, Wilcoxon test, $n = 42$, $P = 0.08$). By contrast, females did not disperse further than males after the death of their mate (medians 31 m in both sexes, $n = 60$ males, 52 females).

What mattered: partner or site characteristics?

While attention is usually focused on phenotypic characteristics of mates when searching for causes of divorce, it is likely that territory quality plays a significant role in the decision to change mates. To know whether an individual may leave another for a better site (or leave a poor one), we needed to assess whether there was consistency in the quality of territories from one year to the next and whether 'quality' can be assessed by potential dispersers (Desrochers and Magrath 1993*b*). One way Blackbirds may evaluate territory quality is by determining the fledgling production in the vicinity of their own territory. From a Blackbird's perspective, recently fledged birds are probably best for this exercise, since they are conspicuous (Snow 1988) and remain in the territory of their parents, thus giving a clue to the quality of that territory. Thereafter, fledglings' movements may make it difficult to know from which territory they originate.

We assessed site quality indirectly by measuring the mean production of fledglings in each 60-m grid in the Garden (mean 12 nests per grid). Although they do not correspond to true territories, these sites are of a

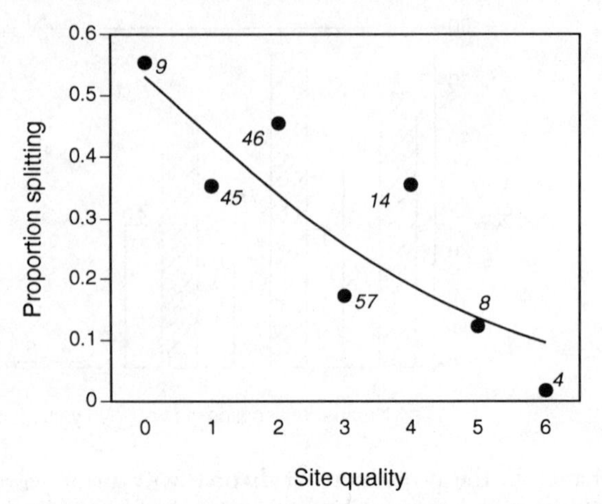

Fig. 9.3 Change in the probability of divorce in sites of varying quality, defined here as the mean annual number of fledglings per pair, for all pairs that have nested on that site. Curve fitted by logistic regression (χ^2 = 8.6, *df* = 1,181, *P* < 0.004). Numbers of pairs in italics. Adapted from Desrochers and Magrath (1993*b*).

similar size to them, and, unlike territories, their limits do not shift with time, thus corresponding to fixed habitat features. The pairs in these 0.36 ha sites showed considerable and consistent variation in the annual number of fledglings per year they produced: site locations were responsible for 21% of the variance in the fledgling production (all years, yearlings excluded, *F* = 2.2, *df* = 42,103, *P* = 0.005). These consistent differences among parts of our study area did not arise because the same individuals remained in the same areas year after year. For example, after keeping just one record per female per site, significant discrepancies among sites remained (14% of variance explained, F = 1.6, *df* = 39,96, *P* = 0.04).

The consistent differences in output among sites made it possible for Blackbirds to use site specific fledgling production as a cue to assess whether or not to divorce and therefore to change territory. Pairs did respond to territory quality, by divorcing more often when their breeding site had a low expected output of fledglings (Fig. 9.3). If changing partners reflects a female's search for a better territory, then we predict that on average, the quality of her new territory will be higher than that of the previous one. Surprisingly, this idea was not borne out by our data: the expected fledgling production in the new sites occupied by females was not different than that of the sites they left (paired *t*-test, *t* = 0.5 *df* = 29, *P* = 0.6). Similarly, site quality did not increase for divorced males (*t* = 0.3 *df* = 30, *P* = 0.7).

Parental characteristics and reproductive success

In Blackbirds, female age has a large effect on reproductive success, while male age has little or no effect. In a previous analysis, we found

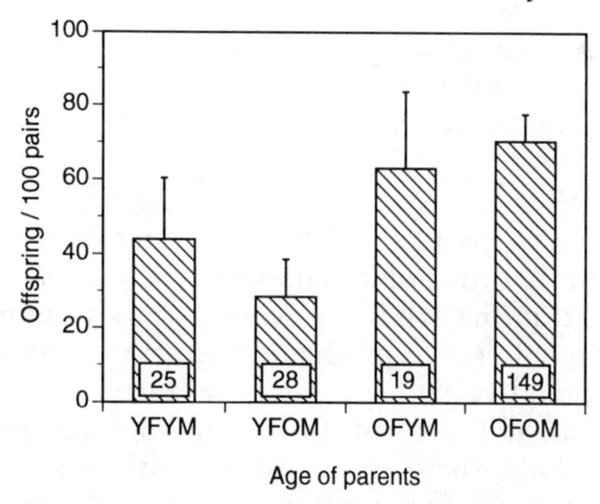

Fig. 9.4 Combined effect of male and female age on the annual number of offspring raised to independence. The female age effect on the probability of getting offspring was significant, after considering male age (logistic analysis, $\chi^2 = 7.9$, df = 1,118, $P = 0.005$), but the reverse was not true ($\chi^2 = 0.4$, $P = 0.5$). Sample sizes and standard errors shown. F = female; M = male; Y = yearling; O = older.

that in their first year, females start breeding later, have shorter breeding seasons and lay smaller clutches than older females (Desrochers and Magrath 1993*a*). Furthermore, these improvements with age were apparently due to changes within individuals, not to selection weeding out poor quality individuals. After controlling for the age of their mate, we found that male age has no effect on fecundity (Desrochers and Magrath 1993*a*). Similarly, by comparing the productivity of different pair combinations, we found that female age-class affected the probability of raising young successfully, but that male age had little or no effect (Fig. 9.4) The improvement in female reproductive success with age stems largely from the poor foraging success of yearlings compared with older birds. Yearlings of either sex are only half as successful at getting earthworms as are 2 year olds, and foraging efficiency reaches a plateau at 3 years of age (Desrochers 1992*a*). The hypothesis that yearling females are constrained to start breeding later because of their poor foraging success was confirmed by a food supplementation experiment. The addition of food to territories before laying led to yearlings laying at the same date as older females (Desrochers 1992*b*). Because of its effect on reproductive success, female age should be an important mate choice criterion. Since correlations between male and female ages occurred within pairs, the preferred option for males (i.e. old females) was achieved mostly by old males.

Discussion

Significance of the pair bond

Unlike the traditional views on pair bond maintenance (e.g. Rowley 1983), we found no advantages of familiarity with partners or sites. We question the importance of familiarity, particularly with mates, because the evidence on divorce from other studies can usually be interpreted in terms of age or site effects.

The role of divorce

Is it adaptive to divorce? Mate losses examined in this study were not accidental in that the population was sedentary so that both mates could have reunited in the following season. Pairs were more likely to divorce if they suffered low reproductive success, suggesting that divorce is indeed an adaptive strategy to move away from poor sites. However, there were possible costs of divorce in our study population, linked with uncertainty. First, there was a risk for males of getting a young replacement mate, second, the risk of losing a good site, and, considering that death rates are low, the risk of not finding a replacement mate. The latter risk seemed more pronounced in males, but our data were insufficient to demonstrate it statistically. These costs could have been sufficient to explain the occurrence of lasting pair bonds.

Who divorced?

Is divorcing a male's or a female's strategy? We have two theoretical reasons for expecting that females will suffer a lower cost of divorce than males and therefore will initiate divorces more often. First, male age has little effect on reproductive success compared with female age, so there is little or no cost to a female of re-pairing with a yearling male. Second, territorial defence is primarily the role of the male, so that any cost of moving to a new site should be smaller for females than males (we did not find a site familiarity effect, but cannot rule it out: Desrochers and Magrath 1993*b*). Our evidence suggests that females may in fact have initiated divorces more than males, because divorcing females tended to move further than females whose mate died, whereas the distances moved by divorcing and widowed males did not differ. On the other hand, females may have dispersed further because they were expelled by their mate or an incoming female, but we have no evidence of this.

Does familiarity with a mate increase reproductive success?

Our data show that familiarity with a mate was largely irrelevant. However, reproductive success *per se* was a significant predictor of divorce. Divorce rates generally increase following poor reproductive performance in many bird species (Chapters 1, 19). As Ens *et al.* (1993) pointed out, that reproductive success and parental age are in general well correlated makes it difficult to determine whether parental age or reproductive success is the real cause of divorce. While the Kittiwake *Rissa tridactyla* data indicate that it is parental age rather than reproductive success that seems to explain why older pairs are more faithful (Coulson and Thomas 1983), we found the opposite result, and question the generality of parental age or pair duration *per se* as a factor directly influencing pair bond maintenance.

Keeping sites rather than mates?

Although the mate choice literature suggests that birds do make choices about partners, we believe there has been too much emphasis in studies of divorce on choosing individual partners, and too little emphasis on choosing sites or avoiding yearling partners. For example, although there is a correlation between female settlement order and male phenotype in the Pied Flycatcher *Ficedula hypoleuca*, Alatalo *et al.* (1986) showed experimentally that females choose nest site characteristics, not male phenotypes.

We have shown that particular areas yield consistently high or low numbers of fledglings per season. Consistent variation between territories is important because if territory quality influences the probability of divorce, then there must be predictable differences in territory quality between years (Oring 1982). There have been demonstrations of consistent variation in bird productivity among sites at different spatial and time scales (e.g. Cody 1985; Ens *et al.* 1992), but only Beletsky and Orians (1987) and Hochachka *et al.* (1989) have shown that apparent reproductive differences among territories were not due entirely to differing quality of site tenacious individuals. Nevertheless, it is still possible that good quality individuals in our study kept the best sites, leaving the poor sites to other individuals, thus exacerbating productivity differences among sites.

Beyond intrinsic site characteristics, site familiarity is often viewed as a reason not to leave a territory. After re-analysing the data from our 1993 paper (Desrochers and Magrath 1993*b*), we conclude that our evidence for benefits of site familiarity in Blackbirds is equivocal, because the presence of a 'site-familiarity' effect depends on the analytical approach used. We cannot rule out a benefit to site familiarity at some spatial scale, because our weak pattern may be due to the small distances moved by the dispersers. Blackbirds were probably familiar with areas much larger than their territories, owing to their daily movements, especially outside the breeding season, and site familiarity should therefore be studied at a larger scale.

Although our evidence suggests that the low quality of certain sites increased the probability of divorce, we cannot rule out an effect of mate quality on divorce. The reason for this is that good quality individuals might be found mainly in good quality sites. However, we suggest that it is more parsimonious to conclude that site quality is the most important influence on divorce. This is because many studies on avian breeding ecology show important effects of the environment on reproductive success (e.g. large seasonal and year effects), yet there appears to be little evidence for large effects of bird quality (apart from age-related differences).

Overview of hypotheses

In this section, we review the different hypotheses for divorce (see Black Chapter 1) in the light of our data on Blackbirds. Although Ens *et al.* (1993) consider 'better options' as defined by the combined quality of mates plus territories, we believe that it is useful to separate choices based on selecting other individuals and selecting other sites.

1. *Better partner hypothesis.* Individuals leave their mate because a bird of higher quality becomes available. If the important differences in 'quality' are genetic, individuals may have the choice of gaining genes via extra-pair copulation rather than divorce (Kempenaers *et al.* 1992; Mulder *et al.* 1994). Nevertheless, mate guarding by males may constrain female choice of genes through extra-pair copulation, thus making divorce a way for females to realize choice on genes.
2. *Better site hypothesis.* In species in which breeding site influences reproductive success, divorce may occur as a side effect of selection of sites. This hypothesis relies on there being consistent territory effects on reproduction but little benefit from remaining with a particular individual. Our data on Blackbirds are most consistent with this hypothesis. A characteristic feature of both this and the previous hypothesis is that only one individual (the initiator) may benefit from divorce (Ens *et al.* 1993).
3. *Incompatibility hypothesis* (Coulson 1966: in Ens *et al.* 1993). This hypothesis suggests that individuals are not intrinsically 'good' or 'bad', but that some individuals are not compatible. In this case both benefit by divorcing (Johnston and Ryder 1987). Our data provide no test of the *incompatibility hypothesis*, but we suspect that behavioural incompatibility is more likely to be a problem in species in which reproductive success is dependent on fine-scale behavioural interactions. For example, the incompatibility hypothesis was proposed initially for the Kittiwake (Coulson and Thomas 1983), a species in which both mates incubate. In Blackbirds, only the female incubates, and we suspect that feeding of chicks does not require as much co-ordination between mates as would incubation.
4. *Errors of mate choice.* The hypothesis suggests that pairs in their first year together should be more likely to divorce, once they discover that they have made the wrong choice for any of the above reasons; we would include a poor choice of site as well as partner. Our data show that, controlling for reproductive success, the duration of the pair bond does not affect the probability of divorcing. Thus the hypothesis is not supported for Blackbirds.
5. *'Non-adaptive hypotheses'.* We believe that it is possible to cast all hypotheses explaining divorce in an 'adaptive' framework. For example,

cases where an individual is evicted from a pair by an intruder (Williams and McKinney Chapter 4) may be explained as the result of an adaptive strategy by the intruder. This may have happened in our study population, but we believe it was infrequent, because it would mean that intruders were more likely to affect pairs that experienced low reproductive success (the other pairs were less likely to divorce). Another 'non-adaptive' hypothesis is that pairs may break up following the *accidental loss* of a partner (e.g. during migration, Owen *et al.* 1988; Black *et al.* Chapter 5). Such a situation may reflect the lack of importance of mate fidelity or a cost of delaying reproduction upon reaching the breeding grounds. Accidental losses were probably rare or nonexistent in our population, because it was sedentary and had a long breeding season. Also, the sedentary status of our study population makes it unlikely that the changing sequence of territorial settlement from year to year (*musical chair hypothesis*, Dhondt and Adriaensen 1994) would greatly affect the pair fidelity patterns.

What is needed to shed more light on the issue is an assessment of the costs and benefits of specific behaviours that can influence the likelihood of separating from or remaining with a partner.

Summary

Divorce in European Blackbirds, was observed in 32% of 183 cases where mates had survived from the previous breeding season. After divorcing, females tended to nest further from their original site than males did, but there was no sex difference in distance moved after the death of a mate. This suggests that females were the most frequent initiators of divorces. The probability of divorce decreased as pair duration increased. However, this pattern was due to the increase in reproductive success with age. The mean annual fledgling production of 60 m × 60 m sites was predictable among years, and divorce rates were greatest in low quality sites. We found no annual fitness cost associated with mating with an unfamiliar bird or in an unfamiliar site. We interpret the maintenance of pair bonds as being primarily a strategy for successful pairs to avoid the risks of losing a good territory, of (for males) gaining a younger, less fecund mate, and the risk of not finding a new mate.

Acknowledgements

We thank Nick Davies for assistance and encouragement throughout this study. Jeff Black, Sharmila Choudhury, André Dhondt, Peter Dunn, David Green, Elsie Krebs, Jan Lifjeld, Penny Olsen, Linda Whittingham, and an anonymous reviewer provided useful comments on a draft of this chapter. We also thank C. D. Pigott and P. Orriss for allowing us to use the Botanic Garden. The authors were members of the Department of Zoology at the University of Cambridge during the collection of the field data, and are grateful for the support and encouragement of all those in the department. This study was funded by the Fonds pour la Formation de Chercheurs et l'Aide à la Recherche (Québec), and the Natural Sciences and Engineering Research Council of Canada to AD, and a Commonwealth Scholarship, Lundgren Research Award, Cambridge Philosophical Society Studentship, and funds from the Company of Biologists (Cambridge) to RDM.

References

Alatalo, R. V., Lundberg, A., and Glynn, C. (1986). Female Pied Flycatchers choose territory quality and not male characteristics. *Nature*, **323**, 152–3.

Beletsky, L. D. and Orians, G. H. (1987). Territoriality among male red-winged blackbirds. I. Site fidelity and movement patterns. *Behavioral Ecology and Sociobiology*, **20**, 21–34.

Cody, M. L. (1985). An introduction to habitat selection in birds. In *Habitat selection in birds* (ed. M. L. Cody), pp. 3–56. Academic Press, New York.

Coulson, J. C. (1966). The influence of pair-bond and age on the breeding biology of the kittiwake gull *Rissa tridactyla. Journal of Animal Ecology*, **35**, 269–79.

Coulson, J. C. and Thomas, C. S. (1983). Mate choice in the Kittiwake Gull. In *Mate choice* (ed. P. P. G. Bateson), pp. 361–76. Cambridge University Press, Cambridge.

Desrochers, A. (1991). Age and reproduction in European Blackbirds, *Turdus merula*. Unpublished Ph.D. thesis. University of Cambridge.

Desrochers, A. (1992*a*). Age and foraging success in European blackbirds: variation between and within individuals. *Animal Behaviour*, **43**, 885–94.

Desrochers, A. (1992*b*). Age-related differences in reproduction by European blackbirds: restraint or constraint? *Ecology*, **73**, 1128–31.

Desrochers, A. and Magrath, R. D. (1993*a*). Age-specific fecundity in European Blackbirds: individual and population trends. *Auk*, **110**, 255–63.

Desrochers, A. and Magrath, R. D. (1993*b*). Environmental predictability and re-mating in European blackbirds. *Behavioral Ecology*, **4**, 271–75.

Dhondt, A. A. and Adriaensen, F. (1994). Causes and effects of divorce in the blue tit *Parus caeruleus. Journal of Animal Ecology*, **63**, 979–87.

Edwards, P. J. (1985). Brood division and transition to independence in Blackbirds *Turdus merula. Ibis*, **127**, 42–59.

Ens, B. J., Kersten, M., Brenninkmeijer, A., and Hulscher, J. B. (1992). Territory quality, parental effort and reproductive success of oystercatchers. (*Haematopus ostralegus*). *Journal of Animal Ecology*, **61**, 703–15.

Ens, B. J., Safriel, U. N., and Harris, M. P. (1993). Divorce in the long-lived and monogamous oystercatcher, *Haematopus ostralegus*: incompatibility or choosing the better option? *Animal Behaviour*, **45**, 1199–217.

Hochachka, W. M., Smith, J. N. M., and Arcese, P. (1989). Song Sparrow. In *Lifetime reproduction in birds* (ed. I. Newton), pp. 135–52. Academic Press, London.

Johnston, V. H. and Ryder, J. P. (1987). Divorce in larids: a review. *Colonial Waterbirds*, **10**, 16–26.

Kempenaers, B., Verheyen, G. R., Van den Broeck, M., Burke, T., Van Broeckhoven, C., and Dhondt, A. A. (1992). Extra-pair paternity results from female preference for high-quality males in the Blue Tit. *Nature*, **357**, 494–6.

Magrath, R. D. (1989). Hatching asynchrony and reproductive success in the Blackbird. *Nature*, **339**, 536–8.

Magrath, R. D. (1991). Nestling weight and juvenile survival in the blackbird, *Turdus merula*. *Journal of Animal Ecology*, **60**, 335–51.

Mulder, R. A., Dunn, P. O., Cockburn, A., Lazenby-Cohen, K. A., and Howell, M. J. (1994). Helpers liberate female fairy-wrens from constraints of mate choice. *Proceedings of the Royal Society, London,* **255**, 223–9.

Newton, I. and Marquiss, M. (1982). Fidelity to breeding area and mate in sparrowhawks *Accipiter nisus*. *Journal of Animal Ecology*, **51**, 327–41.

Oring, L. W. (1982). Avian mating systems. In *Avian biology*, Vol. VI (ed. D. S. Farner and J. R. King), pp. 1–92. Academic Press, New York.

Owen, M., Black, J. M., and Liber, H. (1988). Pair bond duration and timing of its formation in Barnacle Geese (*Branta leucopsis*). In *Waterfowl in winter* (ed. M. W. Weller), pp. 23–38. University of Minnesota Press, Minneapolis.

Perrins, C. M. and McCleery, R. H. (1985). The effect of age and pair bond on the breeding success of Great Tits *Parus major*. *Ibis*, **127**, 306–15.

Rowley, I. (1983). Re-mating in birds. In *Mate Choice* (ed. P. Bateson), pp. 331–60. Cambridge University Press, Cambridge.

Snow, D. W. (1988). *A study of blackbirds*, (2nd edn). British Museum (Natural History), London.

10 *Mate fidelity and divorce in ptarmigan: polygyny avoidance on the tundra*

SUSAN HANNON AND KATHY MARTIN

Introduction

Ptarmigan are small, primarily herbivorous grouse that breed in alpine or circumpolar arctic tundra during the short summer season. Although strongly territorial during the breeding season, they spend the winter in large sex-segregated flocks sometimes numbering in the thousands. The sexes appear similar in their white winter plumage but during summer they have sexually dimorphic breeding plumages and vocalizations. Monogamy is the norm for all species, but up to 20% of males may be polygynous in any year. Pairs are together in spring and summer for 2–4 months, spend the winter apart, and if both members of the pair survive, usually reunite the next spring. The duration of the pair bond and the extent of male parental care vary among species, but in all species the male accompanies the female during the prelaying and laying periods and appears to provide vigilance against predators while she forages.

Our previous work focused on costs and benefits of monogamy to females within a season (Hannon 1984; Martin and Cooke 1987; Hannon and Martin 1992). Here we examine mate fidelity and divorce within and between seasons to assess the costs and benefits of longer-term monogamy in two species of ptarmigan. Using a sample of 1396 pair-years for Willow Ptarmigan *Lagopus lagopus* collected over 16 years from two populations (see drawing above) and 212 pair-years for White-tailed Ptarmigan *L. leucurus* collected over 8 years we examine why ptarmigan divorce their mates and compare the reproductive success of birds that divorce with those that reunite with their mates.

Background, study sites, and procedures

Average clutch sizes for ptarmigan vary with species and location from 8 to 11 for Willow Ptarmigan and 4 to 6 for White-tailed Ptarmigan (Braun *et al.* 1993). Most individuals first breed at 1 year of age. Ptarmigan are single-brooded but may produce up to four clutches in one breeding season if their clutches are depredated (Martin *et al.* 1989). Clutch predation by mammalian and avian predators is the most important factor limiting annual reproductive output for ptarmigan (Martin *et al.* 1989; Braun *et al.* 1993). There is about a 40% reduction in brood size from hatch to fledging caused by predation, starvation, and hypothermia. Family groups break up in late autumn when birds migrate from 2 to 88 km away from the breeding areas, with females moving further than males (Hoffman and Braun 1975; Gruys 1993). Mortality of females is higher than that of males in both species (Braun *et al.* 1993).

The studies on Willow Ptarmigan were conducted at two sites in northern Canada: Chilkat Pass (CP) in north-western British Columbia (59° 50′N, 136° 30′W) from 1980 to 1982 and from 1984 to 1992, and La Perouse Bay (LPB) in northern Manitoba (58° 24′N, 94° 24′W) from 1981 to 1986. CP is located in subalpine tundra and LPB in subarctic tundra. Both areas are dominated by willow *Salix* spp. and Dwarf Birch *Betula glandulosa* with a herb layer of grasses and sedge (Hannon *et al.* 1988). White-tailed Ptarmigan were studied on an alpine site, 3500–4200 m in elevation, on the front range of the Colorado Rocky Mountains (Colo); (39° 34–40′N, 105° 35–53′W; Braun *et al.* 1993; Martin *et al.* 1993) from 1987 to 1994. Over 90% of all birds were individually colour-banded and at CP and Colorado we placed radio transmitters on most females in order to locate nests and broods. At capture, sex and age (adult or yearling) were determined and morphometric measurements were taken. We searched study areas regularly and recorded the location of each bird and the identity of each bird it was associated with. We considered birds to be paired if they were seen consistently with the same mate prior to laying, or if the female nested on the male's territory or the pair was observed with newly hatched young on the territory.

We compartmentalized reproductive success into three stages allowing us to examine potential mechanisms relevant to mate fidelity and divorce. The first measure, clutch size of the first nest, gives an estimate of the potential fecundity of a pair. Second, we determined whether or not a pair produced any hatchlings in a season. This measures a pair's ability to evade nest predation and to produce a replacement clutch if the first clutch is depredated. The third measure, the number of fledged chicks observed/brood at 15–20 days after hatching for successful nests, indicates parental ability to raise hatched young. Information on how we determined clutch size and brood size at fledging can be found in Martin *et al.* (1989).

Demographics

Willow Ptarmigan numbers in North America often cycle with a periodicity of 10–11 years. At CP, ptarmigan numbers were high (about 40 males/km^2) from 1979 to 1981, declined to a low of about 20–25 males/km^2 from 1982 through 1989, increased to 30 males in 1990 and declined again in 1991. Ptarmigan density at LPB varied from 5 to 10 pairs/km^2 and White-tailed Ptarmigan varied in density from 1 to 2 pairs/km^2. Since the sex ratio is slightly skewed in favour of males, and polygyny occasionally occurs, unpaired territorial and nonterritorial males may comprise up to 26% of the total number of males.

Pair formation

After spending the winter in sex-segregated flocks, birds return to the breeding areas in spring to establish a territory and choose a mate. Males arrive first and set up territories, advertising their presence with territorial calls and aerial flights. Females arrive singly or in small groups and move across territories of males. Males display to returning females by calling, wing-dragging, tail fanning, and head bobbing. Early in the pairing process, females sometimes take flight and they are actively pursued by the male who attempts to guide them back to his territory. After pairs have formed, the birds stay in close proximity often giving soft contact calls. Copulations begin well before females are fertile and may help to cement the pair bond.

Pair maintenance

The male remains in close proximity to the female before, during, and after her fertile period (Martin 1984*b*; Braun *et al.* 1993); while she feeds, he scans for predators (Artiss and Martin 1995). Males are more conspicuous than females (Bergerud and Mossop 1984), spend more time alert and give more alarm calls than females (Hannon and Martin 1992). The fact that males continue to guard after the female's fertile period suggests that guarding may also have the function of enhancing the foraging efficiency of the female by reducing predation risk or harassment from other males.

Paternity and parental care

Seasonal pair duration and parental care vary with species: in White-tailed Ptarmigan pair bonds dissolve in mid summer prior to hatch, while in Willow Ptarmigan pairs remain together into autumn and both sexes attend the chicks and defend them against predators. Only females incubate, but males attempt to deflect predators from the nest. Attempted extra-pair copulations have been observed in Willow Ptarmigan (Martin and Hannon 1988) and chicks produced from extra-pair fertilizations occur in about 7% of broods produced by socially monogamous pairs (Freeland 1993). Egg dumping has been observed in Willow Ptarmigan (Martin 1984*a*), but egg laying patterns consistent with egg dumping have only been recorded in 6% of nests (Schieck and Hannon 1993).

Results

The pair bond

Monogamous pair bonds were most common in both species (91%, n = 665, and 95% n = 279, of Willow Ptarmigan pairs at CP and LPB, respectively; and 93%, n = 203 for White-tailed Ptarmigan), but some polygyny occurred in most years in all populations. Most pair bonds lasted for only 1 year in both species, primarily because of high adult mortality. The longest pair duration for Willow Ptarmigan was 5 years at CP and the longest for White-tailed Ptarmigan was 4 years. The number of mates in a lifetime varied between the sexes for both species: most females had only one mate whereas proportionately more males had two or more mates. Males had more mates in a lifetime than did females because males lived longer and some males were polygynous whereas polyandry was extremely rare. Up to 26% of males may remain unmated in some years, whereas resident females were always mated. Mean pair duration remained the same with each subsequent mate for females of both species and for males at CP and Colorado (Table 10.1). Males at LPB, however, had longer pair bonds with their second mates than with other mates.

Mate familiarity effect

Breeding experience may be an important factor that increases an individual's reproductive success, particularly if a bird breeds with a former mate. However, increases in reproductive success may also accrue with increasing age as yearling birds are often smaller and/or physiologically less developed than adults (Hannon *et al.* 1979). To remove the confounding effects of female age (male age had no significant effect, see below) when examining the effect of pair duration, we excluded yearling females from the analysis and controlled for adult female age in the analysis. Pair duration did not significantly influence clutch size or number of chicks fledged for either species (Fig. 10.1). Willow Ptarmigan pairs at CP that had been together for 3 years had a higher probability of hatching chicks than pairs with shorter pair bonds.

Table 10.1 Mean number of years together for male and female ptarmigan with their first and subsequent mates (M=male, F=female)[1]

No. years together	First mate M	First mate F	Second mate M	Second mate F	Third mate M	Third mate F	Fourth mate M	Fourth mate F
Site CP								
Mean	1.28	1.24	1.13	1.39	1.38	1.30	1.11	1.25
SE	0.05	0.03	0.05	0.08	0.19	0.15	0.11	0.25
n	162	344	64	80	24	23	9	4
Site LPB								
Mean	1.08	1.14	1.48	1.22	1.15	1.25	1.00	1.00
SE	0.04	0.04	0.10	0.09	0.08	0.13	0.00	0.00
n	60	77	40	32	20	12	5	4
Site Colo								
Mean	1.23	1.21	1.50	1.33	1.31	1.00	1.12	1.00
SE	0.08	0.06	0.14	0.11	0.18	0.0	0.12	0.00
n	61	89	26	18	13	4	8	1

[1] One-way ANOVA for pair duration with each mate: CP males: $F = 1.56$, $df = 3$, $P = 0.20$; CP females: $F = 1.61$, $df = 3$, $P = 0.19$; LPB males: $F = 7.09$, $df = 3$, $P = 0.0002$; LPB females: $F = 0.62$, $df = 3$, $P = 0.60$; Colo males: $F = 1.41$, $df = 3$, $P = 0.24$; Colo females: $F = 0.55$, $df = 3$, $P = 0.65$.

Mate change

Mate change occurred in one of four ways:

(1) the partner of the previous year died;
(2) for males, they acquired a second mate (i.e. became polygynous);
(3) both partners survived, but one or both of them changed to a new partner (i.e. 'divorce'); and
(4) the pair produced a first clutch that was depredated, then the female changed to a new male and renested in the same season.

In both species and all populations, the majority of mate changes were due to the death of the previous year's partner (Table 10.2).

Costs of mate change

Pairing with a new mate did not influence reproductive success for females of either species (Table 10.3). Clutch size for male White-tailed Ptarmigan that paired with a new mate was lower than for males re-uniting with the same mate and male Willow Ptarmigan at CP showed the same trend (Table 10.3). When males changed mates they had about a 49% chance of pairing with a yearling female, whereas if they reunited with their previous mate they had a 100% chance of being with an adult. Yearling females had smaller clutches than adults (see below), thus it appears that males would do better to try to reunite with previous mates. At LPB, males did not have a lower clutch size when they paired with a new mate (Table 10.3).

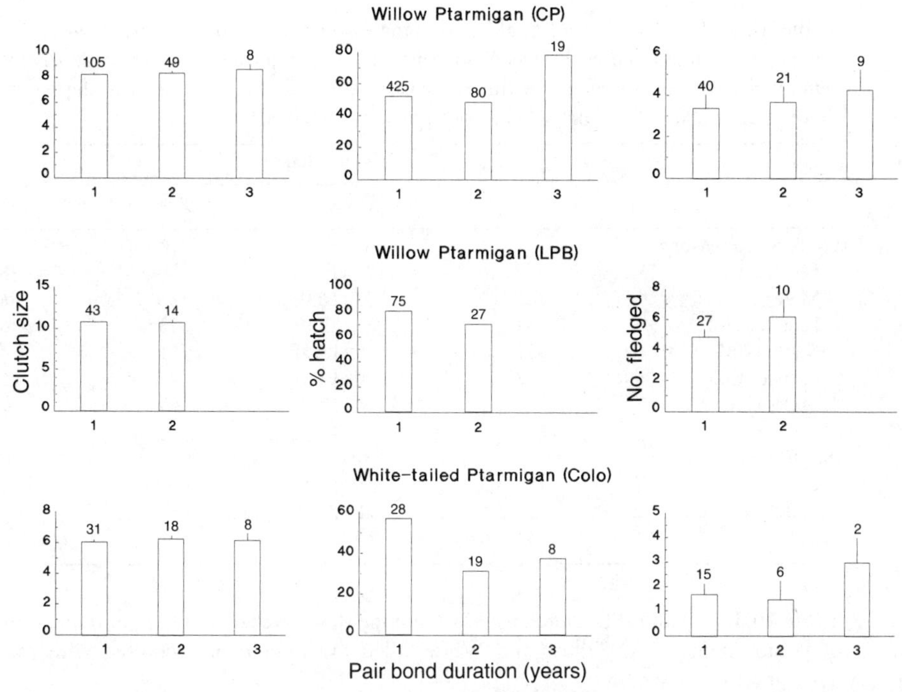

Fig. 10.1 Adjusted mean clutch size, % of pairs hatching a clutch, and number of young fledged from successful nests for Willow and White-tailed Ptarmigan paired with the same mate for 1, 2, or 3 years (Sample size above bars). Clutch size and young fledged adjusted in a three-way ANOVA with year, female age, and pair duration as factors (no interactions, type III sums of squares used); % hatch for CP analysed with GLIM (year, age, and pair duration as factors); for LPB and Colorado cell sizes too low to control for year and age; so data combined for *G*-test. (CP: clutch size $df = 2$, $P = 0.61$, % hatch $P = 0.01$, number fledged $P = 0.74$; LPB: clutch size $df = 1$, $P = 0.41$, % hatch $P = 0.30$, number fledged $P = 0.29$; Colo: clutch size $df = 2$, $P = 0.72$, % hatch $P = 0.22$, number fledged $P = 0.39$).

Divorce between and within seasons

For both species of ptarmigan, only 14–20% of pairs divorced between seasons (Table 10.2). Within season divorce was rare in Willow Ptarmigan, but 21% of White-tailed Ptarmigan females switched partners to renest after their first clutch was depredated (Table 10.2). In the majority of cases, females appeared to initiate the divorce since they moved to another territory while their previous mate stayed on the former territory (Table 10.4). If females divorce because of mistakes in mate choice, we might expect yearling females to have made more mistakes than older ones. However, yearlings were not more likely to divorce the next year than older females.[1]

[1]Likelihood of divorce in young and older females: Willow Ptarmigan CP: 19.7% of yearlings ($n = 86$) and 13.0% of adults ($n = 92$) divorced ($G = 1.45$, $df = 1$, $P = 0.23$); LPB: 15.4% of yearlings ($n = 13$) and 13.0% of adults ($n = 23$) divorced ($G_{ran} = 0.04$, $df = 1$, $P = 0.88$); White-tailed Ptarmigan: 26.3% of yearlings ($n = 19$) and 13.6% of adults ($n = 22$) divorced ($G_{ran} = 1.05$, $df = 1$, $P = 0.35$).

Table 10.2 (a) Number of male and female ptarmigan that reunited with the same mates or changed mates between seasons and (b) number of pairs that divorced or remained together to renest within a season after their first nest was depredated (% divorce in brackets for pairs where both partners survived)

	Same mate	Mate change		Total
		Divorce	Partner dead	
(a) Between seasons				
Females CP	152	31 (16.9)	104	287
Males CP	152	31 (16.9)	183	366
Females LPB	31	5 (13.9)	33	69
Males LPB	31	5 (13.9)	51	87
Females Colo	33	8 (19.5)	27	68
Males Colo	33	8 (19.5)	49	90
(b) Within seasons				
CP	115	5 (4.2)		120
LPB	38	1 (2.6)		39
Colo	52	14 (21.2)		66

Table 10.3 Mean (±SE) clutch size, percentage that hatched a clutch, and mean number of young fledged for Willow and White-tailed Ptarmigan that reunited with the same mate or had a new mate[1]

		Same mate (*n*)	New mate (*n*)	*P*
CP Willow				
Females	Clutch size	8.5 ± 0.18 (54)	8.4 ± 0.14 (80)	0.64
	% Hatched	55.2 (96)	53.6 (110)	0.89
	Young fledged	4.3 ± 0.59 (29)	3.5 ± 0.58 (31)	0.24
Males	Clutch size	8.3 ± 0.33 (70)	8.0 ± 0.31 (173)	0.08
	% Hatched	54.6 (97)	51.8 (112)	0.93
	Young fledged	5.1 ± 0.80 (29)	4.7 ± 0.75 (66)	0.58
LPB Willow				
Females	Clutch size	10.7 ± 0.27 (15)	10.9 ± 0.24 (25)	0.42
	% Hatched	72.4 (29)	77.8 (36)	0.62
	Young fledged	5.9 ± 1.2 (9)	4.7 ± 0.7 (13)	0.28
Males	Clutch size	10.7 ± 0.27 (15)	10.9 ± 0.20 (34)	0.34
	% Hatched	72.4 (29)	77.4 (53)	0.62
	Young fledged	5.9 ± 1.2 (9)	4.7 ± 0.64 (18)	0.39
White-tailed				
Females	Clutch size	6.3 ± 0.19 (26)	6.1 ± 0.14 (27)	0.48
	% Hatched	33.3 (27)	59.3 (27)	0.06
	Young fledged	1.9 ± 0.64 (8)	1.7 ± 0.51 (12)	0.80
Males	Clutch size	6.3 ± 0.19 (26)	5.8 ± 0.12 (50)	0.04
	% Hatched	33.3 (27)	57.1 (42)	0.05
	Young fledged	1.9 ± 0.64 (8)	2.0 ± 0.43 (22)	0.88

[1]Clutch size and young fledged adjusted in a three-way ANOVA with year, female age, and same mate/new mate as factors, no interactions, type III sums of squares used; % hatch for CP analysed with GLIM (year, age, and same mate/new mate as factors); for LPB and Colorado cell sizes too low to control for year and age; so data combined for *G*-test. $df = 1$ in all tests.

Table 10.4 Distances moved from the previous year's territory by divorced feale ptarmigan[1]

	Stay on territory	Move to adjacent	Move <3 territories	Move >1 km
Females CP	0	21	2	8
Males CP	27	0	0	0
Females LPB	0	4	0	1
Males LPB	2	1	0	0
Females Colo	1	5	0	2
Males Colo	5	0	0	2

[1]Females were more likely to move away from the territory than males (stay *vs.* move; CP: G_{ran} = 80.13, $P < 0.001$; LPB: $G_{ran} = 5.18$, $P = 0.056$; Colo: $G_{ran} = 5.78$, $P = 0.04$). $df = 1$ in all tests.

Why did females divorce?

We examine four hypotheses to explain divorce by females:

(1) they had poor breeding success the previous year or within a season when they changed to renest;

(2) their previous territory declined in quality;

(3) to mate with an older male with more breeding experience; or

(4) to avoid having to mate polygynously.

Poor breeding success

Poor breeding success the previous year did not appear to influence whether or not females divorced their mates (Fig. 10.2). Within season divorce was preceded by clutch predation; however, not all females deserted their mates to renest (Table 10.2).

Decrease in territory size

Females may have divorced their mates because territory quality declined from one season to the next. Here, we use change in territory size as an

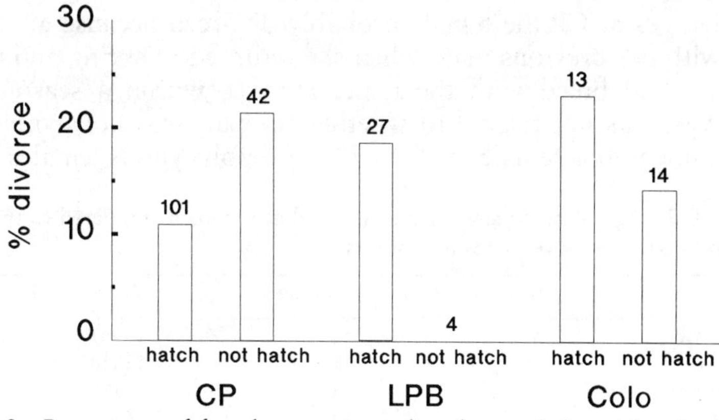

Fig. 10.2 Percentage of female ptarmigan that divorced that either hatched or did not hatch any young the previous season (CP: $G = 2.49$, $P = 0.11$; LPB: $G_{ran} = 1.52$, $P = 0.32$; Colo: $G_{ran} = 0.35$, $P = 0.62$; sample size above bars). $df = 1$ in all tests.

index of change in territory quality at the CP study site. Males showed very high site fidelity to their territories and often the territorial boundaries were similar from year to year. Sometimes males increase their territory size by expanding into a neighbouring territory when the owner dies. Alternatively, if a new male inserts or a neighbour expands his territory, a male's territory might shrink or, more rarely, he might lose his territory. For Willow Ptarmigan at CP, one divorce was preceded by a male losing his territory, eight after their previous mate's territory decreased in size, six after their previous mate enlarged his territory and six when territory size did not change (within 0.5 ha). Ten females changed to a male that had a larger territory than their previous mate, but five changed to a smaller territory, and seven to a territory of similar size (within 0.5 ha). Clearly, territory size was not a major predictor of whether a female divorced a male or not, except in the obvious case where a male lost his territory.

To change to an older male

In general, females did not choose new mates that were older than their former mates. For between season divorce, the majority of females changed to males that were the same age or younger than their previous mate (Table 10.5); however, yearling males were avoided. For within season divorce, the age of the original mate appeared to influence whether a female changed mates to renest for White-tailed Ptarmigan: 33% of 18 yearling males were deserted for the renesting attempt compared to 12% of 49 adult males ($G = 3.64$, $df = 1$, $P = 0.05$). Two of 11 females changed to yearling males and the remainder to males the same age or older than their original mate.

Avoidance of polygyny

Several females may have changed mates to avoid polygyny: in 20 out of 30 divorces at CP the female probably divorced because another female was with her previous mate when she returned in spring and that female remained to breed with the male. Divorce within a season to renest, however, was not related to whether the pair was polygynous (17% of 58 monogamous females and 16.7% of 6 polygynous females changed to

Table 10.5 Age of new mate for ptarmigan that divorced compared to the age of their previous mate (no. of yearlings in brackets)

	Older	Same age	Younger	No mate
Females CP	5	12	13 (4)	0
Males CP	3	8	12 (11)	5
Females LPB	1	4	0 (0)	0
Males LPB	1	0	2 (0)	2
Females Colo	2	2	2 (0)	0
Males Colo	0	3	4 (4)	1

Table 10.6 Pairing status (% polygynous) of female ptarmigan that divorced or reunited with their mates in the year of the divorce

	Divorced	Reunited	G (P)[1]
Willow Ptarmigan CP	3.7% (27)	22.6% (137)	6.59 (0.01)
Willow Ptarmigan LPB	20.0% (5)	9.7% (31)	0.40 (0.81)
White-tailed Ptarmigan Colo	0% (6)	15% (39)	1.85 (0.26)

[1] df = 1 in all tests.

renest, $G = 0.001$, $df = 1$, $P = 0.99$). After divorce, females at CP were less likely to pair polygynously than females that reunited with their mates (Table 10.6). Small sample sizes at LPB and Colorado clouded this relationship, but when all populations were combined 5% ($n = 38$) of divorced females and 19% ($n = 207$) of females that reunited with their previous mates were polygynous ($G = 5.6$, $df = 1$, $P = 0.03$). This may explain why some females at CP divorced when their mate of the previous year expanded his territory: males on larger territories were more likely to become polygynous (Hannon 1984).

Reproductive performance in relation to divorce or mate fidelity

There may be a number of costs associated with changing mates. First, it may be difficult to find a new mate; this would be particularly likely to happen to the bird that was divorced. Second, a bird might suffer decreased reproductive success with a new mate.

Divorce between seasons: females that divorced always found a new mate (Table 10.5) and they appeared to avoid yearling males, despite the fact that yearlings comprised about 40% of the males in the populations. Divorced males, on the other hand had a high probability of pairing with a yearling female and several remained unmated (Table 10.5). Overall, reproductive success (controlled for female age) was similar for females and males that divorced and those that reunited with their previous mate, with the exception of divorced female White-tailed Ptarmigan which were more likely to hatch their clutches with new mates than hens with previous mates (Table 10.7). Thus, in general, divorce between seasons was not costly for female ptarmigan, and in some cases it appeared to be beneficial. The major cost for males was that they ran the risk of remaining unmated.

Divorce within a season: we had sufficient data for White-tailed Ptarmigan only. The clutch size of the renest of females that divorced within the season did not differ from that of females that renested with the same mate (divorce: mean 4.5, SE 0.27, $n = 13$; same mate: mean 4.8, SE 0.12, $n = 42$, $F = 1.35$, $df = 1$, $P = 0.25$ (two-way ANOVA with status (divorce/same mate) and female age as factors). Females that changed mates were no more likely to hatch a clutch than were females that

Table 10.7 Adjusted mean (±SE) clutch size, percentage that hatched a clutch, and mean number of young fledged from successful nests for female and male Willow and White-tailed Ptarmigan that divorced or reunited with their mate of the previous year[1]

		Divorced	Reunited	*P*
CP Willow				
Females	Clutch size	8.4 ± 0.33 (17)	8.5 ± 0.19 (70)	0.77
	% Hatched	47.6 (21)	55.8 (95)	0.30
	Young fledged	3.7 ± 1.22 (5)	5.2 ± 0.56 (29)	0.25
Males	Clutch size	8.1 ± 0.30 (17)	8.5 ± 0.17 (69)	0.20
	% Hatched	66.7 (18)	55.7 (97)	0.14
	Young fledged	4.7 ± 1.14 (8)	5.0 ± 0.63 (29)	0.78
LPB Willow				
Females	Clutch size	10.3 ± 0.33 (3)	10.7 ± 0.27 (15)	0.53
	% Hatched	80 (5)	72.4 (29)	0.78
	Young fledged	4.0 ± 1.4 (4)	6.3 ± 0.92 (12)	0.23
Males	Clutch size	10.3 ± 0.33 (3)	10.7 ± 0.27 (15)	0.53
	% Hatched	66.6 (3)	72.4 (29)	0.87
	Young fledged	2.5 ± 0.5 (2)	6.3 ± 0.92 (12)	0.14
White-tailed				
Females	Clutch size	6.4 ± 0.51 (5)	6.4 ± 0.18 (24)	0.97
	% Hatched	83.3 (6)	32.1 (28)	0.04
	Young fledged	0.25 ± 0.25 (4)	1.9 ± 0.64 (8)	0.12
Males	Clutch size	6.2 ± 0.31 (6)	6.4 ± 0.18 (24)	0.53
	% Hatched	50 (6)	32.1 (28)	0.44
	Young fledged	1.0 ± 1.0 (3)	1.9 ± 0.64 (8)	0.49

[1]Clutch size and young fledged adjusted in a three-way ANOVA with year, female age, and divorce/reunite as factors, no interactions, type III sums of squares used; % hatch for CP analysed with GLIM (year, age, and divorce/reunite as factors); for LPB and Colorado cell sizes too low to control for year and age; so data combined for G-test. *df* = 1 in all tests.

remained with their mates to renest (divorce: 50%, $n = 14$; same mate 30.9%, $n = 55$; $G = 1.73$, $df = 1$, $P = 0.18$). The number of chicks fledged from successful nests did not differ between the two groups (divorce: mean 1.7, SE 0.67, $n = 6$; same mate: mean 2.0, SE 0.50, $n = 15$; $F = 0.14$, $df = 1$, $P = 0.71$).

Pair characteristics and reproductive success

Age

The majority of pairings for both species of ptarmigan were adult–adult. Adult females appeared to avoid yearling males (CP: $G = 8.35$, $df = 1$, $P = 0.004$; LPB: $G = 10.5$, $df = 1$, $P = 0.001$; Colo: $G = 20.7$, $df = 1$, $P < 0.001$); whereas yearling females paired with yearling males at a higher (CP: $G = 4.7$, $df = 1$, $P = 0.02$; LPB: $G = 3.9$, $df = 1$, $P = 0.05$) or similar (Colo: $G = 1.8$, $df = 1$, $P = 0.18$) proportion than their availability. For both Willow Ptarmigan at CP and White-tailed Ptarmigan, clutch size increased for females from 1 to 3 years of age and declined for females of 4 years and older. Clutch size for LPB Willow Ptarmigan did not vary with female age (Wiebe and Martin 1994). Date of first egg varied significantly with female age but not male age for both species. Number of young fledged from nests that hatched did not vary with female or male age, except for Colo females (S. Hannon and K. Martin, unpublished data). In pairs of varying age structure, pairs including adult females for Willow Ptarmigan at CP and White-tailed Ptarmigan at Colorado had slightly higher clutch sizes, whereas pairs with yearling females had lower clutch sizes (Fig. 10.3). So it would appear that adult females should be preferred by males. Pairs with adult males at LPB, on the other hand, fledged more chicks, suggesting that adult males should be preferred by females at least at this site.

Discussion

Why divorce?

For ptarmigan of both species, we believe that divorce was usually initiated by females, thus we examine hypotheses to explain divorce from the female's perspective. Ens *et al.* (1993) discussed two main hypotheses to explain divorce in Oystercatchers *Haematopus ostralegus*: the 'incompatibility' and the 'better option' hypotheses. The incompatibility hypothesis states that traits of the male and female are not compatible leading to poor reproductive success. Thus divorce should result in benefits to both members of the pair. Our data do not support this hypothesis as, contrary to a number of other studies (e.g. Harris *et al.* 1987), females that divorced did not have poorer reproductive success the previous year than females that reunited with former mates. In an earlier paper (Schieck and Hannon 1989) we reported that CP females were more likely to change mates if they were unsuccessful in the previous year; however, that relationship did not hold with the larger sample size reported here. In addition, males that divorced did not fare better than those that reunited with their mate and some remained unmated. In some measures of reproductive success, divorced females in some populations did better than those that reunited with their previous mate, but this was not a consistent result across species and populations. One might expect younger females to make more mistakes in mate choice, yet we found that younger females were not more likely to divorce their mates than older birds.

The better option hypothesis states that birds should change mates or territories if the benefits of divorce outweigh the costs. Ens *et al.* (1993) outlined four types of evidence that would support this hypothesis:

(1) that divorce was by choice, not by coercion;

(2) that there are differences in quality among mates and territories;

(3) that vacancies are available; and

(4) that birds that choose to divorce benefit from that choice.

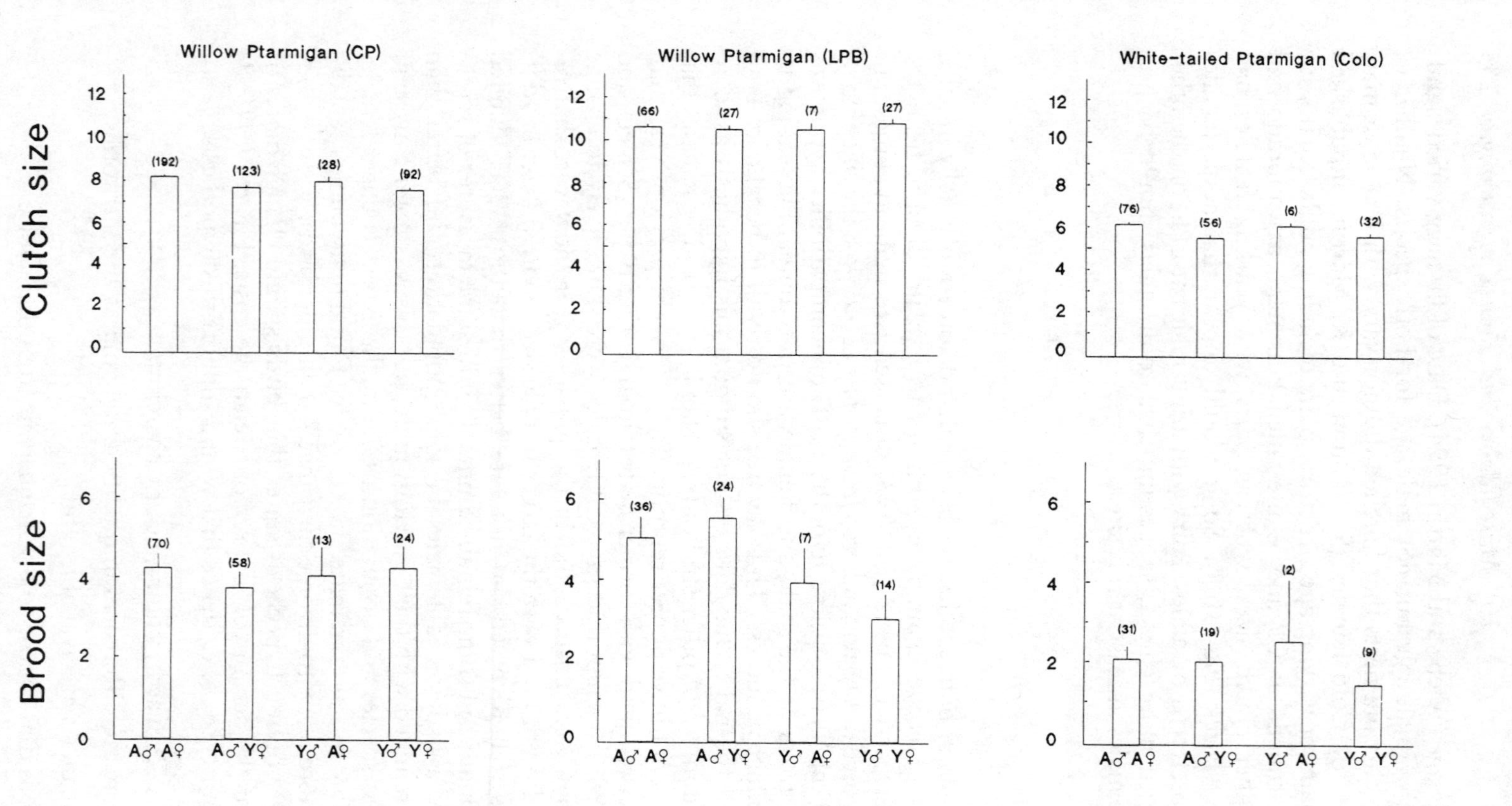

Fig. 10.3 Effect of the age of the female and male in a ptarmigan pair on clutch size and brood size at fledging (A = adult, Y = yearling); (two-factor ANOVA with year and pair age as factors; CP clutch size, $F = 6.5$, $P = 0.0002$; brood size, $F = 0.43$, $P = 0.73$; LPB clutch size, $F = 0.56$, $P = 0.64$; brood size, $F = 2.32$, $P = 0.08$; Colorado clutch size, $F = 4.10$, $P = 0.008$; brood size, $F = 0.33$, $P = 0.81$; posthoc comparisons with Fisher's protected LSD). $df = 3$ in all tests.

Again, since males rarely changed territories, we examine these four points for females.

Divorce was by choice

Although we have no behavioural data to support this contention, we think that females were not forced to divorce their mates. First it is unlikely that their previous mate forced them to divorce, as males benefit from polygyny (Hannon 1984). Second, the other female on the territory (in cases where the male paired with a new mate), would be unlikely to expel a female that had previous tenure on the territory. Third, in a few cases the previous mate remained unpaired after the divorce. A few females, however, changed mates because their male partner of the previous year had lost or moved his territory.

Quality differences among males and territories

Schieck (1988) found that for Willow Ptarmigan at Chilkat Pass, quality of mate, other than that due to age or familiarity, did not influence survival or reproduction. However, more than half of the variability in date of laying and clutch size was due to differences among territories. This suggests that territories vary in their quality, but mates do not.

Availability of vacancies

Female mortality is higher than male mortality in both species (see below), and in any year there are usually unmated territorial males in the population. Thus, territories and mates are available for females to change to.

Females benefit from divorce

As discussed above, females did not consistently appear to benefit from divorce. In a repeated measures analysis using the same individuals before and after divorce, Schieck and Hannon (1989) found that after divorce, female Willow Ptarmigan had a higher probability of losing their clutch to predators, although other measures of reproductive success did not change. It is possible that females divorce to avoid a polygynous relationship with their previous mate. Female Willow Ptarmigan at CP that changed mates between seasons had far fewer polygynous relationships than females in the general population. Female Willow and White-tailed Ptarmigan appear to prefer to pair monogamously; polygynous females fight and chase each other prior to laying (Hannon and Martin 1992; K. Martin unpublished data) and in both species monogamous females repel potential secondary hens from their territories (Martin *et al.* 1990). Hannon and Martin (1992), however, failed to detect any reproductive cost to pairing polygynously. In some years at CP, however, females that mated polygynously had lower survival over the winter. Thus, if females have the option of moving to a neighbouring adult male whose previous mate has not returned, they should do so.

Costs of divorce

Divorce may be costly to birds for four main reasons (Ens *et al.* 1993 and Chapter 19):

(1) costs of searching for a new mate;
(2) loss of breeding status;
(3) costs of fighting with a rival for the new mate; and
(4) costs associated with lower reproductive success with a new and unfamiliar mate.

For most female ptarmigan that divorced, these costs are likely to be low. First, most females changed to an adjacent territory, thus costs of mate searching were low. Second, no females remained unpaired, and many of those that divorced improved their breeding status (i.e. became monogamous). Third, most females that changed moved to a territory where the hen of the previous season had not returned, thus there was no rival to fight. Those females that divorced to avoid polygyny may have gained by avoiding fighting with the female on their previous territory. Finally, females that divorced had similar reproductive success to females that reunited with their original mate and females that divorced showed no significant change in reproductive success between seasons. Females that divorced paired with experienced adult males, most of which would have been familiar to them because they held adjacent territories the previous season. Thus, for females, divorce was not costly, and for those that avoided polygyny divorce may have been preferable to remaining with their old mate in a polygynous arrangement.

Males, because they were the ones left behind, did suffer some costs of divorce. For all populations combined, 20% of divorced males did not gain a new mate in the year of the divorce. Of those that did attract a new mate, 45% of these were yearling females: males paired with yearling females produced lower clutch sizes than males paired with adults; however, this did not translate into lower numbers of chicks fledged from successful nests.

Mate fidelity in ptarmigan

Pair bonds in ptarmigan usually last for 1 year, mainly because annual adult mortality is high: 54% for female and 44% for male Willow Ptarmigan at CP (Hannon 1995); 61% for female and 43% for male Willow Ptarmigan at LPB (Martin *et al.* 1989; Martin 1991); and 50% for female and 43% for male White-tailed Ptarmigan in Colorado (Braun *et al.* 1993). Based on these mortality estimates, the probability of both members of a pair surviving to the next season is between 0.22 and 0.28.

Divorce accounts for only 9–24% of the terminations of pair bonds. Thus, mate fidelity is the norm for ptarmigan and we now discuss potential benefits of mate fidelity.

Pairs may remain together for three main reasons:

(1) to increase reproductive success because of previous experience with the partner;
(2) to benefit from a high quality territory or familiarity with the territory; or
(3) because the current mate is of equal or better quality than other available males.

Schieck and Hannon (1989) argued that pairs stayed together because of the benefits of pairing with a familiar partner. In our analysis, however, we found that this benefit only accrued to birds with a pair bond of 3 years or more as they had a higher probability of hatching a clutch. This effect may have been due to experienced pairs being better able to evade clutch predation and/or the increased ability of older females to replace depredated clutches (K. Martin and S. Hannon unpublished data). Remaining with one's mate may be beneficial to males in particular, since clutch size increases in females as they age up to 3 years. Age of the male, however, did not appear to influence reproductive success, so these age-related benefits would not accrue to females that remained with older males.

Familiarity with the territory does not appear to influence mate fidelity in ptarmigan. Schieck and Hannon (1989) found that females breeding on unfamiliar territories had similar reproductive success to those breeding on familiar territories. We do not have data on mate quality to address hypothesis (3); however, one would have to include the costs of reuniting in this analysis. In other words, the gain from the difference in quality of two potential mates would have to be greater than the cost of moving to a new mate. If the potential new mate is already paired or at a distance from the previous territory, then the costs of divorce may be too high. Thus, low cost options for divorce will be rare as the potential new mate must be close by (within one or two territories) and unpaired. Mate fidelity in ptarmigan may be a default situation and divorce may only occur if it is forced or a better option becomes available close by.

Is longer-term monogamy advantageous?

In our previous work on Willow Ptarmigan, we examined potential costs and benefits of monogamy to males and females within a breeding season (Hannon 1984, 1995; Martin and Cook 1987; Hannon and Martin 1992). Monogamy appeared to benefit females at CP, as females that paired polygynously had poorer annual survival in 2 of 9 years (Hannon and Martin 1992), although reproductive success did not differ between monogamous and polygynous females. Our results may support this contention, as some females appeared to divorce mates to avoid polygyny. Males, on the other hand, appeared to benefit from polygyny as they produced more young than monogamous males in years of high clutch predation (Hannon 1995), although some polygynous males suffered kleptogamy (Freeland 1993). Monogamy appears to be forced on most males by female–female aggression and the male-biased sex ratio (Martin *et al.* 1990; Hannon and Martin 1992). Longer-term mate fidelity may reduce clutch predation, a major component of breeding success. Long-term monogamy is rare though, because annual mortality is high and females can change to a new mate if males attempt to become polygynous.

Comparisons among species and populations

Our results were remarkably consistent among species and populations of ptarmigan. The only major difference was for divorce within a season; it was rare for Willow Ptarmigan but 20% of female White-tailed Ptarmigan that had a clutch depredated changed mates to renest.

Summary

Pair durations are relatively short in ptarmigan, primarily because of high annual mortality of both sexes. Of the pairs that survive intact to the next year, 14–20% do not reunite. Females appear to initiate divorce and they do not suffer reproductive loss, and in some cases, do better than those that remain with their mates. Males, on the other hand, risk remaining unpaired after the divorce. We suggest that mate fidelity is the norm if both members of the pair return to breed, and that divorce occurs only in unusual circumstances, such as the loss or desertion of the territory by the male, high clutch predation combined with high availability of unmated males within a season, or the potential costs to females of pairing polygynously.

Acknowledgements

The study was financed by grants from Natural Sciences and Engineering Research Council of Canada, University of Alberta, Université de Sherbrooke, University of Toronto and Queens University, Canadian Wildlife Service, Colorado Division of Wildlife and the Circumpolar Institute (BAR and NSTP grants). We thank R. B. Bennett, C. McCallum and K. Field for technical help, A. Desrochers, R. Moses, and K. Wiebe for commenting on the manuscript, and the many graduate students and summer students who have worked on ptarmigan with us.

References

Artiss, T. and Martin, K. (1995). Male vigilance in paired white-tailed ptarmigan, *Lagopus leucurus*: mate guarding or predator detection? *Animal Behaviour*, **49,** 1249–58.

Bergerud, A. T. and Mossop, D. H. (1984). The pair bond in ptarmigan. *Canadian Journal of Zoology*, **62,** 2129–41.

Braun, C. E., Martin, K., and Robb, L. A. (1993). White-tailed Ptarmigan. In *The birds of North America* (ed. A. Poole and F. Gill), pp. 1–24. The Academy of Natural Sciences, Washington, DC.

Ens, B. J., Safriel, U.N., and Harris, M. P. (1993). Divorce in the long-lived and monogamous oystercatcher, *Haematopus ostralegus*: incompatibility or choosing the better option? *Animal Behaviour*, **45,** 1199–217.

Freeland, J. R. (1993). The evolution and maintenance of monogamy in Willow Ptarmigan. Unpublished M.Sc. thesis. Queen's University, Canada.

Gruys, R. C. (1993). Autumn and winter movements and sexual segregation of Willow Ptarmigan. *Arctic*, **46,** 228–39.

Hannon, S. J. (1984). Factors limiting polygyny in willow ptarmigan. *Animal Behaviour*, **32,** 153–61.

Hannon, S. J. (1995). Ecological and behavioural constraints on monogamy in the Willow Ptarmigan In *Proceedings of the International Symposium on Grouse*, (ed. D. Jenkins), pp. 43–7. Reading, UK and Instituto Nazionale per la Fauna Salvatica, Ozzano dell 'Emilia, Italy.

Hannon, S. J. and Martin, K. (1992). Monogamy in willow ptarmigan: is male vigilance important for reproductive success and survival of females? *Animal Behaviour*, **43,** 747–57.

Hannon, S. J., Simard, B. R., and Zwickel, F. C. (1979). Differences in the gonadal cycles of adult and yearling blue grouse. *Canadian Journal of Zoology*, **57,** 1283–9.

Hannon, S. J., Martin, K., and Schieck, J. O. (1988). Timing of reproduction in two populations of Willow Ptarmigan in northern Canada. *Auk*, **105,** 330–8.

Harris, M. P., Safriel, U. N., Brooke, M. L., and Britton, C. K. (1987). The pair bond and divorce among Oystercatchers *Haematopus ostralegus* on Skokholm Island. Wales. *Ibis*, **120,** 45–57.

Hoffman, R. W. and Braun, C. E. (1975). Migration of a wintering population of White-tailed Ptarmigan in Colorado. *Journal of Wildlife Management*, **39,** 485–90.

Martin, K. (1984*a*). Intraspecific nest parasitism in Willow Ptarmigan. *Journal of Field Ornithology*, **55,** 250–1.

Martin, K. (1984*b*). Reproductive defence priorities of male willow ptarmigan *Lagopus lagopus*: enhancing mate survival or extending paternity options? *Behavioral Ecology and* Sociobiology, **16,** 57–63.

Martin, K. (1991). Experimental evaluation of age, body size, and experience in determining territory ownership in Willow Ptarmigan. *Canadian Journal of Zoology*, **69,** 1834–41.

Martin, K. and Cooke, F. (1987). Bi-parental care in willow ptarmigan: a luxury? *Animal Behaviour*, **35,** 369–79.

Martin, K. and Hannon, S. J. (1988). Early pair and extra-pair copulations in Willow Ptarmigan. *Condor*, **90,** 245–6.

Martin, K., Hannon, S. J., and Rockwell, R. F. (1989). Clutch size variation and patterns of attrition in fecundity of Willow Ptarmigan. *Ecology*, **70,** 1788–99.

Martin, K., Hannon, S. J., and Lord, S. (1990). Female–female aggression in White-tailed Ptarmigan and Willow Ptarmigan during the pre-incubation period. *Wilson Bulletin*, **102,** 532–6.

Martin, K. Holt, R. F., and Thomas, D. W. (1993). Getting by on high: ecological energetics of arctic and alpine grouse. In *Life in the cold III: ecological, physiological and molecular mechanisms* (ed. C. Carey, G. L. Florant, B. A. Wunder, and B. Horwitz), pp. 33–41. Westview Press, Boulder, Colorado.

Schieck, J. O. (1988). Territory selection and site fidelity in Willow Ptarmigan: the importance of quality and familiarity with territory and partner. Unpublished Ph.D. thesis. University of Alberta, Canada.

Schieck, J. O. and Hannon, S. J. (1989). Breeding site fidelity in willow ptarmigan: the influence of previous reproductive success and familiarity with partner and territory. *Oecologia*, (Berlin), **81,** 465–72.

Schieck, J. O. and Hannon, S. J. (1993). Clutch predation, cover, and the overdispersion of nests of the Willow Ptarmigan. *Ecology*, **74,** 743–50.

Wiebe, K. L. and Martin, K. (1994). Growing old in the cold: age and reproduction in two ptarmigan. *Journal für Ornithologie*, **135,** 385.

11 *Causes and consequences of long-term partnerships in Cassin's Auklets*

WILLIAM J. SYDEMAN, PETER PYLE, STEVEN D. EMSLIE, AND ELIZABETH B. McLAREN

Introduction

Cassin's Auklets *Ptychoramphus aleuticus* are wing-propelled diving seabirds which feed extensively on krill and small fish of various species. Cassin's Auklets are generally migratory; however, our study population on Southeast Farallon Island, California, is sedentary (Ainley *et al.* 1990). Here, its breeding biology, demography, foraging ecology, and diet have been studied since 1969. On Southeast Farallon, these birds nest in burrows or natural rock crevices. Birds will also use nest boxes which greatly facilitates research and monitoring. Cassin's Auklets are nocturnal in their visits to breeding colonies owing to predation on adults by diurnally active Western Gulls *Larus occidentalis*.

This paper evaluates the causes and consequences of monogamy and long-term partnerships in Cassin's Auklets based on studies conducted from 1982 through 1993. Drawing on a sample of 303 pairs, we ask whether this sedentary auklet benefits from maintaining partnerships over several years. In particular, we evaluate the effects of mate change and pair duration on reproductive performance while controlling for the confounding effects of year (representing inter-annual changes in food availability) and previous breeding experience of the parents.

Background, study site, and procedures

Cassin's Auklets are small seabirds (roughly 180 g). Both sexes are greyish-brown above and dusky-white below; the only conspicuous characteristics are bluish feet and an obvious white eyebrow fleck over and under each eye. Among pairs, males are larger than females, and can be reliably sexed in the hand by bill depth (Nelson 1981).

Southeast Farallon Island (37°42′N, 123°00′W) is located 42 km west of San Francisco in the central region of the California Current marine ecosystem. This environment is characterized by substantial inter-annual variation in physical oceanographic processes and biological productivity. Moderate to extreme El-Niño Southern Oscillation events and other oceanographic anomalies occur often, disrupt food web development, and reduce productivity of seabirds in this system (Ainley *et al.* 1995).

During the study period (1982 through 1993), we monitored the reproductive activities of breeding birds in 44 nest boxes originally installed in 1978 (see also Emslie *et al.* 1992). We initiated a banding programme of breeding birds in 1982. Beginning in early March each year, we checked boxes every 5 days for egg laying. When an egg was found we banded and measured the incubating adult, and then returned the next day to band and measure its mate. We define a pair based on catching the male and female in the same nest box. After banding we left the site undisturbed for 35 days and then checked for hatch. After hatching, chicks were left undisturbed for 35 days, then banded and weighed daily to ascertain fledging weight.

Pair duration (the number of years a pair remained together) was recorded beginning with a score of 0 (indicating no previous experience as a pair). A few cases where members of the pair divorced and were subsequently found together in a later year were considered new pairs. We also scored whether or not the pair had been together in the previous year and called this pair fidelity. Divorce was tallied when both birds were known to be alive in a following year, yet breeding with other individuals. Experience was scored based on the number of previous records for each individual in our data set. Birds skip breeding during years of poor food supply or may move into burrows where their presence is unrecorded; hence experience is a minimal estimate. We use experience as an index to age, although we caution readers that experience and age are not interchangeable, especially for seabirds. Experience is equivalent to *age* minus *age at first breeding* minus *the number of years when breeding was skipped* (or missed if birds were in burrows). Therefore, experience and age, although correlated, provide different measures of parental characteristics.

We restricted analyses of experience and pair duration to data from 1985 through 1993 in order to avoid biases towards low values of these variables in the early years (1982–1984, see also Emslie *et al.* 1992). Unbiased estimates of mate fidelity and divorce, however, could be accurately determined for the early years of study (although not in 1982, the first year of banding). We analysed both first and subsequent (re-lay and second brood) reproductive attempts for each pair. Double-brooding is known only for our population of Cassin's Auklets, and only occurred during years of early reproduction and high food availability (Ainley *et al.* 1990). First, re-lay, and second attempts were added to estimate the total number of chicks fledged per pair per year. We restricted analyses pertaining to hatching success (the proportion of eggs that hatched) and fledging success (the proportion of chicks that fledged) to first attempts.

We used multiple logistic regression and ANCOVA to analyse (1) the probability of divorce in relation to experience and reproductive success, and (2) factors affecting reproductive success, hatching success, fledging success, date of laying of first eggs, and weight of fledglings. Statistical analyses were performed using *STATA* version 3.1 (STATA 1993). ANCOVA was used to investigate differences in date of egg laying and fledging weight. Cassin's Auklets lay one egg per clutch; therefore, logistic (which is equivalent to binomial) regression was used to investigate factors affecting breeding success (see Cox and Snell 1989, cf. Sydeman *et al.* 1991). Because Cassin's Auklets double-brood (i.e. they may raise 0, 1, and 2 chicks in a year), we analysed annual chick production via ordered logistic regression. We used a backward modelling procedure. A likelihood ratio statistic (*LRS*), which is equivalent to the difference in deviance between a model containing all variables and a model without the variable of interest, is reported for mate change in Table 11.1, and pair duration and experience in the text. Regression coefficients, $b \pm SE$, are also presented. We tested whether curvilinear patterns between experience and reproductive performance and experience and divorce were best represented by quadratic (x and x^2, $df = 2$), inverse ($1/x$, $df = 1$), or logarithmic ($\ln x$, $df = 1$) relationships (cf. Sydeman *et al.* 1991; Emslie *et al.* 1992). We chose the best model based on the transformation that produced the largest *LRS* statistic and associated coefficient of determination (or pseudo r^2 in logistic regression analyses).

Demography and population trends

Southeast Farallon Island is about 49 ha. Manuwal (1974*a*) estimated 105 000 breeding birds in 1972. The population has apparently undergone a dramatic decline since that time. Carter *et al.* (1992) estimated 46 000 nesting birds in 1990. The decline in the numbers, especially in deep-soil areas on the island, has continued in recent years as evidenced by decreases in burrow numbers in a series of plots. Predation by Western Gulls and/or long-term oceanographic changes (e.g. increasing sea temperature, Sydeman and Ainley 1994) may be responsible for the recent population decline; some gulls are known to specialize on auklets (Emslie and Messenger 1991). Annual adult survival values have been estimated from 80 to 90% on Southeast Farallon (Speich and Manuwal 1974) and other locations in the Pacific (Gaston 1992). Occupancy rates of nest boxes increased during the study, averaging about 80%.

Process of pair formation

Based on returning birds banded as chicks, some individuals breed at age 2 years, but the mode is usually age 3 or 4. Pair formation probably occurs on nesting colonies and perhaps within nest cavities. Cassin's Auklets visit the colony year-round, although activity decreases during moonlit nights (Nelson 1989). Mate and site selection are likely to occur just prior to the breeding season, in January and February each year. Little is known about mate choice.

Paternity and parental care

Nothing is known concerning extra-pair copulation or fertilization in Cassin's Auklets. There are also no data concerning mate guarding. Both sexes incubate the egg, relieving each other on a 24-h schedule (Manuwal 1974*b*). Incubation lasts approximately 40 days. The nestling period lasts approximately 42 days; chicks are brooded alternately by both parents for the first week after hatching. Partially digested food, carried in sublingual gular pouches, is delivered nightly to chicks by both parents. Both birds spend the evening in nest cavities with their chick, before departing the colony to forage during daylight. Estimates of food intake, nestling growth, and the nestling period indicate that biparental care is necessary to fledge chicks. The death of one parent during reproduction always leads to abandonment of the egg and/or death of the chick.

Results

The pair bond

Cassin's Auklets are 100% socially monogamous; there is no evidence of polygyny, polyandry, or communal or co-operative nesting. The modal number of mates in our data set was 1. The average number of mates was 1.46 (SD 0.83, n = 303) with a range of 1–5. The modal pair duration was 1 year. The average pair duration was 1.45 (SD 0.93) with a range of 1–7 years.

Mate change

The overall rate of mate change between years was 58.6% (SE 2.6). Mate change between breeding attempts within a season was 3.7% (n = 327). The frequency of mate change varied significantly between years ($\chi^2 = 33.14$, $df = 10$, $P < 0.001$). There was a relationship between pair fidelity (the opposite of mate change) and annual reproductive success of the population (Fig. 11.1). Pair fidelity increased with increasing annual reproductive success. Given the low divorce rate in this population (see below), these data imply that mortality and reproductive success are also correlated. The data suggest, in particular, that female mortality is high during years of poor reproduction.

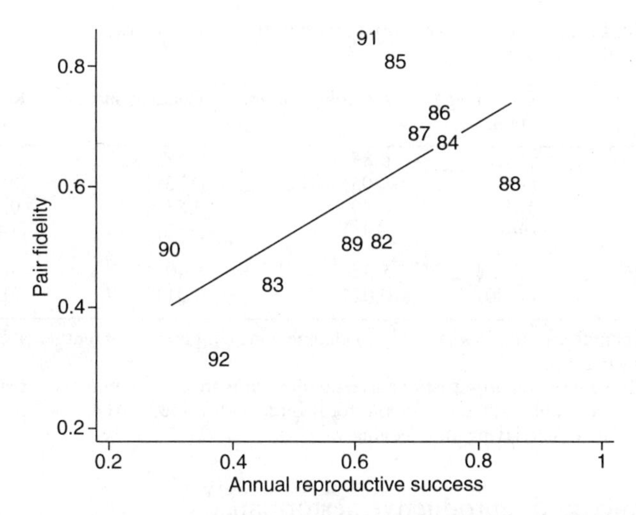

Fig. 11.1 The relationship between pair fidelity (the opposite of mate change) and annual reproductive success for the population in Cassin's Auklets, 1982–1993 ($y = 0.22 \pm 0.61\ (RS)$, $T = 3.37$, $P = 0.042$, $r^2 = 0.38$).

Divorce

Divorce in Cassin's Auklets was relatively infrequent owing to the frequent disappearance of one member of the breeding pair. The probability of divorce was recorded at 7.3% (16 events in 220 pair-years in 108 pairs). Taken from the viewpoint of re-mating, there was about a 92% chance that members of the pair would reunite if both survived to a future year. There were no time trends or patterns to divorce between years. There were, however, relationships between experience of the focal bird, successful or unsuccessful reproduction, and the probability of divorce. Divorce was more common among inexperienced pairs. Of 77 pairs with inexperienced males, 5 (6.5%) divorced. For pairs with males with 1 year of experience 9 out of 57 (15.8%) divorced. By comparison, of 86 pairs with males having > 1 year of previous breeding experience, only 2 (2.3%) divorced. A similar pattern was found with female experience. With respect to reproductive success, of 64 pairs that experienced reproductive failure and survived to the subsequent year, 7 (10.9%) divorced; by comparison, of 156 pairs that did not fail, only 9 (5.8%) divorced. Experience was significantly related to the probability of divorce in a parabolic fashion.[1] After controlling for experience, however, reproductive success was not significantly related to divorce ($LRS = 2.17$ $df = 1$, $P > 0.10$).

[1]Controlling for reproductive success; $LRS_{quadratic} = 8.86$, $df = 2$, $P < 0.01$, $b_1 = 2.22$ (SE 1.45), $b_2 = -0.55$ (SE 0.33); $LRS_{log} = 2.34$ and $LRS_{inverse} = 0.86$, $df = 1$, both $P > 0.10$.

Table 11.1 The relationship between reproductive performance and mate change for Cassin's Auklets

Mate status	Date of egg laying	Hatching success	Fledging success	Reproductive success	Fledging weight (g)
Same	99.71	0.85	0.78	0.76	152.47
	(157)	(205)	(173)	(204)	(99)
New	105.92	0.80	0.65	0.58	151.67
	(96)	(146)	(112)	(142)	(42)
[1]*LRS/F*	10.56	5.35	6.40	14.11	0.23
P	<0.001	0.021	0.011	<0.001	0.629

Results indicate that it was costly to change mates. *n* given in parentheses. See text for definitions of other variables.
[1]*LRS* (likelihood ratio statistic) corresponding to mate change in logistic regression analyses controlling for year and experience of the focal bird, or *F*-value in ANCOVA model for continuous variables (date of egg laying and fledging weight).

Mate change and reproductive performance

As measured from several reproductive variables, auklets performed significantly less well with new mates than with previous mates (Table 11.1); these include measures of annual reproductive success, hatching success, and fledging success. Established pairs also began to breed about 6 days sooner than new pairs. However, there was no difference in the fledging weights of chicks between established and new pairs.

Pair duration and reproductive performance

After controlling for the confounding effects of annual variation in food supply and breeding experience of *both* members of the pair (including quadratic terms for males and an inverse term for females [see below]), reproductive performance improved significantly with increasing pair duration (Fig. 11.2; $LRS = 18.55$, df = 1, $P < 0.001$, $b = 0.65$, SE 0.16). Analyses of hatching success revealed no influence of pair duration ($P > 0.25$). Fledging success, however, increased with pair duration ($LRS = 11.10$, $df = 1$, $P < 0.001$, $b = 0.60$, SE 0.19). Date of egg laying and fledging weight were not influenced by pair duration (Fig. 11.3; ANCOVA, $P > 0.10$). There were no curvilinear effects of pair duration on any measure of reproductive performance (all $P > 0.25$).

Pair characteristics, experience, and reproductive performance

Experience values of mated birds were correlated, although not as highly as could be expected (males: $r = 0.514$; females: $r = 0.503$; both $P < 0.05$). Roughly 60% of the inexperienced males were paired with inexperienced females. Similarly, approximately 50% of the males with 5 or more years of experience were also paired with females with 5 or more years of experience.

Controlling for the confounding effects of year, pair duration, and breeding experience of an individual's partner (using quadratic terms for

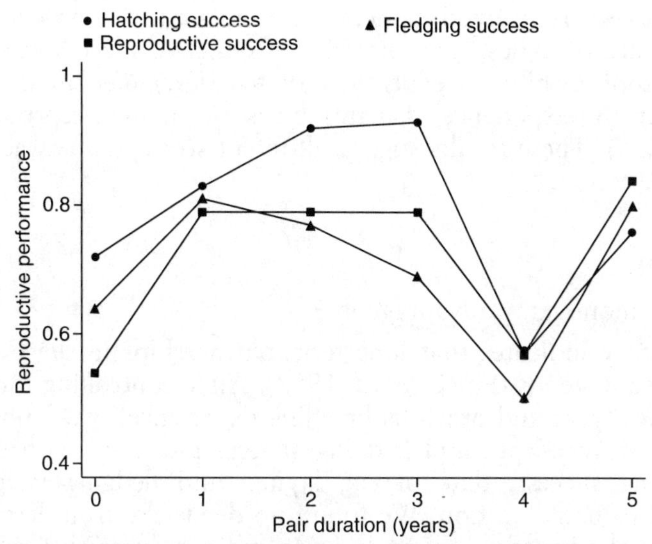

Fig. 11.2 Unadjusted means in reproductive performance with pair duration in Cassin's Auklets. Performance variables include hatching success, fledging success, and annual reproductive success. Data for pair bonds > 4 years were pooled. Sample sizes vary by reproductive parameter. Hatching success: n = 211, 77, 38, 14, 7, and 13 for 0, 1, 2, 3, 4, and 5+ years of pair duration, respectively. Fledging success: n = 150, 63, 35, 13, 4, and 10. Annual reproductive success: n = 209, 76, 38, 14, 7, and 13.

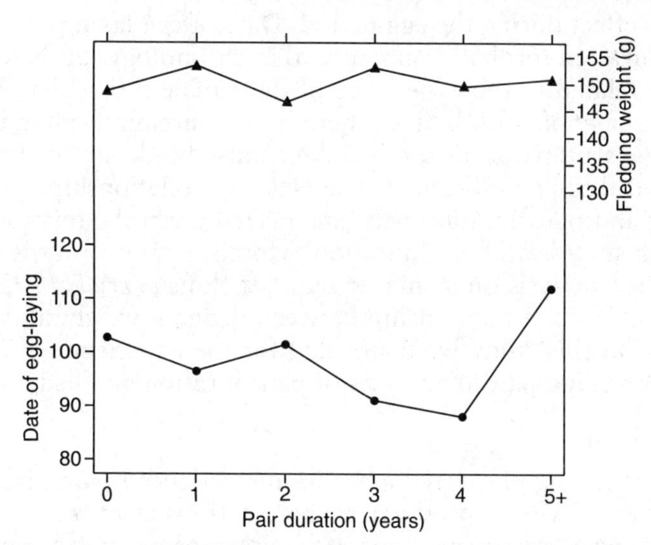

Fig. 11.3 The relationships between date of egg laying and fledgling weight and pair duration in Cassin's Auklets. Date of egg laying is expressed as the number of days after 1 January. Data for pair bonds > 4 years are pooled. Sample sizes varied with parameter. Date of egg laying: n = 209, 76, 38, 14, 6, and 13 for 0, 1, 2, 3, 4, and 5+ years of pair duration, respectively. Chick weights at fledging: n = 97, 50, 27, 9, 2, and 8.

both sexes), reproductive success was related in a parabolic fashion to experience of males.[1] In contrast, reproductive success was related in an asymptotic fashion to experience of females.[2] We found no interaction between the experience of mated birds and annual reproductive success ($P > 0.25$). These results suggest different strategies between the sexes in mate selection.

Discussion

Is long-term monogamy advantageous?

This study indicates that long-term partnerships in the Cassin's Auklet are advantageous (Emslie *et al.* 1992). After controlling for confounding variables (year and previous breeding experience), we found that annual reproductive output and fledging success increased with pair duration. Hatching success, date of egg laying, and fledging weight were not related to pair duration. We found no departure from linearity, indicating that the benefits of pair duration did not diminish even for the most experienced pairs.

The long-term advantage of pair bonds in Cassin's Auklets was important during chick rearing but not during incubation, suggesting that parental co-ordination in feeding young ultimately influences reproductive success. Egg neglect in Cassin's Auklets apparently has little effect on hatchability (Ainley *et al.* 1990) which may explain why pair duration had no effect during the egg period. Date of egg laying was not related to pair duration probably because the chronology of breeding is most closely related to the experience of the female rather than the male (also see Emslie *et al.* 1992). Since there is no courtship feeding in this species, a female's preparation for breeding must be determined exclusively by her own foraging efficiency. The lack of a relationship between fledging weight and pair duration was unexpected given the relationship between fledging success and pair duration. Moreover, this result is different from an earlier analysis on a smaller data set (Emslie *et al.* 1992). Perhaps we failed to detect a relationship between fledging weight and pair duration because in this study we controlled for the experience of *both* members of the breeding pair in analyses of pair duration and fledging weight.

The role of divorce

Divorce was significantly higher in inexperienced pairs, but did not vary significantly with reproductive success (Emslie *et al.* 1992). Although divorce was infrequent, this behaviour provides an opportunity for

[1]Effect of experience on reproductive success in males, $LRS_{quadratic} = 20.82$, $df = 2$, $b_1 = 0.24$ (SE 0.17), $b_2 = -0.10$ (SE 0.03), $P < 0.001$; $LRS_{log} = 0.02$ and $LRS_{inverse} = 0.48$, $df = 1$, both $P > 0.25$.

[2]Effect of experience on reproductive success in females, $LRS_{inverse} = 5.00$, $b_1 = -0.03$ (SE 0.01), $P < 0.05$; $LRS_{quadratic} = 2.28$, $df = 2$, $P > 0.25$ and $LRS_{log} = 4.03$, $df = 1$, $P < 0.05$.

incompatible pairs to choose new partners. However, the relatively low rate of divorce (about 8%) also indicates that this option is not often used for change.

The cost of divorce and mate change is long-term. Manuwal (1974*a*) found that birds that lost mates (to removal) generally obtained a replacement partner within a few days. However, these results were obtained in the early 1970s when the Cassin's Auklet population on Southeast Farallon was markedly larger than it is today. Under current demographic constraints, there could be greater problems with finding a replacement partner after divorce or death of a mate. We also documented a cost of mate change in terms of future fecundity. Egg laying was delayed and measures of productivity were reduced for birds with new mates (Table 11.1). The cost of divorce would also become apparent through the long-term effects of pair duration. Another cost of mate change is the probability of obtaining an inexperienced bird in a subsequent partnership. In this study, from the male perspective 72% (70/97) of the replacement females were inexperienced. From the female perspective, 69% (68/99) of the replacement males were also inexperienced (Emslie *et al.* 1992). Thus the chances of finding an experienced bird for a replacement mate are low. Even when the population was considerably larger, Manuwal (1974*a*) concluded that there were few available and experienced birds for new partnerships. Thus mate change through divorce or death leads to pairing with an inexperienced bird. Even if more experienced birds were available, it is difficult to imagine how auklets could assess the experience levels of potential partners.

Preferred options

Reproductive success was significantly influenced by male and female previous breeding experience, which provides an indication of how age would also affect the breeding biology of Cassin's Auklets. An unusual aspect of this study, however, is that we statistically controlled for the experience of an individual's partner while simultaneously testing for the effects of experience of the focal bird. We found that reproductive success (i.e. number of chicks produced per pair) was related to male experience in a parabolic fashion, indicating a decrease in productivity for the most experienced individuals in our population. This result might indicate senescence, but it could also reflect variation in life history as determined by age at first breeding and/or intermittent reproduction (Sydeman *et al.* 1991). For females, reproductive success increased asymptotically with experience, indicating that the benefits of female experience neither increase nor decrease for the most experienced individuals in our population. Moreover, we found no interaction between the experience of mates and annual productivity which indicates that the benefits (or detriment) of experience do not depend on an individual's partner's experience; this greatly simplifies interpretations of preferred options. Given these relationships, the preferred mate for males is a female with at least one previous season as a breeder. For females, the preferred male has an intermediate level of experience as both inexperienced and highly experienced males showed reductions in productivity. As indicated earlier, however, obtaining mates with previous experience is difficult and most birds are not successful in establishing new partnerships with experienced parents.

The evolution of monogamy—why one partner?

Monogamy may maximize fitness by improving annual reproductive success (and offspring survival), and enhancing the residual reproductive value and survival of the parents (Pianka and Parker 1975). Proximately, monogamy in Cassin's Auklets is closely related to the need for biparental care during chick rearing and incubation (see Mock *et al.* Chapter 3). Single-parent nests always fail, thereby demonstrating the necessity of biparental care.

Ultimately, monogamy in Cassin's Auklets also may be related to a life history characterized by variable reproductive effort. Although Cassin's Auklets typically lay only one egg per year, incubation and the nestling period are long when compared with alcids and other seabirds of similar size. Also unlike many other alcids, Cassin's Auklet chicks depart the nest at or near adult body mass (Ainley *et al.* 1990). Moreover, our population double-broods in years of substantial food supply. These traits indicate substantial energetic demands for reproduction, especially in certain years, which may be reduced by spreading the effort over a long period of time. Mortality of Cassin's Auklets on Southeast Farallon in the late 1980s and early 1990s also was apparently higher (Emslie *et al.* 1992) than other populations (Gaston 1992) or the Farallon population in the early 1970s (Speich and Manuwal 1974).

Selection will favour monogamy not only if it leads to improvements in the co-ordination of parental duties, covering for an ill, injured, or poorly foraging partner (Bart and Tornes 1989; Wolf *et al.* 1990), but also if there is a reduction in the energetic requirements of individual parents. We have not directly assessed these ideas, but monogamy and long-term pairing may reduce the cost of reproduction and thereby increase future fecundity of the parents. For long-lived species with variable reproductive effort, such as the Cassin's Auklet, in which small changes in survivorship often lead to large changes in demography (Croxall and Rothery 1991), understanding energetics and survival may provide the key to understanding the evolution of mating systems as well as other behavioural traits.

Summary

We investigated the causes and consequences of mate change and pair duration in a sedentary and declining population of Cassin's Auklets in central California from 1982 to 1993. Divorce in Cassin's Auklets was rare; most mate change was related to the disappearance of one of the members of a pair. In our data set, the modal number of mates was one, and the modal pair duration was 1 year. We demonstrate that mate change and pair duration influence annual reproductive success and various components of reproductive performance including the date of egg laying, the probability of successfully hatching an egg, and the probability of successfully fledging hatched chicks. There was no interaction between experience of the male and experience of the female. Thus, upon losing a mate, males should seek females with at least 1 year of previous breeding experience, while females should seek males with an intermediate level of breeding experience. It is likely that selection has favoured monogamy in Cassin's Auklets owing to the need for bi-parental care and a life history characterized by variable reproductive effort.

Acknowledgements

We thank the US Fish and Wildlife Service—San Francisco Bay National Wildlife Refuge for support of this project. J. Black, B. Ens, M. Harris, and N. Nur provided comments on earlier drafts of this paper. This is PRBO contribution no. 661.

References

Ainley, D. G., Boekelheide, R. J., Morrel, S. H., and Strong, C. S. (1990). Cassin's Auklet. In *seabirds of the Farallon Islands: ecology, dynamics, and structure of an upwelling-system community* (ed. D. G. Ainley and R. J. Boekelheide), pp. 306–38. Stanford University Press, Palo Alto.

Ainley, D. G., Sydeman, W. J., and Norton, J. (1995). Upper-trophic level predators indicate negative and positive anomalies in the central California marine food web. *Marine Ecology Progress Series*. **118**, 69–79.

Bart, J. and Tomes, A. (1989). Importance of monogamous male birds in determining reproductive success. *Behavioral Ecology and Sociobiology*, **24**, 109–16.

Carter, H. R., McChesney, G. J., Jaques, D. L., Strong, C. S., Parker, M. W., Takekawa, J. E., Jory, D. L. and Whitworth, D. L. (1992). *Point Reyes Bird Observatory and Channel Islands National Park. Breeding populations of seabirds in California, 1989–1991*. Unpublished Report, U.S. Fish and Wildlife Service, Dixon, California.

Cox, D. R. and Snell, E. J. (1989). *Analysis of binary data*. Chapman and Hall, London.

Croxall, J. and Rothery P. (1991). Population regulation in seabirds and implications for their demography and conservation. In *Bird population studies: their relevance to conservation and management* (ed. C. M. Perrins, J. D. Lebreton, and G. M. Hirons), pp. 272–96. Oxford University Press, Oxford.

Emslie, S. D. and Messenger, S. (1991). Pellet and bone accumulation at a colony of Western Gulls (*Larus occidentalis*). *Journal of Vertebrate Paleontology*, **11**, 133–6.

Emslie, S. D., Sydeman, W. J., and Pyle, P. (1992). The importance of mate retention and experience on breeding success in Cassin's auklet (*Ptychoramphus aleuticus*). *Behavioral Ecology*, **3**, 189–95.

Gaston, A. J. (1992). Annual survival of breeding Cassin's Auklets in the Queen Charlotte Islands, British Columbia. *Condor*, **94**, 1019–21.

Manuwal, D. A. (1974*a*). Effects of territoriality on breeding in a population of Cassin's Auklet. *Ecology*, **55**, 1399–06.

Manuwal, D. A. (1974*b*). The natural history of Cassin's Auklet. *Condor*, **76**, 421–31.

Nelson, D. A. (1981). Sexual differences in the measurements of Cassin's Auklet. *Journal of Field Ornithology*, **52**, 233–4.

Nelson, D. A. (1989). Gull predation on Cassin's Auklet varies with the lunar cycle. *Auk*, **106**, 495–7.

Pianka, E. and Parker (1975). Age-specific reproductive tactics. *American Naturalist*, **109**, 453–64.

Speich, S. and Manuwal, D. A. (1974). Gular pouch development and population structure in Cassin's Auklet. *Auk*, **91**, 291–306.

Stata corporation (1993). *Stata reference manual: release 3.1, 6th edition*. College Station, Texas.

Sydeman, W. J., and Ainley, D. G. (1994). Marine birds in the California current ecosystem: contributions to US GLOBEC's goals. *US GLOBEC News*, **7**, 4–7.

Sydeman, W. J., Huber, H., Emslie, S. D., Ribic, C., and Nur, N. (1991). Age-specific weaning success of northern Elephant Seals in relation to previous breeding experience. *Ecology*, **72**, 2204–17.

Wolf, L. Ketterson, E. D., and Nolan V. (1990). Behavioral response of female dark-eyed juncos to the experimental removal of their mates: implications for the evolution of male parental care. *Animal Behaviour*, **39**, 125–34.

12 *Monogamy in a long-lived seabird: the Short-tailed Shearwater*

RON WOOLLER AND STUART BRADLEY

Introduction

Short-tailed Shearwaters *Puffinus tenuirostris* are 500 g burrow-nesting seabirds which breed colonially on islands and headlands in south-eastern Australia. On average, their breeding season occupies 195 days (53%) of each year (Warham 1990). For the rest of the year they are pelagic, making an extensive annual transequatorial migration within the Pacific basin. All 23 million Short-tailed Shearwaters are extremely synchronous in their laying, most of which occurs between 23 and 28 November every year, over the entire breeding distribution of the species. Only one egg is laid each year and lost or unsuccessful eggs are not replaced. Up to one million of the large young of Short-tailed Shearwaters (muttonbirds) have been harvested annually for over a century. Concern about this exploitation resulted in the establishment in 1947 of a study that still continues, allowing analysis of the complete reproductive careers of all individuals in one population.

Direct observations of a species whose breeding activity occurs only underground at night are inevitably limited. In addition, the sexes differ little, if at all, in size, colour or much of their behaviour. Our account, therefore, seeks to explore the value of monogamy to individuals in a species with high annual survivorship (87–95%), a delay of 4–15 years before first breeding, and in which long sequences of breeding with the same partner are often recorded. None the less, low but measurable rates of divorce occur in breeding Short-tailed Shearwaters and we shall explore some possible costs and benefits of this. The results of pairing with various aged partners are examined. We also suggest that reproductive success increases not only with breeding experience overall but, additionally, as a result of familiarity with a particular partner.

Background, study site, and procedures

Each year all burrows and shearwaters on Fisher Island (40°10′S, 148°16′E), a small island between Tasmania and mainland Australia, have been individually marked. Durable metal leg bands and a programme of double-banding have ensured consistency in identification of breeding adults (Bradley *et al.* 1991). Pairs were determined by capturing the birds at the burrow. Mate change occurred when one or both mates were recorded as dead, i.e. missing for 3 years or more (Bradley *et al.* 1989). Divorce occurred when two individuals paired in one year were both alive but not recorded as paired with each other in the following season; usually, one or both had paired with a new mate.

Demographics

Since 1975 the breeding population has remained constant at around 100 individuals. Natal recruits of known age form a relatively constant 45% of the breeding population with no obvious differences between the sexes in their return rates. There is some movement of young birds between neighbouring islands but, despite extensive checks, no shearwater that had bred on Fisher Island has ever been found breeding elsewhere (Bradley *et al.* 1991). The mean age at which males (7.3 years) and females (7.0 years) first breed is similar (Wooller *et al.* 1989). Thereafter, males live for 9.2 years and females for 9.4 years after breeding for the first time; all known shearwaters died within 27 years of first breeding. Year effects on reproductive success were controlled for, as offspring production varied considerably between years due to intermittent flooding of burrows (Wooller *et al.* 1988).

Pair formation process

Short-tailed Shearwaters come ashore only at night and depart before sunrise, often *en masse*. Their colonies are silent by day but can be very noisy at night. They exhibit high fidelity to their breeding sites and displays do not occur outside the burrow. Copulations are most frequent at the end of October, when spermatogenesis peaks (Marshall and Serventy 1956). Pairs then vacate their burrows for 20–22 days before returning to lay synchronously. Partners meet only briefly during the next 3–4 months as they alternate incubation and chick-feeding duties. Unfortunately, the basis of mate selection and the behaviour involved in pair maintenance are unknown in the species.

Results

The pair bond

Instances when three individuals were recorded in one burrow were rare and in no case were there more than two regular occupants (100% social monogamy). No birds formed more than one pair bond in each season (Bradley *et al.* 1990). About half of the shearwaters had only one partner during their complete reproductive careers; the mean was 1.9 mates (SE 0.07, n = 328) and the maximum was seven mates in a lifetime. About 40% of all partnerships lasted only 1 year, 15–20% lasted 2 years and 10% 3 years, with very few lasting more than 15 years (Table 12.1). Pair duration increased with age. Thus, half of the pair bonds with first mates, but only one-third of the pair bonds with third mates, lasted 1 year (Bradley *et al.* 1990). There was no difference between the sexes in either number of mates or pair duration.

Mate change

Overall, 74% of 2114 males and 2136 females reunited with their partners of the previous year, 10% took a new partner following the disappearance of their previous partner and 16% changed mates even though their previous partner was still alive (Bradley *et al.* 1990). The pattern of mate change due to death of a partner started high at younger ages, fell somewhat in the middle years, then peaked in the older birds (Fig. 12.1) when age-dependent mortality again rose (Bradley *et al.* 1989). Mate change due to divorce showed a different pattern; initially, the rates were very high, then fell, and remained low throughout life.

Table 12.1 Percentage frequency distributions of the durations of the first, second, third, and subsequent pair bonds in Short-tailed Shearwaters

	Pair duration (years)										Mean (±SE) duration (years)	Sample size
	1	2	3	4	5	6	7	8	9	≥10		
First partner												
Males	50	19	9	7	4	1	1	3	1	8	2.7 ± 0.2	196
Females	48	21	11	6	2	1	3	3	1	4	2.6 ± 0.2	159
Second partner												
Males	44	12	15	10	5	4	3	3	2	2	2.9 ± 0.3	96
Females	42	16	15	4	9	4	2	1	0	7	3.0 ± 0.3	81
Third partner												
Males	33	21	16	5	7	3	3	3	2	7	3.5 ± 0.4	58
Females	36	11	11	5	7	11	11	5	0	4	3.7 ± 0.4	56
Fourth and subsequent partners												
Males	34	29	11	4	7	5	2	2	2	2	3.0 ± 0.3	56
Females	47	17	12	6	4	4	4	1	4	4	2.9 ± 0.3	83

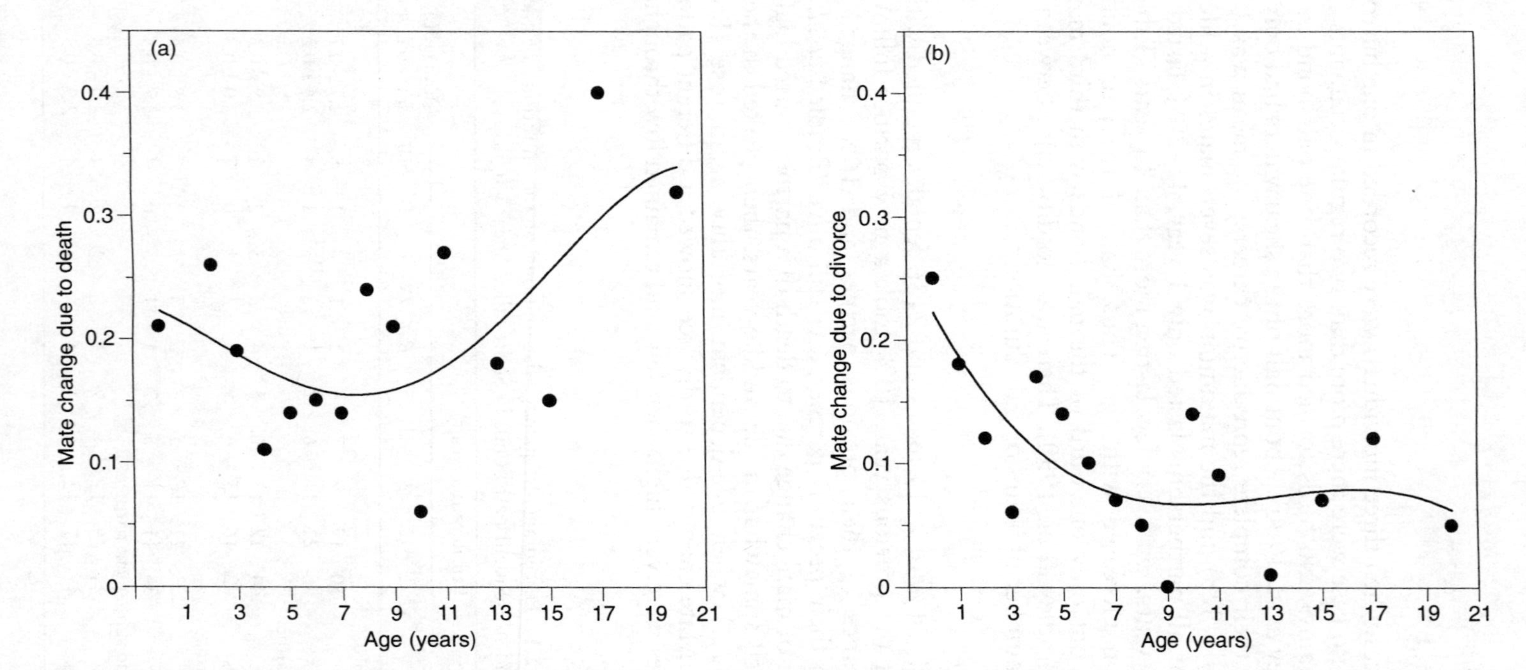

Fig. 12.1 Pattern of mate change in Short-tailed Shearwaters at different ages; mate change due to (a) death of a partner, and (b) divorce. Whereas mate change due to death was greatest in the oldest age-class, divorce occurred more in young birds. Sample sizes for each point were between 50 and 201 pairs. Adapted from Bradley *et al.* (1995).

Mate change was associated with reproductive success in the previous year and divorce was significantly less frequent if a fledgling had been produced (Table 12.2), although the poorer reproductive success of younger birds might account for this trend. However, logistic regression analysis showed that the probability of divorce significantly decreased if a fledging had been produced in the previous year, even when controlled for age and pair duration ($\chi^2 = 11.6$, $df = 1$, $P < 0.001$), although age and pair duration did not have a significant effect on the probability of divorce in the 162 cases examined.

Mate fidelity and reproduction

Reproductive success of Short-tailed Shearwaters increases with increasing breeding experience, except in the most experienced birds (Wooller *et al.* 1990). It is also influenced by chronological age early in their breeding careers and, later, possibly by residual reproductive effort (*sensu* Pianka & Parker 1975). Longer-lived individuals are also more successful reproductively than shorter-lived birds, especially early in their lives (Bradley *et al.* 1989; Wooller *et al.* 1990).

Reproductive success appears to increase with pair duration in the first three pair bonds but not thereafter (Fig. 12.2). Indeed, mate change resulted in decreased reproductive success early in a new partnership before building up again. This effect and the progressive improvement with successive partners (Fig 12.2) may be due to accumulated breeding experience rather than mate familiarity. However, a logistic regression controlling for year and age showed that the probability of producing young in their second and third partnerships did indeed vary significantly with pair duration (female data set, $\chi^2 = 13.7$, $df = 1$, $P < 0.001$; male data set, $\chi^2 = 4.8$, $df = 1$, $P < 0.05$).

We repeated the analysis to determine which variable was more important, age or pair duration. Reproductive success increased with age in females (but not males) when year and pair duration were controlled ($\chi^2 = 6.0$, $df = 1$, $P < 0.05$). Based on relative delta deviance values pair duration explained more variation in reproductive success than age.

The power of these tests was severely limited by the use of only one observation per individual, selected at random, to obviate repeat measures effects. The age of the partner had no significant effect upon reproductive success when the age of the individual, pair duration, and year were controlled but this is not surprising given the correlation between the ages of partners.

Partner characteristics and reproduction

Shearwater partners have very similar 'ages', defined as the number of years since an individual first bred. Overall, 30% of breeding shearwaters have partners of the same age, and 60% have partners that differ in age by 2 years or less (Bradley *et al.* 1995). Most of this similarity

Table 12.2 Fate of pair bonds after reproductive failure

Preceding reproductive event	Males				Females			
	Mate died (%)	Divorced (%)	Reunited (%)	Sample size	Mate died (%)	Divorced (%)	Reunited (%)	Sample size
No eggs laid	11	22	67	246	14	22	64	258
Egg did not hatch	10	21	69	789	8	21	71	779
Fledgling produced	9	10	81	1079	10	11	79	1099

The effect of previous reproductive success on the probability of divorce was significant (see text); the least successful birds divorced more.

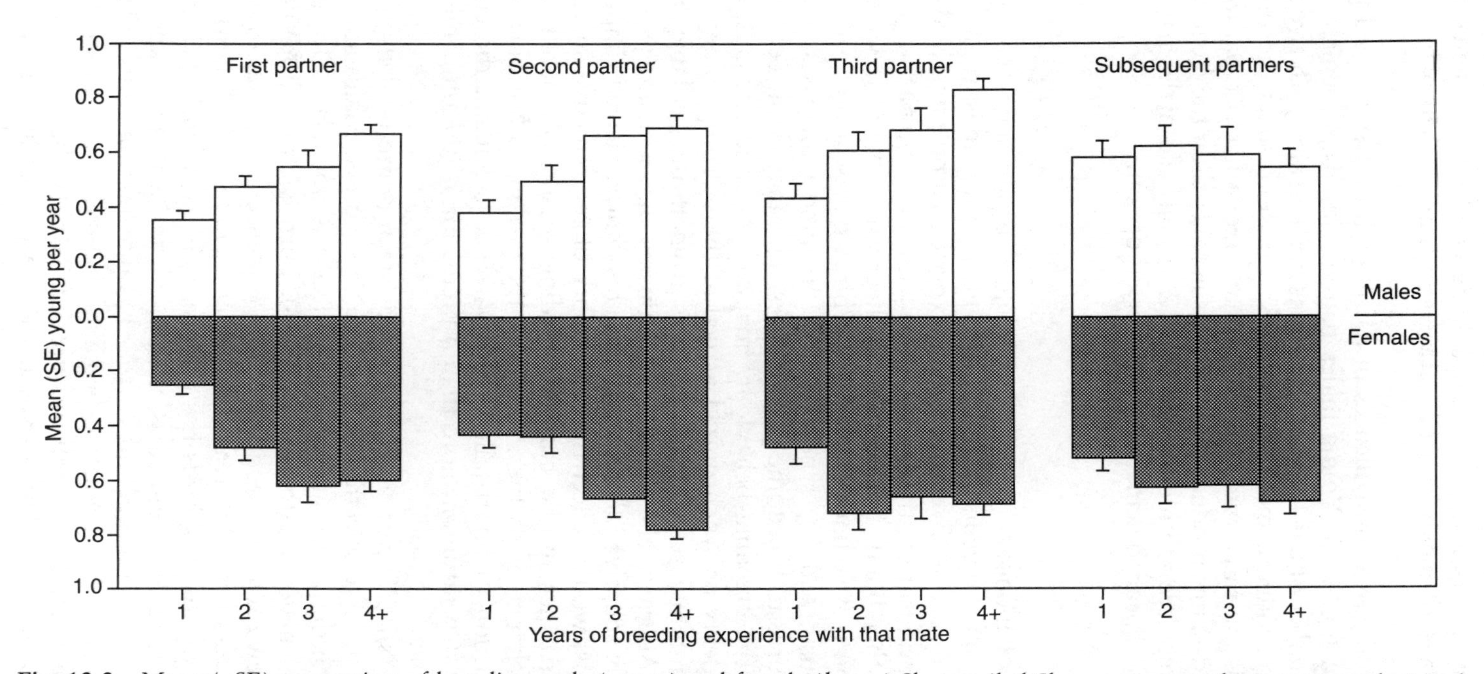

Fig. 12.2 Mean (±SE) proportion of breeding male (*upper*) and female (*lower*) Short-tailed Shearwaters producing young during the first, second, third, and subsequent years with their first, second, third, and subsequent partners (adapted from Bradley *et al.* 1990). Sample sizes for each bar were between 22 and 218 individuals.

results from prolonged partnerships between the same mates and the large numbers of inexperienced partners available when pair formation occurs.

The reproductive consequences of pairing with different aged partners is shown in Fig. 12.3. Young males paired with young females fledged the least number of chicks (0.33), whereas partnerships with two middle-aged mates produced most chicks (0.83). For both sexes the success of young birds increased with increasing mate age, otherwise pairings with similar aged mates produced most offspring. Based on the frequencies of these pairings (see sample sizes in Fig. 12.3), only old males paired with old females appear to achieve their preferred option.

Discussion

The value of monogamy to an individual depends upon the increment that continuous breeding with the same partner adds to its lifetime reproductive success. Every year, each experienced breeding bird is faced with the option of continuing to breed with the same partner or establishing another bond. In Short-tailed Shearwaters the time window for this decision is very limited because all birds that will breed are present simultaneously and lay synchronously. In almost all cases where the partner returns to breed, the pair reunites. This may result because some aspects of the breeding behaviour, adaptive for other reasons, such as high site fidelity or synchronized arrival at the breeding colony, bias an otherwise random selection of partners from those available, towards reuniting. Alternatively, specific mechanisms may encourage reuniting because it is adaptive in its own right. This choice, viewed simply on an annual basis, would reflect the probability of reproductive success with the same partner in the current breeding season compared with an attempted change of partner. However, the prolonged breeding histories of a proportion of Short-tailed Shearwaters suggest that the choice, if made, may depend upon more than a prediction of the outcome from a single season.

In Short-tailed Shearwaters, phenotypic factors are unlikely to play a major role in mate selection or reproductive success. Males and females differ very little in size and breeding success is not related to the body size of the female (Meathrel *et al.* 1993). Burrows, too, differ little and are dug anew each year, and there is no evidence of differences in reproductive success between areas (Bradley *et al.* unpublished data). Rather, the quality of breeding birds has often been equated with their breeding experience, reproductive success often increasing with such experience. A bird reuniting with a previous partner has information from at least one season as a basis for predicting breeding quality. Such a prediction may contain information not only on previous success but the expectation that, with additional shared experience, the probability of success may

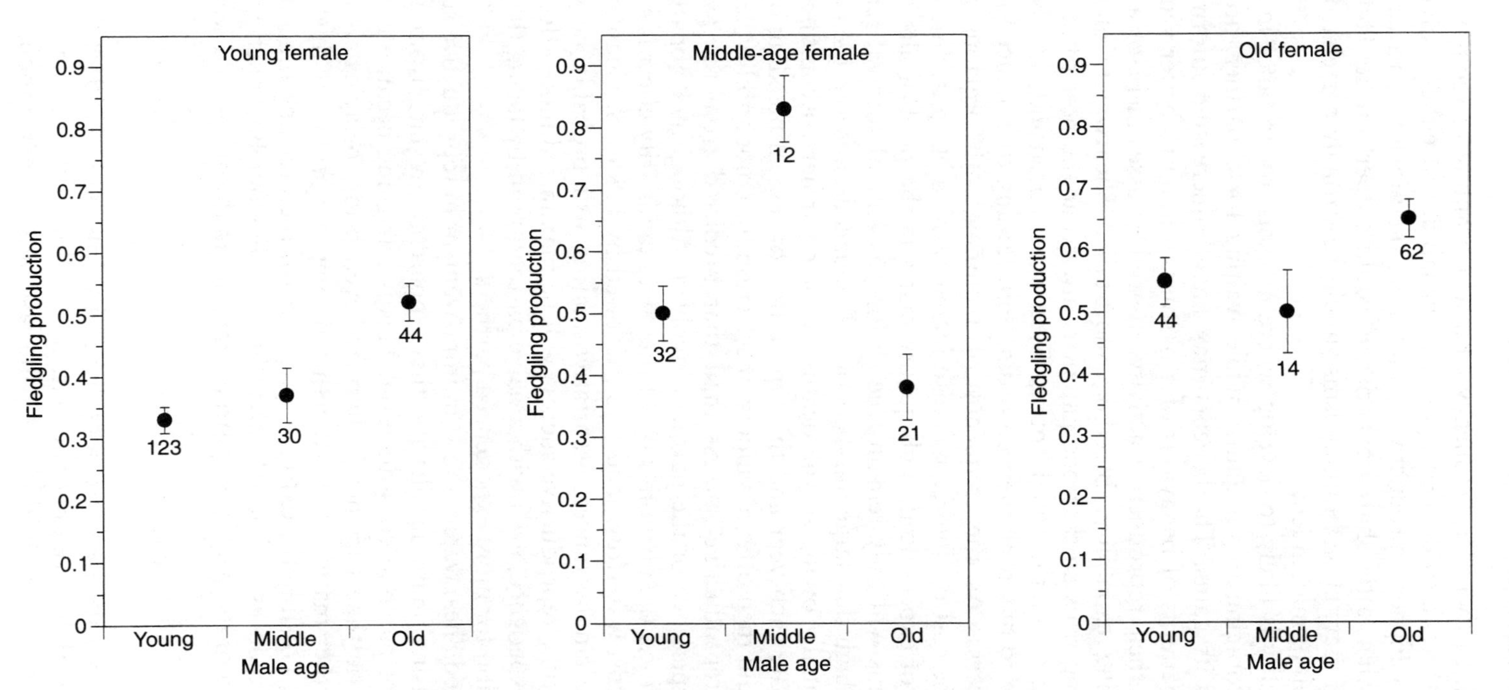

Fig. 12.3 Fledging success in relation to partners' ages. Except for the youngest birds, partnerships with similar aged mates were the most productive options for Short-tailed Shearwaters. Age-classes consisted of the following: young (0–4 years as breeding birds), middle-aged (5–9 years), and old (10+ years). Sample sizes accompany each point. Since reproductive success increases with pair duration, data were limited to the first year of the pair bond. Data from poor breeding years were omitted: 1966, 1969, 1970, and 1976.

increase still further. The frequency with which failure to reunite occurs could, therefore, also depend upon the availability of either individuals of inherently higher reproductive quality than the previous partner or of individuals that, after shared experience, could outperform it. Thus, the evolution of mate choice mechanisms should be strongly correlated with the evolution of mate fidelity.

Mate fidelity is likely to develop where its costs, for instance in terms of loss of foraging opportunities while waiting for a partner, are outweighed by its gains. The benefits may include increasing information about the patterns of behaviour of a partner, leading to better synchronization of their reproductive activity, as well as assessment of a partner's foraging capability. Breeding Short-tailed Shearwaters alternate their foraging trips so that one parent is always at the nest to incubate the egg or protect the small young. Even experienced individuals may require time to fine-tune their co-ordination, leading to enhanced reproductive success. As the frequency of reuniting with previous mates increases, the cost of divorce will also increase, owing to a decline in the proportion of experienced, high-quality mates in the pool of alternative partners. This will only remain true so long as annual survival remains sufficiently high that large numbers of individuals do not join the pool of recruits annually owing to the deaths of their partners. Reuniting with the same mate each year and the expectation of several breeding opportunities would then offer a window of selection for compatibility, leading to greater reproductive success than that predicted from the average breeding experience of the partners involved. Although not apparent in most passerines, the phenomenon of this mate familiarity effect is seen in other seabirds, waterfowl, and jays (reviewed by Ens *et al.* Chapter 19). Under these circumstances, assortative mating will probably be weak, because lifetime reproductive success will be maximized more effectively by shared experience, even with a naive partner, than by facing the costs of competition for more experienced partners.

Short-tailed Shearwaters show an improvement in reproductive success with increasing mate familiarity that apparently results from subtle refinement of parental co-ordination. This is intriguing because partners spend so little time together in burrows even prior to laying. It is not known whether mates remain together during migration but this seems improbable. Possibly the extreme synchrony of the breeding timetable in this species reduces individual variability in behaviour but much has yet to be learned of the ways that partners interact in this long-lived seabird.

Summary

Short-tailed Shearwaters are long-lived seabirds with 87–95% annual survivorship. Half have only one mate in their lifetimes but some take up to seven mates. Overall, 40% of pairs last only 1 year although some pairs bred together for over 15 years. Divorce is most common among younger individuals. Divorce also occurs more often after low reproductive success, even allowing for the poorer reproductive success of young birds. Pairs consisting of middle-aged partners produce most fledglings. Reproductive success improves not only with accumulated breeding experience but, additionally, with growing mate familiarity. We suggest that this would be expected in a long-lived species in which precise coordination of parental activities is vital to reproductive success and in which the parents seldom meet.

Acknowledgements

We thank Catherine Meathrel and Irynej Skira for their roles in the shearwater study, which was started by Dom Serventy. The work was supported by the Australian Research Council (A19332808 and A19031054).

References

Bradley, J. S., Skira, I. J., and Wooller, R. D. (1991). A long-term study of Short-tailed Shearwaters *Puffins tenuirostris* on Fisher Island, Australia. *Ibis*, **133**, S55–S61.

Bradley, J. S., Wooller, R. D., Skira, I. J., and Serventy, D. L. (1989). Age-dependent survival of breeding short-tailed shearwaters *Puffinus tenuirostris*. *Journal of Animal Ecology*, **58**, 175–88.

Bradley, J. S., Wooller, R. D., Skira, I. J., and Serventy, D. L. (1990). The influence of mate retention and divorce upon reproductive success in short-tailed shearwaters *Puffinus tenuirostris*. *Journal of Animal Ecology*, **59**, 487–96.

Bradley, J. S., Wooller, R. D., and Skira, I. J. (1995). The relationship of pair-bond formation and duration to reproductive success in short-tailed shearwaters *Puffinus tenuirostris*. *Journal of Animal Ecology*, **64**, 31–8.

Marshall, A. J. and Serventy, D. L. (1956). The breeding cycle of the Short-tailed Shearwater, *Puffinus tenuirostris* (Temminck), in relation to trans-equatorial migration and its environment. *Proceedings of the Zoological Society of London*, **127**, 489–510.

Meathrel, C. M., Bradley, J. S., Wooller, R. D., and Skira, I. J. (1993). The effect of parental condition on egg-size and reproductive success in short-tailed shearwaters *Puffinus tenuirostris*. *Oecologia*, **93**, 162–4.

Pianka, E. R. and Parker, W. S. (1975). Age-specific reproductive tactics. *American Naturalist*, **109**, 453–64.

Warham, J. (1990). *The petrels*. Academic Press, London.

Wooller, R. D., Bradley, J. S., Skira, I. J., and Serventy, D. L. (1988). Factor contribution to reproductive success of Short-tailed Shearwaters *Puffinus tenuirostris*. *Proceedings International Ornithological Congress*, **19**, 848–56.

Wooller, R. D., Bradley, J. S., Skira, I. J., and Serventy, D. L. (1989). Short-tailed Shearwater. In *Lifetime reproduction in birds* (ed. I. Newton), pp. 405–17. Academic Press, London.

Wooller, R. D., Bradley, J. S., Skira, I. J., and Serventy, D. L. (1990). Reproductive success of short-tailed shearwaters *Puffinus tenuirostris* in relation to their age and breeding experience. *Journal of Animal Ecology*, 59, 161–70.

13 *Between- and within-population variation in mate fidelity in the Great Tit*

ANDRÉ A. DHONDT, FRANK ADRIAENSEN, AND WERNER PLOMPEN

Introduction

The Great Tit *Parus major* is a small (18 g) hole-nesting passerine that readily uses nest boxes. It has been extensively studied in Western Europe. In this chapter we present results from two long-term studies carried out in nine different study plots in northern Belgium where we provided a superabundance of wooden nest boxes. The quality of food and shelter varies considerably between study sites and seasons and this may be linked with the birds' social behaviour. The Great Tits in these sites are socially monogamous, but extra-pair nestlings are found in about one-third of the nests. During the breeding season males defend a multi-purpose territory in which females alone build the nest and in which the pair collects all food for the chicks. Outside of the breeding season (July to January) Great Tits often live in flocks in which the pair bond is not maintained.

The field study of Perrins and McCleery (1985), and Lindén's (1991) experimental work, have shown that relatively unsuccessful pairs of Great Tits are more likely to divorce than are successful pairs. But whereas Perrins and McCleery believed that keeping the same partner was adaptive (individuals improve reproductive success by reuniting with the same partner), Lindén concluded that divorce was adaptive (individuals improve their reproductive success by changing partners).

In our nine study sites, divorce rates varied from 0 to 51% in 280 surviving pairs. In this chapter we shall try to determine what caused these extreme differences among the nine study populations. By considering divorce rates in relation to reproduction, survival, and winter social organization, this chapter forms a link to interspecies comparisons, as it attempts to explain why divorce rates differ between populations of a single species. To examine the mechanisms and adaptive value of divorce within one population, we shall compare breeding results and nonbreeding behaviour from colour-banded individuals that reunited with the same partner in successive years with those that did not.

Background, study sites, and procedures

In Western European populations, adult Great Tits are nonmigratory and first-autumn birds make irruptive migratory movements in some years (Van Balen and Speek 1976). In September the birds exhibit a brief period of territorial behaviour (Dhondt 1971*a, b*), but when it becomes colder, they usually show extensive flocking. Adults may move over extended areas during the day, but resident birds usually roost at night in cavities within their domicile. Kluyver (1951) defined a domicile as 'the part of its range which each Great Tit frequents most often by day, and in which it also sleeps and breeds'. Winter domiciles overlap to a large extent with autumn and/or spring breeding territories. In the winter, members of a former breeding pair behave as two solitary individuals (W. Plompen unpublished data), even when they are resident on overlapping domiciles. In Belgium pair formation starts at the end of January (see below). The first clutch of about nine eggs is usually started in April. In successful broods the young fledge about 6 weeks later. If the clutch fails the pair will usually renest together; if it succeeds, a variable proportion of pairs will start a second brood in the same season. Genuine third broods are rare.

Basic information on the study plots is summarized in Table 13.1. The plots covered a wide range of habitats, and the breeding density of the tits varied by an order of magnitude. During the breeding season we recorded/calculated lay date (first egg date in April), clutch size (of first broods), fledglings per egg (from successful first brood), percentage failed broods (the proportion of first broods not producing any fledglings), percentage of second broods, and reproductive rate (mean number of fledglings per pair and per season, including all breeding attempts of all breeding pairs). Adults were trapped and individually ringed after the young hatched. In Table 13.1 we also give an estimate of the proportion of adults that remained within the study plot throughout winter.

To calculate mate fidelity we only included birds that we knew were alive in two successive years. If they bred together in both years we scored this as one case of mate fidelity, and if they bred with a different partner in the second year, then we scored this as one case of divorce.

Table 13.1 Brief description of study plots included in this chapter

Plot name	Habitat type	Size (ha)	Years	Years survival calculated	Mean breeding density (pairs/ha)	Mean lay date (April)	Adult survival rate (%)	Clutch size	Fledglings per egg	Second broods (%)	Residency status in winter[4]
Ghent-Cit[1]	Mixed city park	20	59–78	64–74	0.4	14	58	8.80	0.60	55	1
Ghent-Mp[2]	Mixed suburb park	20	59–78	64–74	1.4	19	55	8.95	0.65	34	2
Ghent-H	Beech	27	64–78	64–74	1.0	21	44	8.34	0.80	30	3
Ghent-Z	Mixed/oak	16	59–78	64–74	1.8	21	48	8.92	0.79	27	3
Ghent-Coo	Pine + deciduous	30	61–78	64–74	1.1	21	52	9.18	0.68	43	5
Antwerp-L	Mixed/oak	7.5	80–88	80–85	2.5	23	44	8.74	0.90	20	3
Antwerp-T	Mixed/oak	12.5	79–83 89–92	79–83 89–92	2.5	22	47	9.02	0.90	20	3
Antwerp-C[3]	Mixed/oak	17	79–92	—	(1.0)	21	52	8.54	0.86	—	2
Antwerp-B	Mixed/oak	12.5	79–92	79–92	3.2	24	43	8.94	0.90	8	5

Note that all plots at Ghent were woodland fragments or parks that were completely studied using nest boxes. Antwerp plots were only parts of larger wooded areas. The distance between Ghent and Antwerp plots is ca. 60 km. Note also that Antwerp-B is well inside a large woodland area, whereas Antwerp-T and Antwerp-L are, on one side, adjacent or close to villa gardens. Antwerp-C is an isolated wooded park. For more details see Dhondt and Hublé (1968), Dhondt and Eyckerman (1980) and Dhondt *et al.* (1990). Average lay date (for Ghent taken from Dhondt *et al.* 1984). Reproductive rate (mean number of fledglings per pair and per season; taken from Dhondt 1987 or unpublished).

[1]Includes some small urban parks covering ca. another 10 ha.

[2]Includes the data from the adjacent similar park 'Maria-Middelares'

[3]In Antwerp C only 10 nest boxes were available for Great Tits during the breeding season. Breeding density, therefore is an estimate including pairs in natural cavities.

[4]We estimated residency status of the breeding birds on a scale of 1 (fully resident: all birds remaining in the study plot during winter) to 10 (fully migratory: all birds leaving the plot during winter).

Survival rates of breeding adults were calculated using SURGE (cf. Lebreton *et al.* 1992). Because we wanted to obtain the average survival estimate for each study population over the study period, we used an s,p_t model (see also Table 13.1). Unless otherwise indicated, we calculated statistics using SAS V6.07 (analyses of variance using PROC GLM). Exact probabilities were calculated using StatXact V2.01 (Cytel Software, Cambridge, US, 1991). For Generalized Linear Models (e.g. Dobson 1990) we used GLIM V3.77 (Royal Statistical Society, London, UK, 1985).

Pair formation process

Starting in January, male Great Tits leave the winter flocks on warm days and sing on favourite perches (Hinde 1952). W. Plompen (unpublished data) observed on plot Antwerp-B that not only males but also females could become established at that period. Pair formation occurs only between birds settled on overlapping domiciles. Furthermore, females develop a specific 'male-visiting' behaviour. Male-visiting females fly towards singing males (or places of conflict between two males) and start foraging close to the male. If this male(s) is already paired, the female leaves the site, otherwise the male stops singing and they forage together. From that moment on the pair is formed and both birds remain together most of the time. It is usual for a female to pair with the first unpaired male she encounters.

Fighting between males can occur if the male who occupied the territory in the previous breeding season returns. These fights sometimes resulted in the returning male re-occupying part of the old territory, but not regaining his former partner. Winter emigrants, who were absent from the plot between mid December and the end of February, could only reclaim their territory upon arrival, so that pair formation started later in winter emigrants than in winter residents.

Results

The pair bond

Great Tits are socially monogamous, although we observed polygyny twice in 2800 nests. The mean pair duration for 1000 pairs was 1.12 years, resulting in an average of 1.30 mates per lifetime for 767 males and 1.46 mates for 687 females. Mate changes within one breeding season were exceptional. Among 46 pairs in plot Antwerp-B no birds changed mates between first and second broods. On Ghent-Mp only two mate changes occurred in 51 possible cases.

Divorce and the pair formation process

Each spring from 1991 to 1993 we followed colour-marked birds in Antwerp-B on an almost daily basis while they were settling and forming pairs. We found that divorce occurred when pair members returned to the former territory at different times (or, if they were settled, entered the pair formation process asynchronously) and the first bird to arrive had already entered the pair formation process with a different partner (see above). Thirteen of 13 pairs in which both partners were winter residents reunited, versus only one of six when at least one pair member was not a winter resident (Fisher exact test: $P < 0.001$).

To test the hypothesis that pairs in which at least one bird was not a winter resident were more likely to divorce we also analysed the entire data set from Antwerp-B. Each winter, we made repeated evening visits to locate roosting birds. A bird was classified as a winter resident if it was observed in January. Of the 33 pairs in which both partners were winter resident, 26 pairs (79%) reunited. Of the 64 pairs in which at least one bird was not a winter resident, only 38 pairs (59%) reunited (one-tailed Fisher exact test: $P = 0.044$).

Most breeders in Ghent-Mp remained on the plot during winter. In this plot we observed seven cases of divorce. In one case an old male was usurped from his territory by a young male in early February. The female bred with the usurper, and the old male did not reproduce until later in the year (cf. Dhondt 1971*b*). In another case a pair of intruders (*sensu* Dhondt and Schillemans 1983) separated, probably because the male was not a winter resident. On three occasions a female shifted to a neighbouring territory leaving her partner behind on his domicile. Finally, two pairs divorced although both partners remained on overlapping domiciles.

Within-population comparisons of reproductive success in relation to divorce

To determine to what extent reproductive success influenced divorce, we compared reproductive success in year A in pairs which subsequently reunited or divorced in year B. We normalized the reproductive parameters in each year (none of the parameters was significantly non-normal), so that the population mean was zero and variance was one. The data from different years could be grouped because each transformed value was expressed as a deviation from the zero mean in units of standard deviations.

The reproductive parameters we examined in Antwerp-B did not differ significantly between pairs that divorced ($n = 26$) and those that reunited ($n = 31$) (lay date for divorced and faithful pairs: mean −0.11, SE 0.16, mean −0.18, SE 0.14; clutch size: mean −0.05, SE 0.19, mean 0.10, SE 0.17; number of fledglings: mean 0.17, SE 0.18, mean 0.26, SE 0.16; one-way ANOVA: all $P > 0.57$). None of the values differed from the population mean. Nor was there a relation between having a second brood and subsequent divorce; 41% of 17 pairs with second broods divorced compared to 34% of 74 pairs without second broods (Fisher exact test: $P = 0.58$).

To determine whether birds improved their reproductive success by divorcing or by reuniting, we compared the *change* in the normalized values in two successive years for lay date, clutch size, and number of young fledged in first broods. Pairing status had no significant effects on the change in breeding performance in either sex (Table 13.2). Since in normalized data the population mean was zero, we also compared the between-year changes in reproductive traits to zero. Pairs that reunited and divorced females had a change in lay date that was significantly smaller than zero, implying that, compared to the average bird in the population, they 'improved' their lay date. This was not found for divorced males (Table 13.2). Divorced females started breeding significantly earlier than divorced males (mean –0.55, SE 0.11, mean –0.04, SE 0.22, n = 52, t = 2.01: P = 0.049). For clutch size and number of fledglings there were no significant differences.

Between-plot variation in divorce rate

The overall probability of divorce in the nine plots was 25% but varied between 0 and 51% (Table 13.3). The proportion of divorces per plot showed a gradual increase from the inner-city park, Ghent-Cit (0% divorce), characterized by early laying, low density, and low reproductive success, to the suburban park, Ghent-Mp, and the isolated plots, Ghent-H, Ghent-Z, and Antwerp-C, to the high density plot Antwerp-B (51% divorce). This last plot, centrally located in a large rural woodland isolated from human habitation, was characterized by late laying and high reproductive success.

By comparing a suite of characteristics of our nine populations, which lie close together in northern Belgium, we were able to use the comparative approach within a single species. Thus we can verify to what extent various adaptive hypotheses of mate fidelity or divorce (which are used to explain differences between individuals in a single population) could explain the between-population differences observed. We tested to see whether divorce was less likely in plots where, on average, reproduction was more successful, or where fewer potential partners were available. The adaptive hypotheses make one-sided predictions (Chapters 1, 19): if variation in divorce rates can be explained by differences in reproductive success (i.e. *incompatibility hypothesis*), then divorce should be lower in plots with early laying, or in plots where clutch size, fledglings per egg, percentage of second broods, or reproductive rate are high. Similarly, divorce should be low in plots where the adult survival rate is high or population density is low, because in these populations, the number of alternative partners would be low (i.e. better options hypothesis).

We tested these hypotheses separately, using stepwise multiple regression, by grouping variables related to reproductive success in one analysis, and those related to the availability of partners in another. All percentages (divorce rate, percentage of second broods, and annual adult

Table 13.2 Change in breeding performance of Great Tits (normalized values for lay date, clutch size, number of fledglings) between two successive years, according to status (divorced/reunited)

Change in	Divorced males (n = 26)	Divorced females (n = 26)	Reunited pairs (n = 31)
Lay date	−0.02 ± 0.19 (NS)	−0.40 ± 0.19 (*)	−0.35 ± 0.17 (*)
Clutch size	−0.13 ± 0.22 (NS)	0.09 ± 0.22 (NS)	−0.07 ± 0.20 (NS)
no. fledglings	−0.29 ± 0.30 (NS)	−0.54 ± 0.30 (NS)	−0.52 ± 0.27 (NS)

For divorced birds the change was measured separately for males and females. Two-way ANOVAs with status and sex yielded no significant main effects or interactions for change in lay date, change in clutch size, or change in number of fledglings (all $P > 0.27$). To the right of each mean ± SE we indicated whether the adjusted mean was different from the population mean of zero (t-test: $\alpha = 0.05$): (*) indicates that the adjusted mean is not equal to zero; (NS) the opposite. Reunited pairs and divorced females laid relatively earlier in year B compared to the population average.

Table 13.3 Mate fidelity of Great Tit pairs in which both partners were known to be alive in at least two consecutive years

Plot	n_p	nD_p	$\%D_p$	n_{py}	nD_{py}	$\%D_{py}$
Ghent-Cit	20	0	0	25	0	0
Ghent-Mp	48	8	16.7	63	8	12.7
Ghent-H	21	3	14.3	24	3	12.5
Ghent-Z	32	75	21.9	36	7	19.4
Ghent-Coo	13	6	46.2	13	6	46.2
Antwerp-L	20	5	25.0	24	5	20.8
Antwerp-T	41	11	26.8	46	11	23.9
Antwerp-B	72	37	51.4	75	37	49.3
Antwerp-C	13	2	15.4	13	2	15.4

The subscript ‘p’ (for pair) indicates that each pair that bred together was counted only once; the subscript ‘py’ (for pair-years) indicates that each pair was counted in each year (n sample size, nD number of cases of divorce, $\%D$ percentage of pairs that divorce). When some pairs stayed together during 3 years the first value gives a higher estimate of the divorce rate. Divorce rate (D_p) varied significantly between plots: $\chi^2 = 31.87$, $df = 6$, $P < 0.001$.

survival rate) were arcsin transformed. The latter analysis showed that breeding density ($r = 0.69$, $P = 0.04$) but not survival rate were correlated with divorce. The stepwise multiple regression analysis using variables related to reproduction showed that lay date and fledglings per egg, but not clutch size, percentage second broods, or reproductive rate, were correlated with divorce (multiple correlation coefficient: $r = 0.95$, $P = 0.0009$[1]).

In an analysis testing the combined effects of lay date, fledglings per egg, and breeding density, density was no longer selected. The partial correlation between divorce rate and breeding density, once lay date and fledglings per egg were entered, was $r_p = 0.52$ ($P = 0.3$). From these analyses we conclude that the differences in divorce rates between our nine plots can be explained by pairs reuniting more often in plots where laying happens early and the number of fledglings per egg is high.

We have described above how the pair formation process takes place, and the likelihood that former partners will breed together again in the following year is greatly influenced by the birds' winter behaviour. We also tested whether there was a correlation between an index of residency and divorce in each plot. The one-sided prediction, that divorce rates should be low in plots in which birds were highly resident, is supported by the data (Spearman rank correlation: $r_s = 0.84$, $P < 0.01$). A combined multiple regression analysis with residency, lay date, and fledgling per egg results in only lay date ($P = 0.035$) and residency ($P = 0.005$) being retained. The multiple correlation coefficient for this analysis was $r = 0.97$ ($P = 0.0002$).

Partner characteristics and reproductive success

Great Tits appeared to pair with partners of similar ages (Table 13.4). In all study plots the observed number of cases of young + young and old + old was higher than expected by chance. This result was less surprising in plots in which mate fidelity was high and birds were relatively long-lived (as Ghent-Cit), but was unexpected in other plots (as in Antwerp-B). Comparison of the breeding performance, over all plots and years, showed that female age (but not male age) had a significant effect on lay date and clutch size (Table 13.4). Divorce rates did not differ significantly between the different age categories but pairs of old birds tended to divorce less (Table 13.4). Although same-age pairs predominated in all study sites, it was not obvious that same-age pairs had higher breeding performance or mate fidelity than mixed-age pairs, probably showing again the strong influence of mechanistic processes on the pair formation process: we suspect that old birds pair with old birds simply because they start the pair formation process earlier, whereas many first-year birds

[1]The regression equation was: divorce rate = −48.2 (SE 12.42; $P < 0.01$) + 6.85 lay date (SE 1.09; $P < 0.01$) −83.7 fledglings per egg (SE 26.83; $P = 0.02$).

Table 13.4 Pair characteristics, divorce events, and reproduction in Great Tits

	F_Y*M_Y	F_Y*M_O	F_O*M_Y	F_O*M_O	χ^2
Ghent-Cit	31 (20.4)	13 (23.6)	20 (30.6)	46 (35.4)	17.11
Ghent-Mp	66 (48.3)	58 (75.7)	42 (59.7)	111 (93.3)	19.13
Ghent-H	73 (54.7)	38 (56.3)	29 (37.3)	67 (48.7)	26.04
Ghent-Z	48 (34.2)	29 (42.8)	23 (36.8)	60 (46.2)	19.40
Ghent-Coo	51 (40.5)	24 (34.5)	24 (34.5)	40 (29.5)	12.93
Antwerp-L	60 (46.9)	22 (35.1)	31 (44.1)	46 (33.0)	15.75
Antwerp-T	93 (72.6)	43 (63.4)	48 (68.4)	80 (59.6)	25.75
Antwerp-B	182 (150.2)	75 (106.8)	98 (129.8)	124 (91.2)	34.89
Divorce rate	0.34	0.38	0.33	0.21	
n	70	32	30	38	
Breeding performance					
Julian lay date	113.89	113.41	112.00	111.13	
Clutch size	8.80	9.04	9.14	9.08	
no. fledglings	7.60	7.71	7.82	7.76	
n	379	160	210	287	

To test if birds mate at random with another partner regardless of age, we counted the number of pairs of different ages in each study plot, whereby all birds of known age were counted once in each breeding season. Frequency of age of partners (expected numbers in brackets): Y = young; O = old; F = female; M = male. All χ^2 values (1 *df*) are highly significant ($P < 0.001$). The second part of the table gives overall divorce rates and breeding performance of the respective pair-types. Neither female nor male age (nor the interaction) was significantly related to divorce rate (all $P > 0.46$). Comparison of the breeding performance, over all plots and years, showed that female age (but not male age) had a significant effect on lay date and clutch size (ANOVA with male, female age, and male*female age: $F = 15.44$, $df = 1{,}861$, $P = 0.0001$ and $F = 6.77$, $df = 1{,}861$, $P = 0.009$). There were no significant effects of pair composition on the number of fledglings ($P > 0.26$). Divorce rates did not differ significantly between the different categories (Fisher exact test: $P = 0.422$).

enter the breeding population relatively late and can, therefore, only find another first-year bird as a mate.

Discussion

Within-population comparisons

Earlier studies of Great Tits that compared individuals within a single population found that pairs that bred early (Perrins and McCleery 1985) or raised enlarged broods (Lindèn 1991) were more likely to remain together. We were not able to confirm these findings, perhaps because our sample sizes were relatively small. In the same study plots, Dhondt and Adriaensen (1994) found that Blue Tits *Parus caeruleus* that laid early in the season were more likely to remain together than late-breeding birds.

Perrins and McCleery (1985) concluded that pairs that remained together bred more successfully than same-age pairs that bred together the first time. We found that Great Tit pairs that remained together tended to lay relatively earlier in the following year. Since pairs that lay early tend to recruit more offspring than pairs that lay late, mate fidelity may be adaptive for both partners. However, females that divorced and found new partners also advanced their lay date in the next year, thus improving their reproductive potential.

On the other hand, Blue Tit pairs that remained together did not improve their reproductive success. In pairs that divorced, females, but not males, improved their reproductive success. This led to the conclusion that female Blue Tits initiated divorce, and that this behaviour was adaptive for females only (Dhondt and Adriaensen 1994). In Great Tits the same might be true.

Comparisons between study plots

Divorce rates in four European Great Tit populations varied between 42% (Winkel and Winkel 1980) and 23% (Kluyver 1951). In our study the overall probability of divorce in nine plots was 25% (range 0–51%). In our populations the main factor correlating with divorce was winter residency of the population: the more resident the population, the lower the divorce rate. Dhondt and Adriaensen (1994) found that between-population variation in Blue Tit divorce rates was explained by differences in winter social behaviour: the less the birds flocked during winter, the lower the divorce rate. They did not consider winter residency, but winter flocking and winter residency are two aspects of winter behaviour that are closely linked, and may be related to winter food availability and shelter. In urban parks, as in plot Ghent-Cit, all breeding birds are winter resident, and birds remain throughout winter on their former territory. In suburban or isolated forest fragments the proportion of winter residents is somewhat lower, and flocking is observed occasionally. In our study plot Antwerp-B, which is well inside the woodland, detailed winter observations have shown that an important portion of the breeding birds does not remain in the plot throughout winter, and that all birds join large, often mixed-species flocks in winter. The extremely high correlation between divorce rates of Great and Blue Tits in the same plots in northern Belgium and elsewhere in Europe (Fig. 13.1), shows that the same factors must be invoked to explain between-plot variation in divorce rate for both species. Winter social behaviour varies in parallel between the two species and is therefore likely to be the common factor.

A similar situation regarding dispersal and food availability occurs for the tit's main enemy, the Sparrowhawk *Accipiter nisus* (labelled the *habitat-mediated hypothesis*, Newton and Wyllie Chapter 14). For this species, individuals disperse during the nonbreeding season in poor quality habitats (e.g. with few Great Tits), thus few pair members reunite in the spring. Dispersal tendency during the nonbreeding season is also thought to influence variation in mate fidelity and divorce across penguin species (see Williams Chapter 15).

Lay date is the other factor that, according to our analysis, explains between-plot variation in divorce rate. Within-population comparisons also show that early-laying birds are less likely to divorce, and this is true

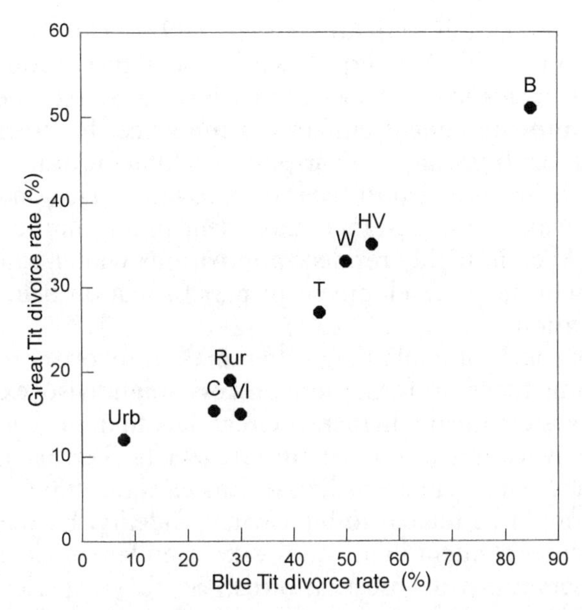

Fig. 13.1 Blue and Great Tit divorce rate in five Belgian, one British and two Dutch plots. The correlation coefficient of $r = 0.98$ is statistically highly significant ($P < 0.0001$). Values for Blue Tit for B (Antwerp-B), T (Antwerp-T), Rur (Ghent rural: mean Ghent-Z and Ghent-H) and Urb (urban: city parks [Ghent-Cit] plus suburban Ghent-Mp) from Dhondt and Adriaensen (1994). The Great Tit value for W (Wytham Woods, Oxfordshire) is taken from Perrins and McCleery (1985). The value for Blue Tits of 50% ($n = 24$) was calculated for us by Perrins and McCleery (*in litt*). The values of the Dutch populations for HV (Hoge Veluwe) and Vl (Vlieland) were calculated for us by Piet Drent (*in litt.*).

both in some Great Tit populations (Perrins and McCleery 1985) and in Blue Tits (Dhondt and Adriaensen 1994). We believe that this correlation probably reflects causation in the within-plot comparison, but does not in the between-plot comparison. Dhondt *et al.* (1984) have shown that in a set of nine study plots between-plot lay date variation for Great and Blue Tits was not significantly correlated. This was mainly because Great Tits lay much earlier in urban plots than they do in suburban and rural plots, whereas Blue Tits do not. Similarly Perrins (1979, personal communication) observed that Great Tits, but not Blue Tits, laid earlier in gardens than in nearby woodlands. If between-plot variation in divorce rates reflected between-plot variation in lay dates, we should not have found such a high correlation between divorce rates in the two species.

Divorce and pair formation

Divorce in Great Tits can simply result from former mates starting the pair formation process asynchronously (the *musical chairs hypothesis* proposed by Dhondt and Adriaensen, 1994). This idea is supported by the situation at plot Antwerp B, where the highest probability of divorce and a high incidence of dispersal in winter were recorded; high levels of dispersal and subsequent emigration influence the precise timing of the settlement on breeding territories. Anecdotal information showed that two more behaviours led to separation in Great Tits: one was usurpation; the other was a short-distance movement in autumn to a new, adjacent, domicile. Even in highly resident populations winter-emigration and perhaps asynchronous development of pair formation behaviour may result in a separation.

Our approach of explaining differences in divorce rates between plots by invoking the pair formation process might also explain why mate fidelity is systematically higher in Great Tits than in Blue Tits (Fig. 13.1). Adult survival rates are higher in Great Tits (Perrins 1979; A. Dhondt unpublished data) and according to Ens *et al.* (1993; Chapter 19) higher survival should be related to higher mate fidelity. Furthermore, Blue Tits are less winter resident in all study sites than Great Tits are (own unpublished observations on roosting birds). So they experience more separations because of the pair formation process. Also Great Tits are more habitat generalists than Blue Tits (Dhondt 1987) and Blue Tits, therefore, might profit more from moving to a better (albeit neighbouring) territory. Males are more attached to their former territory than females, so they are less likely to move. If females have more to gain by moving to a better territory they will split the pair bond by moving. We have shown recently (Dhondt *et al.* 1992) that between years, female Blue Tits do indeed move to sites in which they will lay a larger clutch. This effect was not found in Great Tits. Thus survival, winter residency, and habitat heterogeneity differ between the two species and between study sites and affect the cost/benefit balance of the mechanisms leading to divorce (survival and winter residency) as well as the benefits of the divorce itself (habitat heterogeneity). Because area effects influence the mechanism leading to divorce in the same way, divorce rates of Great and Blue are highly correlated. Because Blue Tits are more sensitive to habitat heterogeneity than Great Tits, Blue Tits are more likely to divorce than Great Tits.

Summary

We studied Great Tits in nine different study plots in northern Belgium. Divorce rate varied between 0 and 51%. Within one study site prior breeding success of the birds did not influence the probability of divorce. Divorced and faithful pairs did not differ in relation to their change in breeding performance. Observations of colour-banded individuals during pair formation in early spring showed that most divorces in the Great Tit probably happen because members of the former pair enter the pair formation process asynchronously (the *musical chairs hypothesis*). In most cases a divorce resulted from differences in wintering behaviour between pair members (resident, flocking, wintering outside the study plot). The distribution of types of wintering behaviour on a plot was strongly dependent on habitat characteristics, which probably explains why differences in divorce rates between study sites could be attributed to general habitat characteristics and why divorce rates of Great and Blue Tits are strongly correlated between study sites.

Acknowledgements

We are grateful to Ken Norris, Marcel Lambrechts, and Jean Clobert for discussions or comments on the manuscript. We thank Chris Perrins, Robin McCleery, and Piet Drent for providing unpublished data. Cindy Berger kindly improved the English. The study was supported by grants from the Belgian National Foundation for Scientific Research (NFWO, FKFO). W.P. was research fellow of the I.W.O.N.L.

References

Dhondt, A. A. (1971*a*). The regulation of numbers in Belgian populations of Great Tits. *Proceedings of the Advanced Study Institute on 'Dynamics of Numbers in Populations'—Oosterbeek 1970* (ed. P. J. Den Boer and G. R. Gradwell), pp. 532–47. Pudoc, Wageningen.

Dhondt, A. A. (1971*b*). Some factors influencing territory in the Great Tit *Parus major* L. *Giervalk*, **61**, 125–35.

Dhondt, A. A. (1987). Blue Tits are polygynous but Great Tits monogamous: does the polygyny threshold model hold? *American Naturalist*, **129**, 213–20.

Dhondt, A. A. and Adriaensen, F. (1994). Causes and effects of divorce in the blue tit *Parus caeruleus* L. *Journal of Animal Ecology*, **63**, 979–87.

Dhondt, A. A. and Eyckerman, R. (1980). Competition and the regulation of numbers in Great and Blue Tit. *Ardea*, **68**, 121–32.

Dhondt, A. A. and Hublé, J. (1968). Fledging date and sex in relation to dispersal in young Great Tits. *Bird Study*, **15**, 127–34.

Dhondt, A. A. and Schillemans, J. (1983). Reproductive success of the great tit in relation to its territorial status. *Animal Behaviour*, **31**, 902–12.

Dhondt, A. A., Eyckerman, R., Moermans, R., and Hublé, J. (1984). Habitat and laying date in Great and Blue Tit. *Ibis*, **126**, 388–97.

Dhondt, A. A., Adriaensen, F., Matthysen, E., and Kempenaers, B. (1990). Non-adaptive clutch-sizes in tits: evidence for the gene flow hypothesis. *Nature*, **348**, 723–5.

Dhondt, A. A., Kempenaers, B., and Adriaensen, F. (1992). Density-dependent clutch size caused by habitat heterogeneity. *Journal of Animal Ecology*, **61**, 643–8.

Dobson, A. J. (1990). *An introduction to Generalized Linear Models*. Chapman and Hall, London.

Ens, B. J., Safriel, U. N., and Harris, M. P. (1993). Divorce in the long-lived and monogamous oystercatcher, *Haematopus ostralegus*: incompatibility or choosing the better option? *Animal Behaviour*, **45**, 1199–217.

Hinde, R. H. (1952). The behaviour of the Great Tit (*Parus major*) and some other related species. *Behaviour*, (Suppl. II), 1–201.

Kluyver, H. N. (1951). The population ecology of the Great Tit, *Parus m. major* L. *Ardea*, **39**, 1–135.

Lebreton, J. D., Burnham, K. P. Clobert, J., and Anderson, D. A. (1992). Modelling survival and testing biological hypotheses using marked animals. A unified approach with case studies. *Ecological Monographs*, **62**, 67–118.

Lindén, M. (1991). Divorce in Great Tits—chance or choice? An experimental approach. *American Naturalist*, **138**, 1039–48.

Perrins, C. M. (1979). *British tits*. Collins, London.

Perrins, C. M. and McCleery, R. H. (1985). The effect of age and pair bond on the breeding success of Great Tits *Parus major*. *Ibis*, **127**, 306–15.

Van Balen, J. H. and Speek, J. B. (1976). Een invasie van mezen (Paridae) in de herfst van 1971. *Limosa*, **49**, 188–200.

Winkel, W. and Winkel, D. (1980). Zum Paarzusammenhalt bei Kohl-, Blau- und Tannenmeise (*Parus major, P.caeruleus* und *P. ater*). *Vogelwarte*, **30**, 325–33.

14 *Monogamy in the Sparrowhawk*

I. NEWTON AND I. WYLLIE

Introduction

The Sparrowhawk *Accipiter nisus* (see drawing above) is a small raptor which breeds in forest and woodland throughout the Palaearctic region, and preys upon small birds. The sexes look much the same, except that females are bigger than males (at twice the weight). Both sexes are short-lived, with an annual mortality of around 33%, an age of first breeding of 1–3 years, and a maximum life span of around 9 years (Newton 1986). Pairs raise no more than one brood per year, containing up to six young. They space themselves through suitable habitat, and tend to nest in the same restricted localities year after year, giving a fairly uniform and stable distribution (Newton 1986). Over a period of years, the same territories may be held by a succession of different individuals, each present for one to several breeding seasons. Individual birds defend the same territory in successive years or they may change territories between one breeding season and the next. Similarly, some individuals retain the same mate in successive years, while others change their mates between breeding seasons, even though the original partner may still be alive. In this chapter, we examine the circumstances underlying mate fidelity and mate change in the Sparrowhawk, and assess the consequences for breeding performance. We also draw comparisons with some other raptors, concluding that local food supply has a major influence on site fidelity and mate fidelity.

Background, study sites, and procedures

The data were derived from three areas, centred on Annandale (55°15′N, 3°5′W) and nearby Eskdale (55°16′N, 3°25′W) in south Scotland, and

Table 14.1 Details of study areas and data available

	Annandale 55°15′N, 3°5′W	Eskdale 55°16′N, 3°25′W	Rockingham 52°30′N, 0°30′W
Area (km^2)	700	200	220
Years of study	1971–1980	1976–1992	1979–1992
Population trend (and annual nest numbers)	Declining (110–70)	Stable (mean 34, range 29–39)	Increasing (3–96)
% Males individually identified in different years	10–36 (mean 24)	0–48 (mean 14)	21–54 (mean 38)
% Females individually identified in different years	18–93 (mean 62)	50–94 (mean 69)	68–86 (mean 78)
Total number of pairs in which both partners were of known age[1]	250	128	346
Total number of known males whose partners were identified in successive years[2]	34	9	73
Total number of known females whose partners were identified in successive years[2]	37	12	65

In other tables data from Annandale and Eskdale are combined under 'South Scotland'.
[1]Years combined, used in analysis of assortative mating; based on trapped birds and others indentified as yearling or adult on moulted feathers.
[2]Used in analyses of territory and mate fidelity and of the costs or benefits of territory and mate changes.

Rockingham Forest (52°30′N, 0°30′W) in east–central England. In all three areas, the landscape was mixed, with patches of woodland scattered through farmland. Each year we searched all the woodland in each area to find the nests, and attempted to catch and ring as many of the breeding birds as possible (Table 14.1); a male and female caught on the same territory in the same breeding season were classed as a pair. Breeders could be aged precisely if they had been ringed as nestlings or if they were first caught in their first or second year of life when they were separable on plumage features. First-year birds had brown dorsal plumage, whereas second year birds had blue dorsal plumage with a few retained brown feathers. From the third year on, the different age groups all had blue dorsal plumage with no retained brown feathers, so at first capture could be ascribed only a minimum age of 3 years. Males were much harder to catch than females, especially in successive years, so sample sizes for some analyses were small.

Demographics

During the study periods, Sparrowhawk numbers in the three areas showed different trends. In Annandale nest numbers declined by 36%

during the 10 years of study; in Eskdale nest numbers remained stable with minor fluctuations during the 21 years of study; while in Rockingham nest numbers increased dramatically during the 14 years of study (Table 14.1). The decrease in Annandale was associated with a reduction in the area of suitable woodland, through tree-felling, while the increase in Rockingham reflected a population recovery, resulting from reduced use of organochlorine pesticides (Wyllie and Newton 1991). In all three areas some unpaired, nonbreeding birds of both sexes were present throughout, especially yearlings. Despite the different population trends, no significant differences in bird behaviour or breeding performance were found between the two Scottish areas, so the data from these areas were pooled for comparison with the Rockingham data, which differed in several respects. For some analyses, the independent unit of observation was taken as one 'bird-year', and refers to whether an individual, caught in successive years, had retained the same territory or mate both years or whether it had changed its territory or mate. Thus, to appear in the data set, a given individual must have been identified along with its supposed mate in at least 2 consecutive years, and an individual identified with its mate in more than 2 successive years appeared more than once, at different ages. A bird that reared young to fledging was counted as successful, whereas one that did not rear young was counted as 'failed', regardless of the stage of nest failure (in practice most failures occurred at the egg stage, Newton 1986).

In this chapter, we use the term 'territory' as synonymous with 'nesting place' because only the occupied nesting place was defended vigorously. Each bird hunted over a wider area, the 'home range'. Particularly among females, home ranges overlapped widely between individuals (Fig. 14.1). The same nesting places were used over several years, with a new nest being built each year near old ones. Thus each nesting place was obvious to the human observer from the cluster of nests of different ages, usually all within a circle of about 50 m radius.

Pair formation process

Pair formation occurred at the nesting place, and the attachment of the birds to the same locality seemed crucial to the maintenance of the pair bond from one year to the next. The process of pair formation, studied by observation and radio tracking (Newton 1986), seemed to involve:

(1) the prior establishment of the male in a home range containing a suitable nesting place;

(2) contraction in the range of the female, from overlapping the ranges of several males to spending most of her time at the nesting place of one male, but occasionally visiting those of other males; and

(3) the provision of food by the resident male, allowing the female to spend more time at his nesting place and to ward off other females (Fig. 14.1).

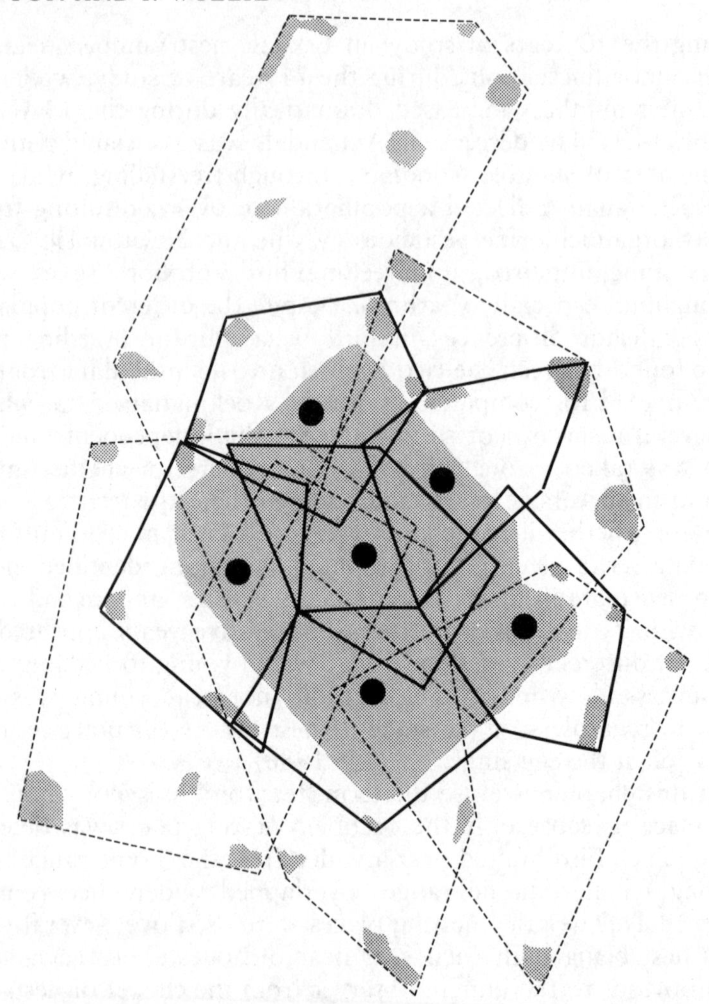

Fig. 14.1 Diagrammatic representation of home ranges (minimum polygons) of male (—) and female (---) Sparrowhawks in early spring, before females settle on the nesting places of males. Potential breeding males are distributed in individual home ranges, evenly dispersed through woodland habitat. Females have much larger and more overlapping home ranges than males. The range of any one female overlaps the ranges of several males, and the nesting place of any one male lies within the ranges of several females. Females each visit the nesting places of several males, but after a time settle on one nesting place for breeding. The success of a resident male in attracting a female for breeding depends largely on his ability to provide enough food (Newton 1986). *Shaded areas* = woodland; *black dots* = nests.

By remaining at the nesting place in the prelay period, the female could also obtain any food provided by the male, for it was only to the nesting place that the male brought prey. The sexes were together only at the nesting place, or when displaying in the airspace above, but they separated for much of each day, as one or both partners left the nesting place to hunt alone elsewhere. As the date of egg laying approached, the partners spent progressively more time together at the nesting place, and the frequency of aerial displays, copulations, and courtship feeding increased. These various findings were later confirmed in another radio tracking study by McGrady (1991), who also observed that 13% of copulations in the prelay period were extra-pair, occurring when a female from one nesting place solicited a male on another. Work is in progress to find whether extra-pair matings lead to extra-pair fertilizations.

Although Sparrowhawks remained year-round in the study areas, the extent to which partners stayed in contact during winter varied with the habitat. Where prey were abundant, one or both partners roosted at the nesting place in winter, and were present on at least some days and nights. Where prey were scarce in winter, local nesting places were much less visited and, as shown by radiotracking, birds roosted in a wider range of localities (Marquiss and Newton 1982). This inevitably meant that former partners had little or no contact in winter. Reuniting depended on both partners settling on the same nesting place as one another in spring, and the chance of this happening was greater if they stayed in the same locality.

Most nestlings raised in the study areas bred at some distance from their natal territory and, among 191 pairs in which both partners could be checked, we found no instances of closely related individuals (brother–sister, parent–offspring) pairing together.

Roles of the sexes in breeding

As in most other birds of prey, in the Sparrowhawk the sexes have different roles in breeding. The male does the hunting and, from before the eggs (usually four to six) are laid until the chicks are about half grown, provides all the food for the female and young. The female remains at the nest, responsible for all incubation and brooding, and for feeding the chicks with food brought by the male. When the young are about half grown they no longer need brooding full time, enabling the female to leave the nest area to hunt. In this, she operates independently of the male. The young are fed for 4 weeks in the nest, and a further 4 weeks postfledging, after which they disperse. Both parents defend the nest contents against potential predators (mainly corvids, squirrels, and owls), acting together if they happen to be there at the same time.

Results

The pair bond

Monogamy seemed to be the usual mating system, but because the species is difficult to observe, this was hard to check at every nest. Over the years, we placed radiotags on about 25 different male Sparrowhawks in spring–summer and none attended or delivered food to more than one nesting territory that year (though most passed through other territories in the course of their hunting). Similarly, the 32 females on which we placed radiotags in spring were resident at only one nesting territory from the date of egg laying. From then on they spent virtually all their time there, defending the nesting place against intruding females. However, we had instances from all three study areas of two females laying simultaneously in the same nest, producing twice the usual number of eggs, and incubating side-by-side. In each case the females were unrelated, in that they were not sisters or mother–daughter pairs (Newton 1986). Overall, the seven records formed about 0.3% of all nesting attempts. Most failed because the eggs were not incubated properly or were pushed out and broken. Because males apparently found it difficult to feed even one female during the early stages of breeding (Newton 1986), it would have been surprising if polygyny were common.

Mating by age and reproduction

Sparrowhawks did not pair randomly with respect to age. The proportion of first-year birds among breeders was greater in Rockingham than in the Scottish areas (Wyllie and Newton 1991). In all three areas, however, the frequency of matings between first-year (brown dorsal plumage) and adult (blue dorsal plumage) birds differed significantly from the frequencies expected in that area if mating were random between age groups. There were more yearling–yearling and adult–adult pairings, and fewer adult–yearling pairings, than expected by chance (Fig. 14.2). This finding confirmed earlier results on smaller samples (Newton 1986).

The advantage of an adult partner was apparent from the nest success (Fig. 14.2). In general, yearling–yearling pairs showed lower nest success than adult–adult pairs, while yearling–adult pairs were intermediate. Moreover, pairs in which the male was adult and the female was a yearling more often produced young than did pairs in which the male was a yearling and the female adult (this category had the lowest success in Rockingham). This difference was not statistically significant, but it held in all three areas and was consistent with the previous finding that males (in their role of food providers) contributed to more of the variation in nest success than did females (Newton 1986). Overall, a much greater proportion of adult females than yearling females achieved the preferred option (i.e. an adult male).

Selective mating was evident not only between yearlings and adults, but in one area also among adults of different ages. Regression analyses of female age (y) against male age (x) for precisely aged birds of 2–9 years gave a slope significantly different from zero, but only for Rockingham (Fig. 14.2; caption). This was because some birds paired together when young and then stayed together in subsequent years, and because mate fidelity was more frequent in Rockingham than in the Scottish

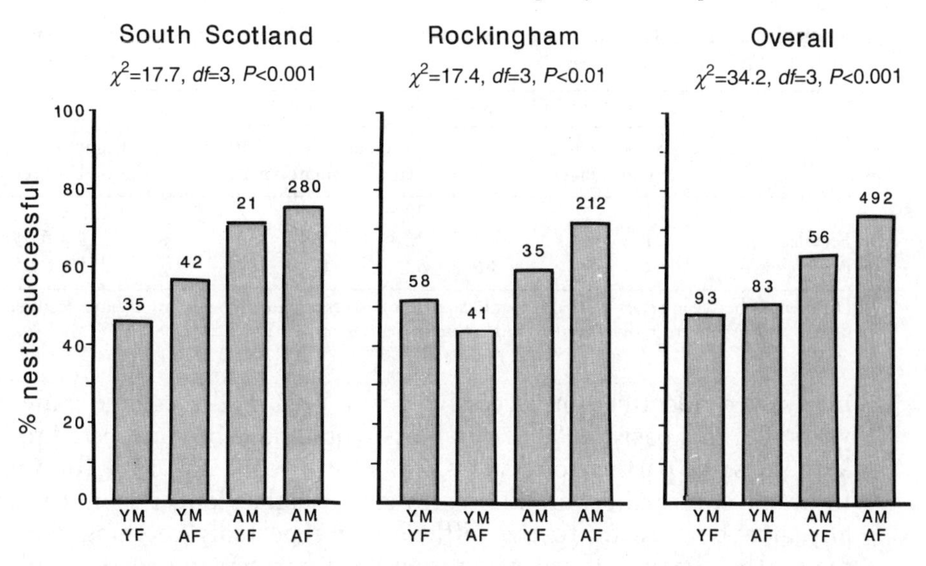

Fig. 14.2 Nest success of pairs involving first-year (Y) and adult (A), male (M) and female (F) Sparrowhawks. Figures show number of nests in each category, and chi-square values show significance of variation in nest success between categories. The mean numbers of young produced per nest in the four categories were 1.43, 1.88, 1.92, and 2.81 for the Scottish areas, and 1.51, 1.41, 1.92, and 2.59 for Rockingham. The observed frequencies of matings in the categories YM–YF, YM–AF, AM–YF, and AM–AF differed significantly from the frequencies expected if mating were random between age groups. For south Scotland, $\chi^2 = 72.0$, $df = 3$, $P < 0.001$; for Rockingham, $\chi^2 = 71.0$, $df = 3$, $P < 0.001$; overall $\chi^2 = 148.2$, $df = 3$, $P < 0.001$. The proportions of the four groups did not differ significantly between the two Scottish areas ($\chi^2 = 8.12$, $df = 3$, NS), but they did differ between the Scottish areas and Rockingham ($\chi^2 = 17.2$, $df = 3$, $P < 0.001$), mainly because of the greater proportion of yearlings in Rockingham. For birds aged 2–9, regression of male age (y) against female age (x) revealed a significant relationship in Rockingham ($b = 0.18$, $r = 0.31$, $P < 0.001$), but not in the Scottish areas ($b = 0.12$, $r = 0.14$, NS).

areas (see later), so such same-age pairs formed a larger fraction of the overall sample for Rockingham. When analysis was restricted to birds that were known to be on their second or subsequent partners, no correlation was apparent in any area between the ages of male and female.

Mate changes

Between one year and the next, mate change was common in all three areas, but was more frequent in the Scottish areas than in Rockingham (Table 14.2). Of 92 female Sparrowhawks in the Scottish areas that were caught with a male, 46% were with the same male in both years and 54% changed males. On this basis, 21% of survivors would have been expected to have kept the same partner for 3 years, 10% for 4 years, and 4% for 5 years. Similarly, of 138 Sparrowhawks in Rockingham whose

Table 14.2 Frequency of mate changes in individual Sparrowhawks whose mates were identified in successive years

	South Scotland		Rockingham		Overall		
	Same mate	Different mate	Same mate	Different mate	Same mate	Different mate	Significance of variation between areas
Males	21	22	43	30	64	52	$\chi^2 = 1.1$, NS
Females	21	28	43	22	64	50	$\chi^2 = 6.2$, $P < 0.05$
Both sexes	42	50	86	52	128	102	$\chi^2 = 6.2$, $P < 0.05$

Differences between sexes were not significant. Some of the mate changes could have resulted from deaths and others from divorces (original mate still alive). $df = 1$ in all tests.

mates were identified in successive years, 62% had the same mate both years. On this basis, 39% of survivors would have been expected to have kept the same partner for 3 years, 24% for 4 years, and 15% for 5 years. In practice, no pairs were known to stay together longer than 4 years in any area but, owing to the difficulty of repeatedly catching the same males, the chances of recording long-term partnerships were small. The actual frequency of mate change may have been slightly higher than recorded, if any birds left the study area, and could not be recaptured.

On the basis of Table 14.2, the average surviving individual in the Scottish areas would be expected to change mates about once every 1.5 years, and in Rockingham about every 1.6 years. As Sparrowhawks in the Scottish areas bred for 1–8 years (mean 2.25 years, Newton 1989), successive age groups would be expected, on average, to have one to five mates in their lifetimes (mean 1.5). The equivalent figures for Rockingham were 1–8 years (mean 2.96 years), giving averages in successive age groups of one to five mates, and an overall mean of 1.9 mates per lifetime.

Of those Scottish Sparrowhawks (sexes combined) that survived from one year to the next, 38% would have been expected to change partners as a result of the death of a former mate (this being the mean annual mortality, Newton *et al.* 1993). In practice, at least 54% of surviving Sparrowhawks changed partners (Table 14.2), so at least 16% must have changed through divorce. In Rockingham, where annual mortality averaged 28% (Newton *et al.* 1993), 38% of surviving birds changed partners between years, so at least 10% must have divorced. Therefore, the overall probability of divorce in Sparrowhawks was 11.3% (26 events/230 pair-years). These figures are minimal, because divorced birds were more likely to change territories, and thus leave the study areas where they would not be recorded.

In all study areas, a change of nesting place usually meant a change of mate. All four possible options were recorded, namely a bird was found in the second year:

(1) on the same territory with the same mate;

(2) on the same territory with a different mate;

(3) on a different territory with a different mate; or

(4) on a different territory with a different mate.

In some instances where a bird had a different mate, the original mate was known to be alive and breeding elsewhere, while in other such instances the original mate was known to be dead, but in most cases the status of the original mate was unknown.

Overall, some 54% of 70 Sparrowhawks in the Scottish areas that occupied the same territory in successive years reunited with the same mate, whereas only 18% of 22 birds that changed territory did so (Table 14.3). The equivalent figures for Rockingham were 73% for 104 birds that kept the same territory, compared with 29% of 34 birds that changed territory. These differences in mate change in relation to site fidelity were statistically highly significant (Table 14.3). Moreover, all 14 moves in which the pair stayed together were to an adjacent territory in the same wood (about 0.5 km), and all 42 individuals that moved to a more distant territory changed their mate. Such longer moves were mostly up to 5 km, but occasionally further. It seemed, therefore, that continued residence in the same restricted locality was a prerequisite to reuniting with the same mate.

Previous analyses revealed that changes of territory were associated with three main factors, namely territory quality (birds were less likely to move from a good territory than a poor one), age (birds became less likely to change territory with increasing age), and previous nest success (birds of a given age were more likely to change territory after a nest failure than after a success) (Newton 1986, 1993; Newton and Wyllie 1992). These findings on territory changes were evident in both sexes, and would have been expected to influence the associated mate changes.

Table 14.3 Year-to-year change of mate in relation to change of territory in Sparrowhawks

	Number of birds that in the second year had				
	Same territory		Different territory		Significance of
	Same mate	Different mate	Same mate	Different mate	variation between areas
South Scotland					
Males	19	13	2	9	$\chi^2 = 5.6, P < 0.05$
Females	19	19	2	9	$\chi^2 = 3.5, P < 0.1$
Rockingham					
Males	38	15	5	15	$\chi^2 = 13.1, P < 0.001$
Females	38	13	5	9	$\chi^2 = 7.4, P < 0.001$
Overall					
Males	57	28	7	24	$\chi^2 = 18.2, P < 0.001$
Females	57	32	7	18	$\chi^2 = 10.3, P < 0.001$

Differences between sexes were not significant. Birds tended to reunite with previous mates when they returned to a previous territory. $df = 1$ in all tests.

Because only birds with known mates were used in the present analysis, samples were small and only previous nest success and mate fidelity emerged as significantly linked with territory fidelity in multiple logistic regression analysis (Table 14.4). In a similar analysis, mate fidelity was associated with previous nest success and with age, independently of the link with territory fidelity, in one or more of the bird-groups considered (Table 14.4).

Mate fidelity and reproduction

To examine the effects of territory and mate fidelity on reproductive success, three types of analyses were performed. The first entailed a comparison of breeding performance in three categories of birds which were:

(1) on the same territory with the same mate as in the previous year;

(2) on the same territory but with a different mate from the previous year; or

(3) on a different territory with a different mate.

The fourth option, in which the pair stayed together but occupied a different territory, was too infrequent (seven cases) to warrant inclusion.

In nest success, birds that had retained both territory and mate (category 1) did better than birds that had retained the territory alone (2), and these in turn did better than birds that had changed both territory and mate (3). Overall, the mean number of young produced in a 2-year period by males that kept the same territory and mate was 6.6, compared with 4.7 males that kept the same territory but changed their mate, and 3.7 for males that changed both territory and mate (Table 14.5). The equivalent figures for females were 6.6, 5.3, and 3.8. These trends were apparent in all three areas but significant only in Rockingham. The mean ages of birds in these three categories differed slightly, but not significantly. However, when the data on individual breeding performance were corrected for age (using information in Newton 1986), differences in mean performance between categories were still apparent, and still statistically significant in Rockingham (Table 14.5). Taking these results at face value, therefore, there seemed to be a cost both in territory change and in mate change, but other interpretations were possible (see below).

The second analysis involved examination of the change in nest success between successive years for the same three classes of birds (Table 14.6, comparison between columns A and B). Those that kept the same territory and mate showed no significant change in nest success between years (despite the suggestion of a slight decline), and the same was true for birds that kept the same territory but changed mate. However, birds that changed both territory and mate showed an improvement in nest success between years (apart from Rockingham females). On the small

Table 14.4 Findings from three logistic regression analyses regarding the relationship between mate change and reproductive success in two successive years, A and B

		South Scotland (*n*)		Rockingham (*n*)		Both areas (*n*)	
Dependent variable	Independent variables	Male (43)	Female (49)	Male (73)	Female (65)	Male (116)	Female (114)
Same or different territory in year B	Success in year A	0.5	3.0	6.1*	4.4*	5.9	5.6*
	Female age in year B	0.0	0.1	0.1	3.3	0.0	2.4
	Same or different mate in year B	6.0*	3.2†	3.1†	6.6**	12.3***	8.7**
Same or different mate in year B	Success in year A	0.0	0.2	12.5**	0.1	4.5*	1.5
	Female age in year B	3.0	0.0	4.3*	2.5	0.3	1.7
	Same or different territory in year B	6.1*	3.2†	2.8†	7.0**	12.3***	8.6**
Success in year B	Success in year A	3.6†	0.0	1.4	0.5	1.6	0.0
	Female age in year B	0.2	0.2	0.4	2.7	0.9	0.9
	Same or different territory in year B	0.0	0.7	2.4†	2.3†	2.0	3.8*
	Same or different mate in year B	0.1	0.3	3.3†	0.2	2.0	0.3

The figures are the delta deviance values, which indicate the contribution of the independent variable concerned when fitted last in a multiple regression. *$P < 0.05$, ** $P < 0.01$, *** $P < 0.001$, † significant at $P < 0.05$ when fitted first in a multiple regression. All results were positive: in one or more of the bird groups considered, retention of the same territory was independently associated with previous nest success and with retention of the same mate; retention of the same mate was associated with previous nest success, increasing female age and retention of the same territory; while successful nesting in the current year was associated with retention of the same territory.

Table 14.5 Total numbers of young (± SE) produced in 2-year periods in relation to territory and mate fidelity; see footnote for *df*.

	Same territory with same mate	Same territory with different mate	Different territory with different mate	Significance of variation
South Scotland				
Males	6.2 ± 0.5	5.6 ± 0.6	6.4 ± 0.8	$F = 0.37$, NS
Females	6.2 ± 0.5	4.8 ± 0.6	4.4 ± 1.2	$F = 1.85$, $P > 0.1$
Rockingham				
Males	6.7 ± 0.4	3.9 ± 0.6	2.3 ± 0.5	$F = 24.1$, $P < 0.001$
Females	6.7 ± 0.4	6.1 ± 0.8	3.2 ± 1.0	$F = 7.3$, $P < 0.01$
Overall				
Males	6.6 ± 0.3	4.7 ± 0.4	3.7 ± 0.6	$F = 13.1$, $P < 0.001$
Females	6.6 ± 0.3	5.3 ± 0.5	3.8 ± 0.8	$F = 8.3$, $P < 0.009$

Sparrowhawks that retained both territory and mate produced most young. When the above data were corrected for age (by adjusting each individual brood size up or down, depending on the age of the individual, to a standard second-year value), the variation between categories was still apparent, although not significant for the Scottish areas (males, $F = 0.13$, $df = 2,38$, NS; females, $F = 1.3$, $df = 2,44$, $P > 0.1$), it was still significant for Rockingham (males, $F = 19.1$, $df = 2,65$, $P < 0.001$; females, $F = 5.4$, $df = 2,57$, $P < 0.01$), and overall (males, $F = 10.1$, $df = 2,106$, $P < 0.001$; females, $F = 6$, $df = 2,104$, $P < 0.01$).

samples available, the improvement was not statistically significant, but was consistent with an earlier result on larger samples, that individuals of both sexes (examined in the absence of knowledge of their mates) showed significantly higher nest success after a change of territory (Newton and Wyllie 1992).

In a third analysis of these data, we used logistic regression to explore whether nest success in year B (the present year, as the dependent variable) was related to:

(1) nest success in year A (the previous year);

(2) age in year B;

(3) same or different territory; and

(4) same or different mate.

Nest success in year B showed a significant relationship with all these variables (except age) in one or more of the bird groups considered, but only with same or different territory after controlling for the other variables (Table 14.4). Present findings gave no indication that a change of mate contributed to an improvement in performance, over and above the effect associated with the change of territory.

Discussion

Differences and similarities between areas

Several findings differed between the two Scottish areas where Sparrowhawk breeding numbers were stable or declining, and Rockingham

Table 14.6 Proportion of nests which produced young in relation to territory and mate fidelity in successive breeding seasons (A and B)

	(1) Same territory with same mate				(2) Same territory with different mate				(3) Different territory with different mate			
	A		B		A		B		A		B	
	n	%	*n*	%	*n*	%	*n*	%	*n*	%	*n*	%
South Scotland												
Males	19	79	19	84	13	85	13	77	9	78	9	89
Females	19	79	19	84	19	74	19	74	9	44	9	67
Rockingham												
Males	38	95	38	82	15	67	15	60	15	27	15	47
Females	38	95	38	82	13	92	13	85	9	67	9	44
Overall												
Males	57	89	57	82	28	75	28	68	24	46	24	63
Females	57	89	57	82	32	81	32	78	18	56	18	56

Sparrowhawks that changed territories and mates tended to improve their nest success. Comparing the three B columns, the variation was significant for Rockingham males ($\chi^2 = 6.89$, $df = 2$, $P < 0.05$) and females ($\chi^2 = 6.20$, $df = 2$, $P < 0.05$) only. In addition to the birds shown, four birds in the Scottish areas changed territory, but retained the same mate, as did 10 birds in Rockingham (seven pairs in all).

where numbers were increasing. The two main differences were the greater proportion of yearlings among breeders and the reduced frequency of territory (and mate) changes in Rockingham. Both could have arisen because the Rockingham population was in rapid growth (from 3 to 96 pairs in 12 years), expanding into vacant habitat where competition for territories was minimal. This could have enabled more yearlings to breed and more birds to retain the same territory from year to year. In contrast, the Scottish areas were occupied nearer to full capacity, with more competition for nesting places. This could have resulted in fewer yearlings breeding and more frequent changes of territory. The habitat in the Scottish areas was also more variable than in Rockingham, and spanned a much greater spread of elevation. Greater variation in the quality of nesting places (tantamount to variation in local food supply) may have promoted more frequent territory and mate changes in the Scottish areas, as birds competed for the better sites. Another difference between areas was that Sparrowhawks, on average, lived slightly longer in Rockingham than in the Scottish areas (Newton *et al.* 1993), and this could also have facilitated longer mate fidelity.

Despite these differences between areas, some general trends emerged. First, assortative mating between the two age-classes was apparent in all three areas, with yearling–adult pairs less frequent than expected by chance. Such selective mating was not necessarily by direct choice, but could have been largely passive, dependent on yearlings settling later than adults, and in generally less favoured places (Newton 1986).

Yearling–yearling pairs showed the lowest nest success, adult–adult pairs the highest, while yearling–adult pairs were intermediate. The implication was that birds of both age groups benefited from an adult partner; perhaps, therefore, birds of both age groups accepted yearlings as partners only when they had no choice. Second, in all three areas failed breeders more often changed territory than did successful breeders (Newton and Wyllie 1992), and change of territory was usually associated with change of mate. Also, in all three areas, birds that stayed on the same territory with the same mate raised more young in a 2-year period than did birds that stayed on the same territory with a different mate, and these in turn raised more young than birds that changed both territory and mate. However, only the latter group (with the poorest success in the first year) improved in nest success between years.

Territory change, mate change, and nest success

Many of the birds that changed territory did so after a breeding failure the previous year, so even if failures fell randomly across the population in each year, individuals that failed in one year would have been less likely to fail the next, leading to an improvement in the success of the group as a whole. Similarly, birds that stayed on the same territory had mostly bred well in the previous year, so on stochastic grounds alone, more would be expected to fail in the second year. Hence, while a change of territory (and usually also mate) may have been stimulated by breeding failure, the following improvement may not have resulted entirely from the move. It was also possible that the tendency to stay or move, and to succeed or fail in breeding, were both dependent on some third (underlying) factor, such as local food supply. In this case, both high site (and mate) fidelity and high breeding success could have occurred in localities that offered a good food supply, while poor site (and mate) fidelity and low breeding success could have occurred in localities that offered only a poor food supply. Many of the birds that moved could then have ended up on a better territory (confirmed for Eskdale, Newton 1991), facilitating an improvement in their breeding success.

Extra-pair copulations

The extra-pair copulations seen by McGrady (1991) were apparently initiated by females visiting nesting places of other males. For participating males the advantage presumably lay in the possibility of producing extra offspring without commitment, and for the females extra-pair copulations could have increased the chance of producing fertile eggs. Since it was those females that were least well fed that most often visited other territories, it was probably the males that had poor territories or were poor food providers that were most often cuckolded. However, all observed extra-pair copulations occurred in the prelay period, and we do not yet know what proportion (if any) of them led to fertilization.

Ecological basis of mate fidelity and change

With an annual mortality of 30–40%, and an age of first breeding of 1–3 years, most Sparrowhawks that survived long enough to breed made only one or two nest attempts in their lives (maximum eight in our study areas). The possibilities for long-term pair bonds were therefore extremely limited. The combined effects of deaths and movements meant that around half the surviving birds had a new mate in any one year. In our study areas, the main factor that kept pairs together seemed to be site fidelity, which was in turn contingent upon a consistent and reliable local food supply. Territory changes were most common in those parts of each study area where prey were least plentiful, especially in winter. Mate fidelity or change could thus be viewed partly as a consequence of habitat variation. As a secondary factor in divorce, some previous occupants may have been forced to leave by newcomers, the change of territory being associated with a change of mate (the *forced divorce hypothesis*). Other birds may have moved from prey-poor localities to take up vacancies in better places (consistent with the *better option hypothesis*). This *habitat mediated* view of mate fidelity and change is similar to that proposed by Desrochers and Magrath (1993, Chapter 9) for the Blackbird *Turdus merula*, and another hypothesis on the same lines has been proposed to explain mating and social systems in general (Rubenstein and Wrangham 1986; Clutton-Brock 1991; Davies 1991). In essence, Sparrowhawks of both sexes that occupied good (prey-rich) habitat and were able to hold it in competition with other individuals, were likely to stay faithful to both nesting territory and mate. Those in poor habitat, by contrast, were unable to remain continuously, and would take any opportunities that arose to settle in better habitat, leading to a change of territory, and hence of mate. It was probably because Sparrowhawk sexes were solitary away from the immediate nesting place, and acted independently of one another, that territory changes for whatever reason usually led to divorce.

A similar situation, regarding winter dispersal and mate fidelity occurs in one of the Sparrowhawk's main prey, the Great Tit *Parus major* (Dhondt *et al.* Chapter 13). For this species, individuals living in dense woodland sites disperse in winter when food availability is low, so the timing of their return in spring seldom coincides and few pair members reunite. On the other hand, tits living in city parks and urban areas show low dispersal (perhaps due to food supplements) and a low incidence of divorce. Dispersal tendency during the nonbreeding season is also thought to influence variation in mate fidelity and divorce among penguin species (Williams Chapter 15).

Comparisons with other raptors

In other raptors, as in the Sparrowhawk, more information is available on territory changes than on mate changes, but mate fidelity is again closely linked with site fidelity. To begin at one extreme, 10 pairs of Greater Kestrels *Falco rupicoloides* remained together on their territories over a 3-year period in the Transvaal, with only one replacement of a male which died (Kemp and Kemp 1977). These birds thus showed great fidelity both to territory and to mate. Several studies of Ospreys *Pandion haliaetus* gave similar results, but divorce occurred occasionally. It usually involved young breeders, followed a breeding failure, and entailed the female moving to another territory and male (Spitzer 1980; Poole 1989; Postupalsky 1989). Four studies of Peregrine Falcons *F. peregrinus* gave a total of 59 instances of males found on the same territory in two successive years and of one on a different territory. The equivalent figures for females were 168 and 16, with no difference between resident and migrant populations (Mearn and Newton 1984; Ambrose and Riddle 1988; Enderson and Craig 1988; Court *et al.* 1989). On these figures, if the sexes acted independently of one another, about 90% of partners that survived to the next year would be expected to remain paired, on grounds of site fidelity alone. In the Peregrine, as in some other species, mate changes were increased by the fact that females showed less site fidelity than males. In two studies of Merlins *F. columbarius* in Canada, 18 males were found on the same territory in two successive years and 11 on different territories, while in females the ratio was 16 to 33 (Hodson 1975; James *et al.* 1989). On these figures, only 20% of surviving Merlins would be likely to remain together from one year to the next, for no pairs were reported to remain intact after a change of territory.

Comparing species, the degree of site fidelity seems to vary partly with the extent of year-to-year variations in food supply. In general, many individuals of raptor species that depend on cyclic rodents, whose abundance varies greatly from year to year and from place to place, move around from year to year, and show low levels of both site fidelity and mate fidelity. Sometimes, behavioural differences are apparent between different populations of the same species, if these populations differ in the stability of their food supply. Thus Hen Harriers *Circus cyaneus* on Orkney (Scotland), whose varied diet was fairly stable from year to year, showed high levels of site and mate fidelity: some 56% of 62 pairs remained intact from one year to the next (Picozzi 1984). In contrast, in the same species in Wisconsin, where the diet was based on cyclic rodents, breeding numbers fluctuated greatly from year to year (Hamerstrom 1969). Only 28% of birds marked as breeding adults were known to return to a 16 000 ha area in later years, and only one female was found to mate with the same male more than once, but she had also mated with another male in the interim.

Among Kestrels *F. tinnunculus* in the Netherlands, ringed adults were often present in the same locality in successive breeding seasons, but not necessarily in the same nest boxes (which were close together). Fidelity to the area was as high as 70% in good vole years when Kestrel numbers were on the increase, but as low as 10% in poor vole years, when Kestrel numbers were declining (Cavé 1968). In another study, male Kestrels showed greater site fidelity than females, and birds that used the same territory in successive years were more likely to retain the same mate than were birds that changed territory (Village 1990). Moreover, pairs that remained intact after a change of territory had moved only a short distance (less than 1 km); longer moves were invariably associated with mate change. As in the Sparrowhawk, Kestrels and other species were more likely to change territory and mate after a breeding failure than after a success, but not every study controlled for age (Picozzi 1984; Postupalsky 1989; Village 1990).

Low levels of site and mate fidelity were also shown by another rodent-eater, the Black-shouldered Kite *Elanus caeruleus* in South Africa (Mendelsohn 1983). Although birds were present in the study area year-round, 86% of individuals stayed for less than 100 days. Of 75 pairs studied, 18 (24%) reunited on the same territory, after one partner had been absent for a time; the majority of birds re-paired with different partners on the same or different territories. Hence, comparing different populations of raptors, the same trend as found in the Sparrowhawk emerges in more extreme form, namely that site fidelity, and the associated mate fidelity, are strongly dependent on year-to-year consistency in food supply. A consistently good food supply (often associated with a varied diet) usually enables birds to breed in the same localities year after year, whereas a poor or variable food supply encourages them to move elsewhere, with a resulting change of mate.

Summary

In three study areas, Sparrowhawk breeding numbers were decreasing (Annandale, south Scotland), stable (Eskdale, south Scotland) and increasing (Rockingham Forest, east–central England). Compared to the Scottish population, the increasing Rockingham population contained a greater proportion of yearlings, and territory and mate changes were less frequent. All three differences were probably linked with reduced competition for nesting territories in Rockingham. In all three areas, Sparrowhawks were normally monogamous but, with an annual mortality of 30–40% and frequent changes of territory, the possibilities for long-term pair bonds were limited. In the two Scottish areas, about 54% of Sparrowhawks changed mates between one year and the next (16% through divorce), and in Rockingham about 38% (10% through divorce). No pairs were known to stay together longer than 4 years. Mate fidelity was largely dependent on site fidelity. Birds of similar age tended to pair together more often than expected by chance. This tendency was especially apparent between first-year and older birds. Birds that kept the same territory and mate raised more young per 2-year period than birds that kept the same territory with a different mate, and more again than birds that changed both territory and mate. Only the latter group improved in breeding success after generally poor success in the first year. All three features (poor territory fidelity, poor mate fidelity, and poor breeding success) could have occurred in response to a poor local food supply, leading to generally improved breeding success after the move. Data from other raptors are consistent with the views that (1) mate fidelity is largely dependent on site fidelity, and that (2) many territory (and mate) changes in these birds are habitat mediated, occurring in response to a poor local food supply.

Acknowledgements

We are grateful to Dr M. Marquiss, who contributed to the study in Annandale, to Mr T. Sparks for statistical help; to the editor, and to Drs N. Fox and R. Payne, in their capacity as referees, for helpful comments on the manuscript.

References

Ambrose, R. E. and Riddle, K. E. (1988). Population dispersal, turnover and migration of Alaska Peregrines. In *Peregrine falcon populations. Their management and recovery*. (ed. T. J. Cade, J. H. Enderson, C. G. Thelander, and C. M. White), pp. 677–84. The Peregrine Fund, Boise.

Cavé, A. J. (1968). The breeding of the Kestrel, *Falco tinnunculus* L., in the reclaimed area Oostelijk Flevoland. *Netherlands Journal of Zoology*, **18**, 313–407.

Clutton-Brock. T. H. (1991). Mammalian mating systems. *Proceedings of the Royal Society*, **235**, 339–72.

Court, G. S., Bradley, D. M., Gates, C. C., and Boag, D. A. (1989). Turnover and recruitment in a tundra population of Peregrine Falcons *Falco peregrinus*. *Ibis*, **131**, 487–96.

Davies, N. B. (1991). Mating systems. In *Behavioural ecology, an evolutionary approach* (ed. J. R. Krebs and N. B. Davies), pp. 263–300. Blackwell, Oxford.

Desrochers, A. and Magrath, R. D. (1993). Environmental predictability and remating in European blackbirds. *Behavioral Ecology*, **4**, 271–5.

Enderson, J. H. and Craig, G. R. (1988). Population turnover in Colorado Peregrines. In *Peregrine falcon populations. Their management and recovery*. (ed. T. J. Cade, J. H. Enderson, C. G. Thelander, and C. M. White), pp. 685–8. The Peregrine Fund, Boise.

Hamerstrom, F. (1969). A harrier population study. In *Peregrine falcon populations: their biology and decline* (ed. J. J. Hickey), pp. 367–85. University of Wisconsin Press, Madison.

Hodson, K. A. (1975). Some aspects of the nesting ecology of Richardson's Merlin *(Falco columbarius richardsonii)* on the Canadian prairies. Unpublished M.Sc. thesis. University of British Columbia, Vancouver.

James, P. C., Warkentin, I. G. and Oliphant, L. W. (1989). Turnover and dispersal in urban Merlins *Falco columbarius*. *Ibis*, **131**, 426–47.

Kemp, A. C. and Kemp M. I. (1977). The status of raptorial birds in the Transvaal Province of South Africa. *Proceedings ICBP World Conference on birds of prey, Vienna, 1975*, 28–34.

McGrady, M. J. (1991). The ecology and breeding behaviours of urban Sparrowhawks *(Accipiter nisus)* in Edinburgh, Scotland. Unpublished Ph.D thesis. University of Edinburgh.

Marquiss, M. and Newton, I. (1982). A radio-tracking study of the ranging behaviour and dispersion of European sparrowhawks *Accipiter nisus*. *Journal of Animal Ecology*, **51**, 111–33.

Mearns, R. and Newton, I. (1984). Turnover and dispersal in a Peregrine population. *Ibis*, **126**, 347–55.

Mendelsohn, J. M. (1983). Social behaviour and dispersion of the Black-shouldered Kite. *Ostrich*, **54**, 1–18.

Newton, I. (1986). *The Sparrowhawk*. T & A. D. Poyser, Calton.

Newton, I. (1989). Sparrowhawk. In *Lifetime reproduction in birds* (ed. I. Newton), pp. 279–96. Academic Press, London.

Newton, I. (1991). Habitat variation and population regulation in Sparrowhawks. *Ibis*, **133**, (suppl.), 76–88.

Newton, I. (1993). Age and site fidelity in female sparrowhawks *Accipiter nisus*. *Animal Behaviour*, **46**, 161–8.

Newton, I. and Wyllie, I. (1992). Fidelity to nesting territory among European Sparrowhawks in three areas. *Journal of Raptor Research*, **26**, 108–14.

Newton, I., Wyllie, I., and Rothery, P. (1993). Annual survival of Sparrowhawks *Accipiter nisus* breeding in three areas of Britain. *Ibis*, **135**, 49–60.

Picozzi, N. (1984). Breeding biology of polygynous Hen Harriers *Circus c. cyaneus* in Orkney. *Ornis Scandinavica*, **15**, 1–10.

Poole, A. F. (1989). *Ospreys. A natural and unnatural history*. Cambridge University Press, Cambridge.

Postupalsky, S. (1989). Osprey. In *Lifetime reproduction in birds* (ed. I. Newton), pp. 297–313. Academic Press, London.

Rubenstein, D. and Wrangham, R. W. (1986). *Ecological aspects of social evolution*. Princeton University Press, Princeton.

Spitzer, P. R. (1980). Dynamics of a discrete coastal breeding population of Ospreys in the northeastern United States during a period of decline and recovery, 1969–1978. Unpublished Ph.D. thesis. Cornell University, Ithica.

Village, A. (1990). *The Kestrel*. Poyser, Calton.

Wyllie, I. & Newton, I. (1991). Demography of an increasing population of sparrowhawks. *Journal of Animal Ecology*, **60**, 749–66.

15 *Mate fidelity in penguins*

TONY D. WILLIAMS

Introduction

Penguins (family Spheniscidae) are a distinctive group of flightless, pelagic seabirds, widely distributed in the cooler waters of the Southern Hemisphere and occupying a wide variety of breeding habitats, ranging from the snow and ice of the Antarctic Continent to the desert-like lava flows of the equatorial Galapagos Islands. The family comprises 17 species in six distinct genera and is most closely related to the petrels, albatrosses, and divers (Procellaridae and Gavidae). All penguins have a very similar body form, which is highly specialized and adapted primarily for swimming and diving underwater. However, they vary considerably in size from the Little Penguin *Eudyptula minor*, which weighs 1.1 kg and stands about 40 cm tall, to the Emperor Penguin *Aptenodytes forsteri*, which weighs over 30 kg and reaches a height of 115 cm. Plumages are generally similar in all species, being characterized by dark, blackish upper parts and white under parts. The major differences between species are largely restricted to the coloured patches and crests on the head (e.g. in *Aptenodytes* and *Eudyptes* spp.) or banding on the head and chest (*Spheniscus* spp.). The sexes are outwardly similar in all species, though males are usually slightly, but significantly, heavier and larger than females.

Penguins have been generally considered a classic example of an avian taxa with a primarily monogamous breeding pattern: birds mate with only one other individual in each year and form a pair bond with both partners contributing more-or-less equally to incubation and the rearing of offspring. Indeed, this equal division of parental care is essential for successful reproduction: at least up to the crèche stage a single parent would be unable to incubate eggs or brood chicks while also foraging at sea (although Yellow-eyed Penguins *Megadyptes antipodes* can successfully rear chicks after losing their mate, Marchant and Higgins 1990). In many penguin species, most birds also retain the same mate for two or more consecutive seasons. In this chapter I compare mate fidelity for 12 different penguin species, and describe the variation in the frequency and pattern of mate fidelity in relation to breeding latitude, nest density, and mortality rates. Particular reference is made to the incompatibility hypothesis (Coulson 1972; Davis 1988), in relation to patterns of incubation and chick rearing, and the pair bond reinforcement hypothesis (Rowley 1983). Finally, I consider whether penguins are in fact truly monogamous, or promiscuously monogamous, within a breeding season.

Background, study sites, and procedures

Breeding biology and demographics

Penguins are relatively easy to capture, to mark with numbered metal or plastic flipper bands, and thus to recapture or re-sight, making them ideal for studies of mate fidelity. Most penguins nest colonially, some colonies comprising up to several hundred thousand pairs at a single site, with nest densities of up to 2–3 pairs per square metre (the exceptions are the Fiordland Penguin *Eudyptes pachyrhynchus* and the Yellow-eyed Penguin which breed only loosely colonially or in solitary pairs). Breeding typically occurs annually, with a two-egg clutch being laid in the austral spring (Sept–Nov) in most species, the exceptions being the two *Aptenodytes* species which lay a single-egg clutch. A few species, such as the Fiordland Penguin and northern races of the Gentoo Penguin *Pygoscelis papua* (see drawing on title page) start breeding during late winter, and the Emperor Penguin is unique in rearing its chick throughout the long, dark winter months on the Antarctic Continent. Incubation lasts about 35 days (65 days in Emperor Penguins), and chicks are reared during the austral summer (Nov–Mar), being brooded and guarded continually by one parent for 2–4 weeks following hatching (the *guard* period). Chicks are then left alone in the colony until fledging, which occurs at 60–80 days of age in most medium-sized species. During this *crèche* period the chicks typically form small, loose groups with both parents foraging at sea and returning every 1–2 days to feed the chick(s). Following fledging, most adults undergo a postnuptial moult, although some Galapagos Penguins *Spheniscus mendiculus* and some African Penguins *Spheniscus demersus* moult at the beginning of their breeding season. All species forage at sea during the breeding season, catching crustaceans, fish, and cephalopods by pursuit diving from the surface. During the nonbreeding period several species (e.g. *Eudyptes* spp.) are migratory and completely pelagic, remaining at sea throughout the winter. Other species, such as the King *Aptenodytes patagonicus*, Little, and Gentoo Penguins are resident or only partially migratory at most breeding locations. Penguins are long-lived, with high annual rates of survival as adults (averaging 80–90%, but reaching 95% in *Aptenodytes* penguins) and individuals have been recorded still breeding at 17–20 years of age. Another characteristic of all species is that they delay onset of breeding until they are at least several years old. Gentoo, Little, and Yellow-eyed Penguins, for example, start breeding at 2 years of age, but the average age of first breeding in Emperor Penguins is 5 years and some birds do not join the breeding population until they are 9-years old (Mougin and van Beveren 1979).

Data were obtained from the literature on rates of mate fidelity for 12 of the 17 extant penguin species, including those with both migratory and resident life cycles, from breeding localities ranging from the equator to 77°S. Several studies have estimated mate fidelity for only a single pair of years, and there have been only a small number of long-term studies of mate fidelity in penguins (e.g. Richdale 1957; Reilly and Cullen 1981; Ainley *et al.* 1983). Mate fidelity is defined as the percentage of birds alive in one year that reunited with their partner of the previous year (thus excluding those cases where one member of the pair died or failed to return); the converse of this is divorce. Several studies present data on *pair* fidelity, which includes all pairs in estimating the proportion remaining together from one year to the next regardless of whether the pair was terminated by divorce, death, or disappearance; the converse of this is mate change. In many penguin studies the fate of chicks is not followed beyond the guard period, at about 21–28 days of age so, unless otherwise stated, successfully breeding pairs have been defined as having reared chicks to this stage.

Pair formation process

Pair formation, and reuniting of established pairs, occurs mainly during the prelaying courtship period in penguins, the duration of which depends on whether the species is sedentary or migratory and on breeding latitude which determines the overall length of the breeding period. In migratory species, there is no evidence that pairs remain together during the pelagic, nonbreeding period. Members of a pair return to the colony separately in spring, males typically arriving from a few days to several weeks before the female (e.g. Warham 1963; Trivelpiece and Trivelpiece 1990; Williams and Croxall 1991). In at least some sedentary species, however, established pairs have been recorded together at their old nest site during the winter period (Reilly and Cullen 1981; T. D. Williams unpublished data). The postnuptial moult, prior to departure at the end of the season, probably also represents an important period for pair formation, in addition to reinforcement of established pairs, in some species. For example, in the Macaroni Penguin *Eudyptes chrysolophus*, 26% of birds that moulted in a pair with a partner different to their mate of that season, bred with the new mate the following year (Williams and Rodwell 1992).

In trying to explain the sexual dimorphism that occurs in penguins (males being larger), Davis and Speirs (1990) proposed a model for the process of mate choice in the Adelie Penguin *Pygoscelis adeliae*, which may be applicable to other penguin species. They tested this model by comparing the observed and predicted sequence of behaviours during pair formation in 23 males and 19 female Adelie Penguins (Davis and Speirs 1990), and showed that although males and females returning to breed appear to adopt different strategies, both sexes maximize the chance of reuniting with their previous year's partner. Davis and Speirs (1990) suggested that male Adelie Penguins benefit most by adopting a single, relatively simple strategy: they should return to last year's nest site, they should return as early as possible (to reduce the likelihood of their being cuckolded), and they should court all females until paired in case their old mate fails to return. This suggests that the primary 'aim' of male Adelie Penguins on returning to the breeding colony is to retain their old nest site and only secondarily to reunite with their old mate. Females should also initially return to their old nest site, but if their previous mate is absent they should pair with the unattached male nearest to the old nest site. This strategy still maximizes a female's chances of successfully reuniting with her mate from the previous year should he return late. This mechanism for females retaining their old mate is supported by the fact that female Adelie Penguins that have already paired with a new partner will leave this bird if their old mate returns late. Later arriving females will also often oust 'new' females that have paired with the arriving female's old mate. A similar pattern of mating appears to occur in Macaroni Penguins (Williams and Rodwell 1992) and may be widespread in penguins because in many species:

(1) nest site fidelity is higher in males than in females;

(2) following divorce, males retain the old nest site, even though they may be unsuccessful in remating for several years; and

(3) females that re-pair nest only a short distance from their old site (e.g. Richdale 1957; Trivelpiece and Trivelpiece 1990; Williams and Rodwell 1992).

Implicit in Davis and Speirs' (1990) model is the assumption that both sexes should aim to reunite with their old mate if at all possible, in order to enhance breeding success through increased familiarity of the pair; benefits (and costs) of mate fidelity are considered further below.

Extra-pair behaviour

As described above, penguins have generally been considered to be typically socially monogamous within each breeding season: forming a pair bond, and mating, and rearing offspring with only a single partner. However, in many species the sex ratio is markedly male-biased within breeding colonies (Richdale 1957; Davis and Speirs 1990). For example, in Macaroni Penguins at South Georgia there are many 'surplus' or unpaired males in the colony during the courtship period and many of these hold territories among the paired birds. If paired males are temporarily removed from nest sites during this period, neighbouring, unpaired males will move on to the vacated nest site, initiate sexual displays with the removed bird's female partner and defend the new territory or partner against other males (T. D. Williams unpublished data). In these situations, if males arrive at the nest later than their female partners, or if they do not guard the female, then there is the possibility that females might copulate with one or more of the unpaired, neighbouring males, that is, female penguins may have multiple 'partners' within a single season. Recent studies have shown that such extra-pair behaviour does occur in a number of penguin species, and that this results in extra-pair copulations and, probably, extra-pair fertilizations. In the only study so far to have used DNA-fingerprinting to estimate extra-pair paternity, chicks in 12 of 13 Royal Penguin *Eudyptes schlegeli* broods were fathered by the male who paried with the female and who subsequently helped to rear the chicks. Only in one brood was there the possibility that an extra-pair fertilization was involved, equivalent to a rate of extra-pair paternity of 7% (C. C. St. Clair personal communication). Davis and Speirs (1990) showed that 27% of female and 28% of male Adelie Penguins had more than one partner in a single season, and suggested that multiple matings by females must be considered normal reproductive behaviour in this species. More recent work on Adelie Penguins has shown that most multiple matings (90%) occur as a result of within-season mate switching (*sensu* Birkhead and Møller 1992), females copulating sequentially with successive male 'partners' (due to the late arrival of their previous year's mate), rather than obtaining 'true' extra-pair copulations (F. Hunter personal communication). Females also copulate most frequently, and always copulate last prior to laying, with the male who subsequently contributes to chick rearing. This would suggest that although extra-pair copulations might be common, the frequency of extra-pair fertilizations might be low, especially if last-male sperm precedence determines paternity (Birkhead *et al.* 1988; F. Hunter personal communication). At South Georgia, copulation behaviour in both Macaroni and Gentoo Penguins was only ever recorded on the nest site following pair formation and no copulations were observed between birds other than the male and female who subsequently reared the resulting chicks (Macaroni, n = 45 copulations, Gentoo, n = 56 copulations; T. D. Williams unpublished data). The constant presence of male partners at the nest site between arrival and incubation, and the generally earlier arrival of the male within pairs, would be predicted to reduce the opportunity for extra-pair copulations by other males in many penguin species.

Mate choice characteristics

The process of mate choice has been little studied in penguins, partly because pair bonds have been considered to be typically long-lasting, birds choosing a new partner only rarely, and partly because there appear to be few obvious differences in size or plumage characteristics within either sex that might be used to assess mate quality. Few studies have considered the relative characteristics of the two birds within pairs and the effect this aspect of mate choice might have on breeding success. Richdale (1957) showed that young female Yellow-eyed Penguins tended to mate with older males and that old females mated more often with young males. However, this may simply be a consequence of the heavily male-biased sex ratio that occurs in this species, particularly among older birds (2 males:1 female in birds ≥ 10 years of age), rather than representing active female choice. Ainley *et al.* (1983), in contrast, showed that Adelie Penguins most often paired with birds of, or close to, their own age. In pairs where there was a disparity in the age of the two birds there was little evidence to suggest that this has a detrimental effect on breeding effort or breeding success (Tables 8.5 and 8.6 in Ainley *et al.* 1983). Similarly, in the Little Penguin birds tended to pair with an individual of similar 'quality' as themselves (defined as length of residence), but fledging success did not vary with residence time (Reilly and Cullen 1981). Finally, in Adelie Penguins, females select for large males (either directly or indirectly) when choosing a mate, but although these larger males breed earlier they again do not have higher reproductive success (Davis and Speirs 1990). There is, thus, little evidence to date that penguins can enhance their probability of breeding successfully by choosing mates based on these 'absolute' measures of quality (although they may still benefit by increasing the 'relative' quality or compatibility of the pair as a whole, see below).

Results and discussion

The pair bond

No mating system other than social monogamy, that is, a single male and female attending eggs\chicks at one nest, has been recorded for any penguin species in the wild (polygynous 'trios' are common in Emperor Penguins but these form only temporarily early in the courtship period, Prevost 1961). Mate fidelity averages 71.8% in penguins (12 species, 17 studies, Table 15.1), that is 28.2% of birds change mates even though their previous year's partner is still alive. However, this varies markedly between different species. About 90% of birds retained the same mate between years, when both partners remained alive, in Macaroni, African,

Magellanic *Spheniscus magellanicus*, Galapagos, and some Gentoo and Little Penguin populations. In Adelie Penguins and northern populations of Gentoo Penguins between 40 and 60% of birds retained the same mate, and only in the two *Aptenodytes* species was mate fidelity very low: King and Emperor Penguins (29% and 15%, respectively), most birds breeding with a new partner each year (see Table 15.1 for references). Mate fidelity also varies within species between different breeding locations, for example, between 49% and 84% in Adelie Penguins (Table 15.1; Trivelpiece and Trivelpiece 1990), and in the same population from year to year. At South Georgia in 1988, 97% of Gentoo Penguin pairs with both partners alive reunited (n = 134), whereas in the previous year (1987) no pairs remained together (n = 13). Over the same period and at the same location, the rate of mate fidelity in Macaroni Penguins remained more or less constant: 86% (n = 21) and 91% (n = 132) in the two years (Williams and Rodwell 1992). Similarly, in his long-term study of Yellow-eyed Penguins, Richdale (1957) showed that mate fidelity varied between 63% and 94% in different years.

Pair durations of up to 11 years have been recorded in Little Penguins (Reilly and Cullen 1981), and up to 10–13 years in Yellow-eyed Penguins (Richdale 1957). However, despite the relatively high rate of mate fidelity, such very long-term pair durations are probably the exception in penguins with most pair bonds being maintained for 3–4 years or less. In Yellow-eyed Penguins pair duration averaged 2.1 years (SD 1.8, 1–13 years, n = 271, Richdale 1957), with 27%, 61%, and 12% of pairs lasting 1, 2–6, and 7–13 years, respectively. Ainley *et al.* (1983) reported that no Adelie Penguin bred with the same mate for more than 4 years, and birds breeding for 3 years (n = 27) had on average one (22%), two (44%), or three (33%) partners. Similarly, in Gentoo Penguins no bird bred with the same partner for more than 3 successive years (n = 13), although in Macaroni Penguins 6 of 28 birds (21%) bred with the same mate over a 4-year period; over 3 years, 36% (n = 99) and 52% (n = 124) of Gentoo and Macaroni Penguin, respectively, retained the same partner (Williams and Rodwell 1992). The reason for these relatively short pair durations, despite very high mate fidelity, is that the main factor leading to mate change in penguins is death (or disappearance) of the bird's previous partner, with divorce only a secondary factor (Fig. 15.1). Pair fidelity varies from 23% in the King Penguin (Barrat 1976) to 83% in the Little Penguin (Reilly and Cullen 1981), and divorce accounted for only 18% (Little Penguin, Reilly and Cullen 1981) to 39% (Macaroni Penguin, Williams and Rodwell 1992) of all mate change.

Mechanisms of divorce and mate fidelity

Divorce may be a 'passive' or nonfunctional event occurring, for example, as a consequence of asynchronous return to the colony, low nest site fidelity or high annual mortality (Rowley 1983), or it may represent an

Table 15.1 Summary of mate and pair fidelity data for penguins (Spheniscidae)

Species	Mate fidelity		Pair fidelity		Nest site fidelity[1]	Mig/Res	Association	Latitude	No. years	Source[2]
	%	n^3	%	n^3	%		%			
King	29	(42)	23	(26)	83	PM	6	46°	1	(1)
	—		9	(11)	—	PM	—	54°	1	(2)
Emperor	15	(81)	—		0	M	17	66°	2	(3)
Gentoo	49	(136)	—		—	R	58	46°	3	(4)
	90	(490)	78	(285)	89–100	R	91	54°	3	(5)
	90	(752)	—		63/60	PM	89	62°	4	(6)
Adelie	62	(916)	—		99/65	M	43	62°	4	(6)
	84	(330)	60	(554)	—	M	—	72°	1	(7)
	56	(56)	—		75	M	—	77°	1	(8)
	49	(256)	—		78	M	—	77°	14	(9)
Chinstrap	82	(804)	—		94	M	53	62°	4	(6)
Rockhopper	79	(134)	59	(90)	++	M	62	53°	1	(10)
	—		62	(25)	++	M	—	54°	2	(11)
Macaroni	91	(305)	75	(367)	69–87	M	54	54°	4	(5)
Yellow-eyed	82	(1078)	63	(737)	30	R	68	45°	12	(12)
Little	97	(238)	83	(139)	68/64	R	70	38°	11	(13)
	—		69	(71)	74	R	—	32°	4	(14)
African	86	(232)	62	(161)	60	R	79	33°	6	(15)
Magellanic	90	(—)	—		80/70	M	35	44°	10	(16)
Galapagos	89	(—)	—		97/98	R	81	0°	1	(17)

++ = 'high'; M = migratory; PM = partly migratory; R = resident. See text for details of index of association, and definition of other terms.
[1]89–100 = range of more than 1 year; 89/100 = value for male/female.
[2](1) Barrat (1976); (2) Stonehouse (1960); (3) Isenmann (1971); (4) Bost and Jouventin (1991); (5) Williams and Rodwell (1992); (6) Trivelpiece and Trivelpiece (1990); (7) Penney (1968); (8) Davis (1988); (9) Ainley *et al.* (1983); (10) P. J. Moors and D. M. Cunningham in Marchant and Higgins (1990); (11) Warham (1963); (12) Richdale (1957), Darby and Seddon (1990); (13) Reilly and Cullen (1981); (14) Wienecke and Wooller (in press); (15) LaCock *et al.* (1987); (16) P. D. Boersma and G. Fowler (personal communication); (17) Boersma (1976).
[3]n = number of birds for mate fidelity, and number of pairs for pair fidelity; sample sizes not given in original paper or unavailable for Galapagos or Magellanic Penguin.

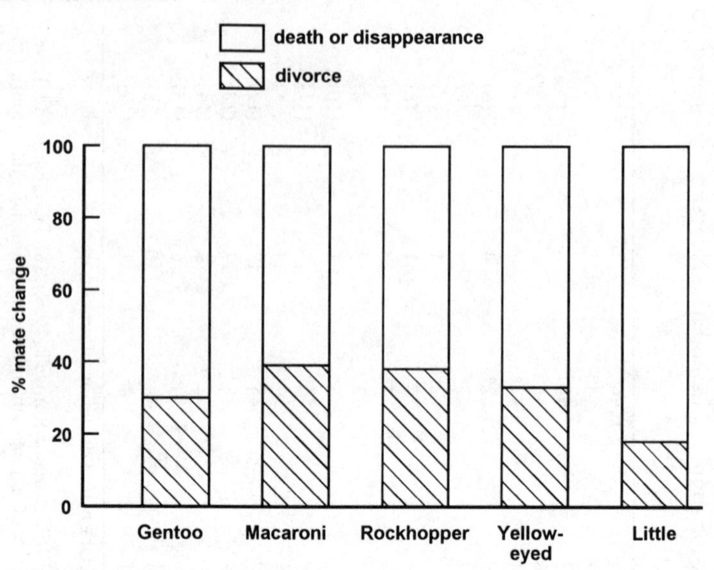

Fig. 15.1 Causes of mate change in Gentoo (n = 65 pairs), Macaroni (n = 99), Rockhopper (n = 37), Yellow-eyed (n = 295), and Little (n = 24) Penguins; data from Richdale (1957), Reilly and Culen (1981), Bost and Jouventin (1991), Williams and Rodwell (1992), and P. J. Moors and D. M. Cunningham (in Marchant and Higgins 1990).

'active' decision by an individual to mate with a different partner (these two mechanisms may not be totally independent, however, as nonfunctional changes such as death of a neighbour or rival may increase the opportunity for 'adaptive' mate change, e.g. the 'better option' hypothesis, Ens *et al.* 1993). Comparing patterns of mate fidelity between species I now assess the applicability of each of these ideas to penguins.

Asynchronous arrival at the colony

In migratory Adelie Penguins, Davis and Speirs (1990) showed that in pairs that changed mates, birds were more asynchronous in their arrival (within the pair) than pairs that reunited (Fig. 15.2). In a second study of Adelie Penguins, Ainley *et al.* (1983) showed that in all cases where mate change occurred the female arrived first before her male partner of the previous year. In contrast in Macaroni Penguins at South Georgia, which are also migratory but which breed at lower latitudes than Adelie Penguins, there was no difference in within-pair synchrony of arrival between birds that divorced and those that reunited (Fig. 15.2). Nor did female Macaroni Penguins that mated with a new partner arrive earlier, in relation to mean arrival date for the population, compared to females that reunited (0.9 days, n = 10 vs. –0.2 d, n = 33, $P > 0.20$). This may suggest that the importance of timing of arrival in determining mate fidelity varies with breeding latitude and with the time available for

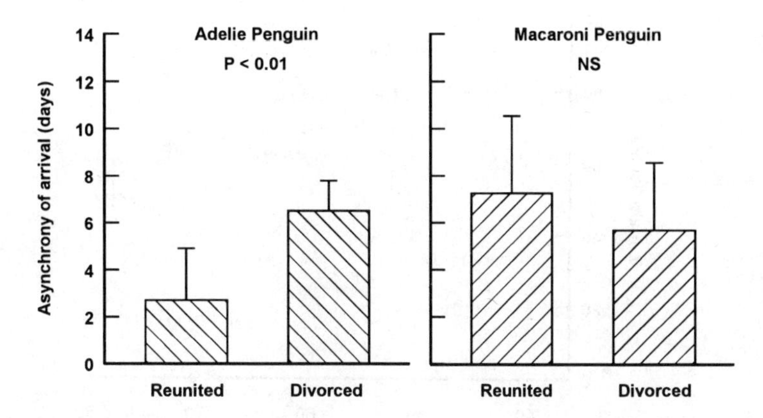

Fig. 15.2 Asynchrony of arrival at the colony at the beginning of the breeding season between the two members of a pair in pairs that subsequently reunited or divorced, in Adelie (n = 19 and 5) and Macaroni Penguins (n = 19 and 4); data from Davis and Speirs (1990) and T. D. Williams (unpublished data).

breeding (i.e. the potential length of the breeding season). At higher latitudes, with very short breeding seasons, there should be a premium on pairing quickly with a new mate, that is, divorcing, rather than waiting for a previous partner (Ainley *et al.* 1983; Davis 1988; Trivelpiece and Trivelpiece 1990). Although mate fidelity is negatively correlated with breeding latitude among species in migratory penguins (Spearman rank correlation, $r_s = -0.66$, $n = 6$; Fig. 15.3) this relationship is not significant ($P = 0.16$). Similarly, within-species mate fidelity is not linearly related to latitude at four different breeding locations in the Adelie Penguin (Trivelpiece and Trivelpiece 1990; see Fig. 15.3). One reason for this might be that shorter breeding seasons select for more synchronous arrival of the population as a whole, thus actually enhancing the probability that a pair will return to the colony over a short period and reunite before either bird has time to mate with a new partner (Williams and Rodwell 1992, cf. Ainley *et al.* 1983).

Nest site fidelity

Mate fidelity will be more likely to occur if nest site fidelity is high: even if both partners are alive they will only reunite if they both return to the same breeding site. Mate fidelity may therefore largely be a function of both mates returning to the same nest site (e.g. Morse and Kress 1984; Cuthbert 1985). Most penguins show a high level of nest site fidelity (60–100%, though as low as 30% in Yellow-eyed Penguins; Table 15.1). However, there is no overall relationship between mate fidelity and nest site fidelity, either including ($r_s = 0.07$, $n = 14$, $P > 0.20$) or excluding ($r_s = -0.16$, $n = 13$, $P > 0.20$) Emperor Penguins (which are anomalous in that they have no nest site or even a 'zone of attachment', cf. King

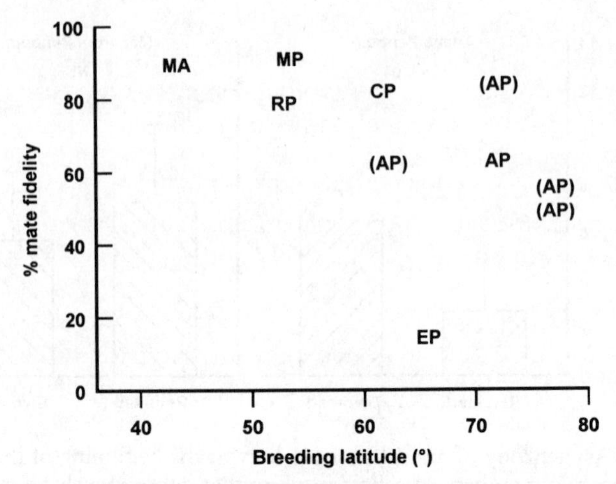

Fig. 15.3 Relationship between mate fidelity and breeding latitude in migratory penguin species; data are from references cited in Table 15.1. AP = Adelie, RP = Rockhopper, CP = Chinstrap, MA = Magellanic, MP = Macaroni and EP = Emperor Penguin. A mean value was used for the Adelie Penguin in the correlation analysis (see text); individual values for the four populations are indicated in parentheses.

Penguins, Barrat 1976). Only 30% of Yellow-eyed Penguins return to and use the previous year's nest site, yet mate fidelity is 82% in these species (Richdale 1957; Darby and Seddon 1990). Conversely, in three populations of Adelie Penguin, nest site fidelity averages 75–82% but mate fidelity is only 49–62% (Penney 1968; Davis 1988; Trivelpiece and Trivelpiece 1990). Furthermore, in Little Penguins which have a high rate of mate fidelity, Reilly and Cullen (1981) showed that on average 50% (n = 113) of birds that retained the same mate were likely to breed in a nesting burrow different from that of the previous year. Thus, although it is obviously important for both members of a pair to return to the same colony or general breeding area, return to a specific nest site does not appear to be a prerequisite for mate fidelity in penguins.

Annual mortality rates

Waiting for a previous partner to return may not be a viable strategy if the rate of adult mortality is high, because birds may subsequently be unable to obtain a new mate in time to breed that year and therefore risk not breeding at all (Ainley *et al.* 1983; Trivelpiece and Trivelpiece 1990; see also below). For this reason mate change should be more common in species with higher annual mortality rates. The *better option hypothesis* also predicts a positive correlation between mate change and mortality rate because of the increased possibility of changing to a better mate without having to incur costs of fighting and eviction of a previous partner (Ens *et al.* 1993). Estimates of annual adult mortality are only available for a few penguin species but these indicate that there is no overall relationship between mate fidelity and mortality. Annual adult mortality is lowest in the *Aptenodytes* penguins, averaging 5–10%, but these species show extremely low mate fidelity (King Penguin 29%, Emperor Penguin 15%; Isenmann 1971; Mougin and van Beveren 1979; Weimerskirch *et al.* 1992). Conversely, the Little Penguin has relatively high annual adult mortality (25%, Dann and Cullen 1990) but a very high rate of mate fidelity (97%, Reilly and Cullen 1981).

The incompatibility hypothesis

Coulson (1966, 1972) suggested that although reuniting with the same mate might carry advantages (in terms of enhanced breeding success), divorce could also be adaptive if this occurred among 'incompatible' pairs. Davis (1982, 1988) presented evidence in support of this *incompatibility hypothesis*, and also a mechanism through which it could operate, in Adelie Penguins. Adelie Penguins, like several other penguin species, have a complex incubation routine, the male and female alternating prolonged incubation shifts with long foraging trips at sea. Incubation shifts can last 10–15 days and failure of pairs to co-ordinate the incubation routine and nest reliefs successfully is a major cause of nest failure in Adelie Penguins (Davis 1982). Davis (1988) suggested that unsuccessful pairs are incompatible or noncomplementary, and that these pairs would show greater asynchrony of arrival (within the pair) the following year, compared to 'complementary' pairs that successfully co-ordinated their incubation routine. Unsuccessful pairs would therefore be more likely to divorce than would successful pairs; this is a general prediction of the *incompatibility hypothesis* (Coulson 1972; Rowley 1983). Unsuccessful breeders should continue to mate with a new individual each year (divorce) until they find a 'compatible' partner (Davis 1988). In penguins, the probability of having bred successfully the previous year was lower among birds that subsequently divorced, but only in species with more complex incubation routines and prolonged incubation shifts (e.g. Macaroni Penguin, 26% vs 61% n = 293, Williams and Rodwell 1992; see also Davis 1988, for Adelie Penguin). In species that have more regular change-overs and shorter incubation shifts (1–2 days) the probability of divorce was unrelated to breeding success the previous year (Little Penguin, 54% vs. 62%, n = 139, Reilly and Cullen 1981; Gentoo Penguin, 74% vs. 68%, n = 304, Williams and Rodwell 1992, and 76% vs. 71%, n = 42, Bost and Jouventin 1991). This supports Davis' (1988) interpretation of the *incompatibility hypothesis* because in species with daily or frequent change-overs it is presumably less important for birds to be 'compatible' or to co-ordinate nest reliefs. The benefit(s) of switching to a potentially better mate will therefore be less, relative to any costs that might be incurred through divorce (though it should be noted that none of these studies controlled for the effect of age).

Pair bond reinforcement

Rowley (1983) suggested that the likelihood of reuniting with the same mate might be dependent on the degree of reinforcement of the pair bond throughout the year (the *keeping company hypothesis*; Black Chapter 1), which in turn would relate to the species' 'lifestyle', for example, whether it was resident or migratory. As discussed above, migratory penguin species do not maintain pair bonds during the non-breeding period whereas resident species at least have the potential to spend time together during winter (in pairs) and thus reinforce the pair bond. Mate fidelity is on average higher in resident penguin species or populations (81.7%, n = 6) compared to migratory species or populations (67.6% n = 9), although this difference is not significant (Mann-Whitney, $P > 0.10$), mate fidelity varying widely between 49% and 97% in resident species and between 15% and 91% in migratory species (Table 15.1). As Williams and Rodwell (1992) pointed out this prediction may, in any case, not be valid. At South Georgia, the Macaroni Penguin is migratory but arrival at the colony is highly synchronous increasing the chance for pair members to reunite over a short period of time. Additionally in migratory species, birds remain in the colony fasting between arrival and egglaying so individuals are less likely to fail to meet up and when they do they will remain together throughout the day maximizing reinforcement of the pair bond. In contrast, the resident Gentoo Penguin has a protracted prebreeding period with both birds continuing to make diurnal foraging trips to sea (during which pairs do not remain together) until clutch initiation. As these foraging trips are not synchronized, this will reduce the amount of time the pair spend together at the nest and may also increase the probability that one or other bird may mate with a new partner.

During the breeding season, prolonged or constant reinforcement of the pair bond can occur only during the courtship period, particularly in species that remain on land fasting between arrival and laying (e.g. *Eudyptes* spp., Adelie, Chinstrap *Pygoscelis antarctica* and Magellanic Penguins), or during moult, in species where pairs moult together at the nest site (e.g. *Eudyptes* spp., Little Penguin). At all other times one or both members of the pair undertake diurnal, or more prolonged foraging trips at sea, and opportunities for pair bond reinforcement are restricted to nest reliefs during change-overs in incubation or brooding, or when feeding visits by both parents coincide during chick rearing. Using information on the timing and pattern of the breeding cycle for different species I have calculated an index of association,[1] defined as the number

[1]In so doing I have made a number of assumptions: (1) during incubation, if shifts last ≥ 2 days birds only meet during nest reliefs and for 1 day (Yellow-eyed, Adelie, *Eudyptes spp.*) or 2 days (Aptenodytes spp.) each change-over; otherwise, for incubation shifts ≤ 1 day, birds meet each day; (2) if chick feeding frequency is > 1 visit/day the pair meet at

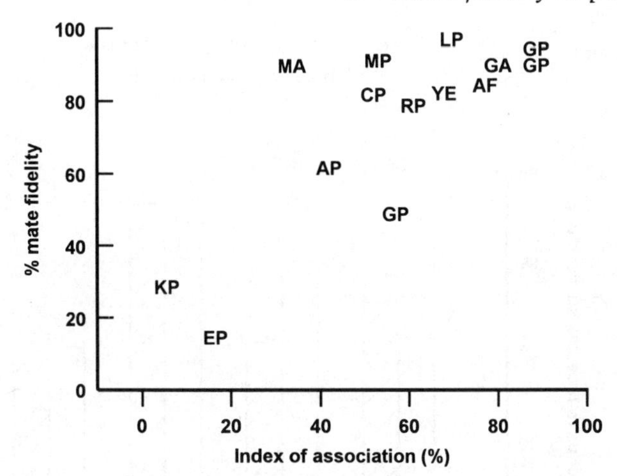

Fig. 15.4 Relationship between mate fidelity and index of association in penguins (see text for details of terms); data are from references cited in Table 15.1. Abbreviations as for Fig. 15.3, and KP = King, GP = Gentoo, LP = Little, MA = Magellanic, MP = Macaroni, AF = African, and GA = Galapagos Penguins.

of days pairs spend together at the nest (for at least part of the day) divided by the total number of days required for breeding and moulting.

Using this index of association, the proportion of the total breeding cycle that pairs spend together varies markedly between species, ranging from 6% and 17%, in King and Emperor Penguins respectively, to 89% and 91% in two populations of Gentoo Penguins (Table 15.1). Consistent with Rowley's (1983) *keeping company hypothesis*, mate fidelity is significantly positively correlated with this index of association (r_s = 0.57, n = 14, P = 0.034; Fig. 15.4). This relationship between index of association and mate fidelity also provides another interpretation for the increased propensity of failed breeders to change mates compared to successful breeders (see above). Pairs that fail early in the breeding attempt may only visit the colony infrequently and/or irregularly during the remainder of the season such that there will be little opportunity for pair bond reinforcement (Davis 1988).

Consequences of mate change

Evidence for improved breeding success in relation to mate retention and increasing pair duration is equivocal in penguins. Several studies, on

---

the nest each day (e.g. Gentoo, Little, Galapagos), if it is < 1 visit/day pairs meet on 50% of days (e.g. Chinstrap, Adelie), except for *Eudyptes* spp. where birds do not meet for the first 10 days of the crèche period (because the male remains at sea during this time); (3) no pairs are together during the premoult period; (4) King, Emperor, Gentoo, Yellow-eyed, and *Spheniscus* spp. do not moult together, Little Penguins are together for 50% of moult (because moult is asynchronous within pairs in this species), otherwise pairs are together for the whole duration of moult.

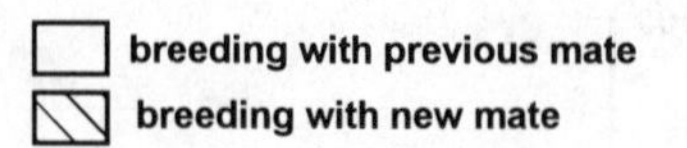

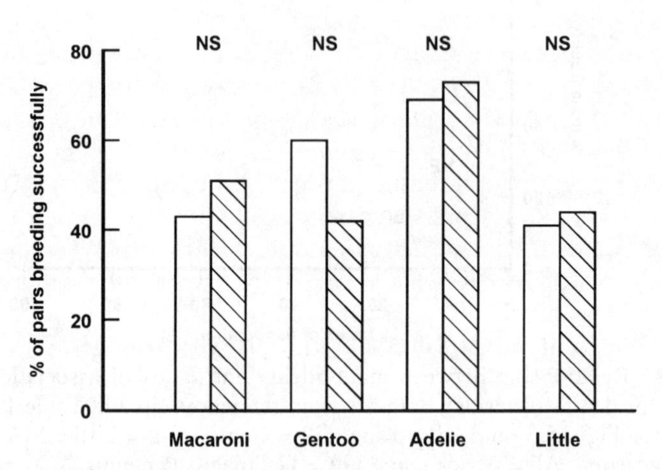

Fig. 15.5 Breeding success (percentage of pairs rearing one chick to fledging) in birds reuniting with their previous partner and those breeding with a new mate in Macaroni (n = 111 and 13), Gentoo (n = 101, 41), Adelie (n = 56 and 161), and Little (n = 115, 24) Penguins; data from Reilly and Cullen (1981), Ainley *et al.* (1983), Williams and Rodwell (1992).

Adelie, Macaroni, and Gentoo Penguins, found no difference in laying date, clutch size, or breeding success between newly mated pairs that bred in the year following mate change and reunited pairs (Fig. 15.5; Spurr 1975; Ainley *et al.* 1983; Bost and Jouventin 1990; Williams and Rodwell 1992). Reilly and Cullen (1981) similarly showed that breeding success of individual Little Penguins was not different in the year before and after mate change, and that females breeding with a new mate actually nested earlier than in the previous year with their former mate (0.78 months earlier, n = 41). However, in Adelie Penguins Penney (1968) showed that even in birds that re-paired with a new mate 27% failed to produce eggs, compared to only 2% of birds that retained the same mate, and that this did lead to an overall lower breeding productivity in birds that changed mates. Evidence for a 'long-term' effect of mate fidelity on reproductive success is also equivocal, or nonexistent, in penguins. Wienecke and Wooller (in press) showed that the total number of chicks hatched was positively related to pair duration in Little Penguins, but the slope of the line (b = 1.96) is no different from a value of two that would be predicted given a two-egg clutch and an increase of one additional breeding attempt per year. Ainley *et al.* (1983) found no overall difference in breeding success in relation to pair duration in their long-term study of Adelie Penguins (considering all pair durations). However, birds breeding for their third season (or more) with the same mate were successful more often (83%) than birds breeding in their third season with a new mate (58%). There is, therefore, little evidence for any significant short- or long-term cost of mate change in penguins in terms of breeding success.

Loss of breeding status, on the other hand, does appear to represent a significant cost of mate change: birds that lose their previous year's mate (either through divorce or the death or disappearance of their partner) often fail to re-pair the following season and miss one or more breeding opportunities. In Macaroni Penguins at South Georgia, 39% of males (n = 18) that divorced remained unmated the following year, and did not breed again for 2–4 years (even though they retained their old nest site). The likelihood of loss of breeding status varies between species and in relation to sex, however, as only one of 18 female Macaroni Penguins (6%) failed to breed the year after mate change. In Gentoo Penguins at South Georgia all males (100%) bred in the year following divorce whereas only 64% of females did so (n = 14 pairs, Williams and Rodwell 1992).

Summary

Mate fidelity, when both birds remained alive in a subsequent season, averaged 72% (29–97%) in 12 different penguin species. Nevertheless, divorce was only a secondary factor in mate change, accounting for 13–39% of all mate change, 61–89% being due to the death or disappearance of a previous mate; overall, pair fidelity averaged 58% (23–83%). Variation in mate fidelity between species appears to be related to: (1) within-pair synchrony of arrival at the breeding colony at the start of the season in species with complex incubation routines, that is, pair 'complementarity' or compatibility: pairs arriving asynchronously are more likely to divorce; and (2) opportunities for pair bond reinforcement during the breeding attempt, defined as the proportion of the breeding cycle that the pair are together at the nest: species in which pairs spend a lot of time together at the nest during breeding have higher rates of mate fidelity. In contrast, variation in mate fidelity appears to be unrelated to the level of annual adult mortality, or to breeding latitude (length of the breeding season); short breeding seasons at high latitudes may select for highly synchronous return to the colony, as well as for rapid onset of breeding, thus maximizing the chance of birds reuniting with their old mate before mating with a new partner. A high rate of nest site fidelity is not a prerequisite for mate fidelity in penguins. There is little evidence in penguins for a cost of divorce in terms of reduced breeding success among birds that attempt to breed following mate change. However, divorce can lead to loss of breeding status for one or more years following mate change. It remains unclear why mate fidelity is so prevalent in penguins. This is partly due to the paucity of the data currently available: more detailed data on variation in mate fidelity and related factors (synchrony of arrival, annual mortality, nest site fidelity, costs of mate change) are still required for most species.

Acknowledgements

Dee Boersma, Colleen Cassidy St. Clair, Fiona Hunter, Barbara Wienecke, and Ron Wooller kindly allowed me to include their unpublished data in this review, and John Croxall and Dirk Briggs provided me with unpublished British Antarctic Survey data. Jeff Black, Bruno Ens, and Simon Pickering provided helpful comments on earlier drafts.

References

Ainley, D. G., LeResche, R. E., and Sladen, W. J. L. (1983). *Breeding biology of the Adelie penguin*. University of California Press, Berkeley.

Barrat, A. (1976). Quelques aspects de la biologie et de l'ecologie du Manchot Royal (*Aptenodytes patagonicus*) des Iles Crozet. *Comité nacional Francaise recherche Antarctique*, **40**, 9–51.

Birkhead, T. R. and Møller, A. P. (1992). *Sperm competition in birds*. Academic Press, London.

Birkhead, T. R., Pellatt, J. E., and Hunter, F. M. (1988). Extra-pair copulation and sperm competition in the zebra finch. *Nature*, **334**, 60–2.

Boersma, P. D. (1976). An ecological and behavioural study of the Galapagos Penguin. *Living Bird,* **15**, 43–93.

Bost, C. A. and Jouventin, P. (1990). Laying asynchrony in Gentoo Penguins on Crozet Island: causes and consequences. *Ornis Scandinavica*, **21**, 63–70.

Bost, C. A. and Jouventin, P. (1991). The breeding performance of the Gentoo Penguin *Pygoscelis papua* at the northern edge of its range. *Ibis*, **133**, 14–25.

Coulson, J. C. (1966). The influence of the pair-bond and age on the breeding biology of the kittiwake gull (*Rissa tridactyla*). *Journal of Animal Ecology*, **35**, 269–79.

Coulson, J. C. (1972). The significance of the pair-bond in the kittiwake. *Proceedings International Ornithological Congress*, **15**, 424–33.

Cuthbert, F. J. (1985). Mate retention in Caspian Terns. *Condor*, **87**, 74–8.

Dann, P. and Cullen, J. M. (1990). Survival, patterns of reproduction and lifetime reproductive output in Little Penguins (*Eudyptula minor*) on Phillip Island, Victoria, Australia. In *Penguin biology* (ed. L. S. Davis and J. T. Darby), pp. 63–84. Academic Press, San Diego.

Darby, J. T. and Seddon, P. J. (1990). Breeding biology of Yellow-eyed Penguins (*Megadyptes antipodes*). In *Penguin biology* (ed. L. S. Davis and J. T. Darby), pp. 45–62. Academic Press, San Diego.

Davis, L. S. (1982). Timing of nest relief and its effect on breeding success in Adelie Penguins (*Pygoscelis adeliae*). *Condor*, **84**, 178–83.

Davis, L. S. (1988). Coordination of incubation routines and mate choice in Adelie Penguins (*Pygoscelis adeliae*). *Auk*, **105**, 428–32.

Davis, L. S. and Speirs, E. A. H. (1990). Mate choice in penguins. In *Penguin biology* (ed. L. S. Davis and J. T. Darby), pp. 377–97. Academic Press, San Diego.

Ens, B. J., Safriel, U. N., and Harris, M. P. (1993). Divorce in the long-lived and monogamous oystercatcher *Haematopus ostralegus*: incompatibility or choosing the better option? *Animal Behaviour*, **45**, 1199–217.

Isenmann, P. (1971). Contribution a l'ethologie et a l'ecologie du Manchot empereur (*Aptenodytes forsteri*) a la colonie de Pointe Geologie (Terre Adelie). *L'Oiseau et Recherche Francais Ornithologie*, **41,** 9–64.

LaCock, G. D., Duffy, D. C., and Cooper, J. (1987). Population dynamics of the African penguin *Spheniscus demersus* at Marcus Island in the Benguela upwelling ecosystem. *Biological Conservation*, **40,** 117–26.

Marchant, S. and Higgins, P. J. (1990). *Handbook of Australian, New Zealand and Antarctic Birds*, Vol. 1A. Oxford University Press, Melbourne.

Morse, D. H. and Kress, S. W. (1984). The effect of burrow loss on mate choice in the Leach's Storm Petrel. *Auk*, **101,** 158–60.

Mougin, J.-L. and van Beveren, M. (1979). Structure et dynamique de la population de Manchot empereur *Aptenodytes forsteri* de l'archipel de Point Geologie, Terre Adelie. *Comptes Rendus Academe Science (Paris)*, **298D,** 157–60.

Penney, R. L. (1968). Territory and social behaviour in the Adelie Penguin. *Antarctic Research Series*, **12,** 83–131.

Prevost, J. (1961). *Ecologie du manchot empereur*. Hermann, Paris.

Reilly, P. N. and Cullen, J. M. (1981). The Little Penguin *Eudyptula minor* in Victoria, II. Breeding. *Emu*, **81,** 1–19.

Richdale, L. E. (1957). *A population study of penguins*. Oxford University Press, Oxford.

Rowley, I. (1983). Re-mating in birds. In *Mate Choice* (ed. P. Bateson), pp. 331–60. Cambridge University Press, Cambridge.

Spurr, E. B. (1975). Breeding of the Adelie Penguin (*Pygoscelis adeliae*) at Cape Bird. *Ibis*, **117,** 324–38.

Stonehouse, B. (1960). The King Penguin (*Aptenodytes patagonica*) of South Georgia, Part 1. *Falkland Islands Dependency Survey Report*, **17,** 1–97.

Trivelpiece, W. Z. and Trivelpiece, S. G. (1990). Courtship period of Adelie, Gentoo and Chinstrap Penguins. In *Penguin biology* (ed. L. S. Davis and J. T. Darby), pp. 113–27. Academic Press, San Diego.

Warham, J. (1963). The Rockhopper Penguin *Eudyptes chrysocome*. *Auk*, **80,** 229–56.

Weimerskirch, H., Stahl, J. C., and Jouventin, P. (1992). The breeding biology and population dynamics of King Penguins *Aptenodytes patagonica* on the Crozet Islands. *Ibis*, **134,** 107–17.

Wienecke, B. and Wooller, R. D. Mate and site fidelity in Little Penguins *Eudyptula minor* at Penguin Island, Western Australia. *Emu*. (In press.)

Williams, T. D. and Croxall, J. P. (1991). Annual variation in breeding biology of macaroni penguins *Eudyptes chrysolophus*, at Bird Island, South Georgia. *Journal of Zoology (London)*, **223,** 189–202.

Williams, T. D. and Rodwell, S. (1992). Annual variation in return rate, mate and nest-site fidelity in breeding gentoo and Macaroni Penguins. *Condor*, **94,** 636–45.

16 *Causes and consequences of mate fidelity in Red-billed Gulls*

JAMES A. MILLS, JOHN W. YARRALL, AND DEBORAH A. MILLS

Introduction

The Red-billed Gull *Larus novaehollandiae scopulinus* is a small coastal breeding gull, which nests colonially in high numbers at high densities. Red-billed gulls are highly philopatric, and usually pair with the same partners each successive breeding season. In the autumn and winter months most adults and young disperse from the breeding colony, but pair bonds are seldom maintained outside of the breeding season. New pairs are formed either at nest sites defended by unattached males, at sites within or adjacent to the colony where unemployed birds gather temporarily. Pairs reunite at the nest site they occupied the previous season, but may not necessarily breed at that site. This study, based on data collected between 1964 and 1993 in a population of known age individuals, examines the causes and consequences of mate change and the advantages of retaining the same partner for a number of breeding seasons. We draw on a sample of 4062 colour-marked individuals that were involved in 34 459 breeding attempts. We also explore the relationship between courtship feeding by males and extra-pair behaviour in their mates. The Red-billed Gull population is rather unique in that there is a large excess of females and this induces major differences in the demography of the sexes.

Background, study sites, and procedures

The study was carried out at Kaikoura Peninsula, New Zealand. Typically 5000–7000 pairs nest in up to seven colonies on the 5 km headland of the Peninsula. Nestlings have been banded annually since 1958 and in total over 90 000 have been marked. Each season all nests of colour-marked gulls were numbered and checked at least every 2 or 3 days to determine laying date, clutch size, and hatching and fledging success. Nestlings from marked nests were usually banded within 4 days of hatching. Very few colour-marked gulls bred without being observed and special searches were made to identify those that were in the Kaikoura region but did not breed. Adults were sexed by a combination of two bill measurements, depth at gonys and bill length (Mills 1971).

In New Zealand, the Red-billed Gull is single brooded, but replacement clutches are laid. At Kaikoura, egg laying extends from late September to early January. Two-egg clutches are the most frequent (77%), but three-egg (9%) and single-egg (14%) clutches are fairly common. Pairs have difficulty in raising three chicks; of those fledging chicks, only 0.4% fledge all three. Both parents share approximately equally in nest building, incubation, and feeding the young, but males have the additional duty of courtship feeding the female (Tasker and Mills 1981).

Red-billed Gulls are potentially long-lived, up to 30 years of age. The annual survival rate of females (mean 0.866, SE 0.003) is significantly higher than that of males (mean 0.846, SE 0.004). This differential survival rate, coupled with higher survival of female nestlings in some seasons, results in an excess of females in the population, which produces major socially induced differences in the demography of the sexes. As a result, males tend to breed more frequently and commence breeding at an earlier age (1–4 years) than females (2–6 years) (Mills 1989, 1991). In a typical season, only 52% of the available adult females breed, compared to 86% of males (Mills 1989). The limited number of males induces the formation of female–female pairings.

Predation is the major cause of egg and nestling failure, accounting for on average 25% of egg and 17% of chick losses (Mills 1989). Predators include Black-backed Gulls *Larus dominicanus*, neighbouring Red-billed Gulls, and introduced Stoats *Mustelia erminea*, Hedgehogs *Erinaceus europaeus*, Cats *Felis catus*, and Ferrets *Putorius furo*.

There is little emigration or immigration of adults and virtually all of the young that survive return to breed (Mills 1973). Whilst most adults disperse from Kaikoura in the nonbreeding period, virtually all return in the breeding season, even as nonbreeders. Pair members were identified from multiple resightings at the nest site. For the purposes of determining whether a pair bond ended due to the death of a partner, the mate had to have been recovered dead, or if colour-marked, was not seen for two seasons. If assumed dead, the death was considered to have occurred during the first year of its disappearance. In the case of divorce, both members of the pair had to be known to be alive but not breeding together. For this paper, the timing of the mate change was assumed to have occurred at the end of the season the pair last bred together. Thus if individuals of a pair did not breed the season after the last time they were seen together, and one or both bred with a different partner the next season after that, the pair was assumed to have split up 2 years before breeding again. Confounding variables such as age and year effects were controlled in a multiple logistic regression process. In less sophisticated analyses breeding data from the 1987–1988 and 1989–1990 breeding seasons were excluded from the data set as these were 'outlier' years due to poor food availability. Age was controlled by including only birds between the ages of 5 and 15 years when reproductive success shows little variation.

Results

The pair bond

Social monogamy was the most common mating strategy in the Red-billed Gull; approximately 94% were female–male and 6% female–female pairs. On occasions two females and one male were observed to be associated at the same nest, but this was very rare making up considerably less than 1% of all pairings. In this paper only male–female pairings are considered.

On average, 68% of pairs had partners with an age difference of only 1 year or less, and 26% of these had mates of the same age (Mills 1973). In 57% of all pairs the female was the older partner (n = 212).

The number of partners an individual had during its lifetime varied from one to nine. There was no significant difference in the mean number of partners of males (1.89, n = 875) or females (1.89, n = 1114). Fifty-four per cent of females and 51% of males had only one partner, and overall, 75% of both sexes had no more than two partners in their lives. The average pair duration of birds ringed between 1967 and 1980 (n = 1365 pairs) was 2.5 years (SE 0.06). The longest known pair duration was 17 years, which terminated with the death of the female at 26 years of age, 1 year older than the male. Fifty per cent of the pairs lasted only 1 year, 18% 2 years, and 10% 3 years. In all, 15% lasted for 5 or more years and 2% for 10 or more years. The probability of divorce, when both members survived from one year to the next, was 10.5% (409 divorce events in 3903 pair-years).

Factors affecting mate change and divorce

Age and pair duration

On average, 83% of pairs (n = 9323) retained mates from one season to the next. Divorce was the most common cause of mate change causing

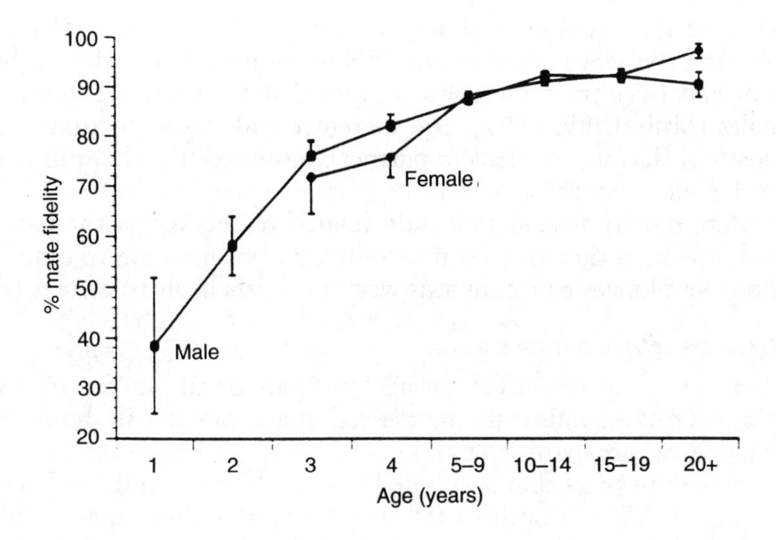

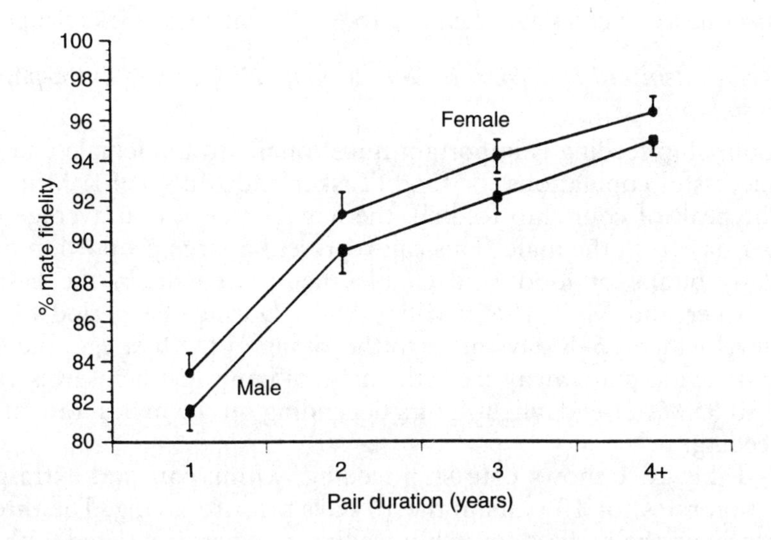

Fig. 16.1 The influence of age and pair duration on mate fidelity and divorce probability in Red-billed Gulls. After controlling for pair duration, the age of both the male (n = 4318) and female (n = 3339) influenced divorce rates; younger birds divorced more (GLIM logistic regression analysis, male $\Delta D = 54$, female $\Delta D = 15$, $\Delta df = 5$, $P < 0.001$, 0.02). Similarly, after controlling for age, pair duration influenced divorce rates; younger pairs divorced more (male $\Delta D = 64$, female $\Delta D = 77$, $\Delta df = 3$, both $P < 0.001$). The factors age and pair duration significantly interacted (male $\Delta D = 8$, female $\Delta D = 4$, $\Delta df = 1$, both $P < 0.05$). Sample sizes for each standardized value varied between 5 and 1681, mean = 634, SE 163.

61% of the break-ups of females and 73% of males. The differences between the sexes arose because there were more females in the population and over their life span they tended to breed less frequently than males (Mills 1989, 1991), and therefore had less opportunity to divorce. Death or disappearance of a partner accounted for 39% of mate changes for females and 27% for males.

Mate fidelity was significantly related to the age of the pair members and how long they had been together; pairs consisting of older birds and those with longer pair durations were much less likely to divorce (Fig. 16.1).

Previous reproductive success

Controlling for the effects of age and pair duration, previous reproductive success significantly influenced mate fidelity in both males and females, where pairs that failed to fledge young were more likely to divorce than pairs that succeeded (for 4359 males and 3382 females: $\Delta D = 5$ and 4, $\Delta df = 1$, both $P < 0.05$). The adjusted divorce probability values increased from 4.2% after two or three chicks were fledged, to 5.4% after just one chick was fledged, to 6.5% when no chicks fledged.

Relationship of courtship feeding to mate change and extra-pair copulations

Courtship feeding is important nutritionally to the female and she trades successful copulations for food (Tasker and Mills 1981; Mills 1994). At the peak of courtship feeding, the female receives on average four feeds per day from the male. This can represent a large proportion of her total daily intake of food, and enables her to remain in the nest territory (Tasker and Mills 1981; Mills 1994). During the period of rapid egg development, 5–8 days prior to the laying of the first egg, the amount of time females are away from the nest foraging for themselves varies from 6 to 95% of the daylight hours depending on the males' rate of courtship feeding.

Table 16.1 shows courtship feeding, within-pair and extra-pair copulation rates for 10 pairs in the 16 days prior to laying. The three females receiving the highest courtship feeding rate per h reunited with their previous mates, but two females that received low rates divorced the next season. Perhaps because females trade copulations for food, the success rate of within-pair copulations amongst poorly provisioned females was also low (Table 16.1). Since high copulation rates probably minimize the risk of extra-pair copulations (see Birkhead *et al.* 1987), males that did not feed their mates adequately were at greater risk of losing paternity.

Extra-pair copulation attempts were common, amounting to 21% of all copulation attempts by males (Mills 1994). The only female actively to solicit or engage in a successful extra-pair copulation during the fertile period was the 'Hide' female which had one of the lowest courtship feeding and within-pair copulation rates. Females that were well provisioned during courtship feeding actively resisted all extra-pair copulation

Table 16.1 Courtship feeding, within-pair (WPC) and extra-pair (EPC) copulation rates in the 16 days prior to laying and the fate of the pair bond the next year

Nest number	Hours of observation (h and min)	Number regurgitations observed	Number regurgitations per h	Number successful WPC	Number successful WPC/h	No. EPC attempts on female	No. EPC attempts per h	Fate of pair bond
5a[1]	18.30	7	0.38	2	0.108	2	0.108	Retained
18	37.40	13	0.35	4	0.106	7	0.186	Retained
12	92.45	28	0.30	18	0.194	3	0.032	Retained
8	48.40	13	0.27	5	0.103	7	0.144	Neither individual seen
13	34.40	6	0.17	4	0.115	1	0.029	Neither individual seen
17	52.20	9	0.17	5	0.096	1	0.019	Neither individual seen
Hide	48.30	8	0.16	4	0.082	1	0.021	Changed
10	103.45	14	0.13	10	0.096	3	0.029	Changed
1[2]	58.20	5	0.09	9	0.155	2	0.034	Unknown
4[2]	46.00	3	0.06	2	0.043	0	0.000	Unknown

After Mills (1994). Spearman rank correlation (number regurgitations/h versus successful WPC/h = 0.655, $P < 0.05$. Pairs only observed from 11 to 16 days prior to laying have been excluded from the analysis because WPCs are significantly less at this time. Spearman rank correlation (number regurgitations/h versus EPC/h) = 0.679, $P < 0.02$.

[1]Pair observed only from 1 to 4 days prior to laying.

[2]Pair observed only from 11 to 16 days prior to laying.

attempts, even though they were potentially at a greater risk of forced copulation attempts by strange males because the incidence of extra-pair copulation attempts was related to the proportion of time females spent on the colony (Spearman rank correlation, $r_s = 0.765$, $n = 10$, $P < 0.001$; Mills 1994).

Although the sample sizes were small, the evidence indicates that males that were good providers were more likely to retain paternity and retain their partner the next season. Conversely, poorly fed females that copulated less frequently with their partner were more likely to have successful extra-pair copulations and to have a different mate the next year.

Consequences of mate change and divorce

A major effect of mate change on potential productivity is the number of breeding seasons individuals miss following a mate change. In all, females were twice as likely never to breed again (32%) as males (16%) following the termination of a pair bond (Table 16.2). Whereas the shortage of males in the population probably ensured that most males (92%), irrespective of their age, were able to obtain a new mate within two seasons, only 69% of females re-paired within the same period of time ($\chi^2 = 194.2$, $df = 1$, $P < 0.001$). Significantly more females than males stayed unpaired for the remainder of their lives after mate change (Table 16.2).

Gulls that paired with a new partner laid on average 4 days later, and had significantly smaller clutches and egg sizes than previously established pairs (Table 16.3). Over the spectrum of ages, established pairs consistently fledged more chicks than those that bred for the first time with a new mate (Fig. 16.2)

If divorce was adaptive in the sense that individuals change to higher quality partners, after replacing a mate we would expect reproductive success to be higher for those that divorce compared with those that re-paired after death of a mate. This expectation was met in two of 10 comparisons (Tables 16.4 and 16.5). Fledging success (from replacement clutches) of those that divorced was significantly higher than for birds that lost their partner through death. Divorced gulls recommenced breeding 3.9 days sooner than those who lost their mate through death.

Mate familiarity effect on reproduction

The effects of pair duration on fledging success were examined on 10 732 pairs using linear and quadratic models. After controlling for year (and not age), fledging success progressively improved for the first 3 years of the pair bond (Fig. 16.3; linear $t = 3.2$, $df = 9149$, $P < 0.001$; quadratic $t = 2.8$, $df = 9149$, $P < 0.001$). However, when age was controlled for as well, the pair duration effect was no longer significant. This was due, in part, to the high correlation between age and pair duration (females $r = 0.44$, $P < 0.001$; males $r = 0.43$, $P < 0.001$); longer lasting pairs were more likely to consist of older individuals. By progressively removing the nonsignificant effects, female age (and female age^2) and male age (and

Table 16.2 Differences in the initiation of breeding by males and females following the termination of a pair bond

Status	Sex	No.	Years since pair bond terminated											
			1	2	3	4	5	6	7	8	9	10	11	12
(a) Bred again	Females	855	59%	22%	11%	4%	3%	1%	<1%	<1%	<1%	<1%		
	Males	953	82%	12%	4%	1%	<1%	<1%	<1%					
(b) Never bred again	Females	400	100%	73%	55%	40%	29%	22%	15%	10%	6%	3%	2%	2%
	Males	185	100%	41%	19%	10%	6%	3%	3%	2%	<1%	<1%	<1%	

The data include 1255 individually colour-marked females which survived at least 1 year following the break-up of a pair bond between the 1975 and 1985 breeding seasons. (a) The number of years it took to resume breeding after a mate change. The difference between the sexes was significant, females taking longer than males after termination of the bond ($G = 117.5$, $df = 7$, $P < 0.001$); (b) the proportion of females and males that never bred again after a mate change; females more than males ($G = 77.0$, $df = 9$, $P < 0.001$).

Table 16.3 Reproductive performance following mate change in Red-billed Gulls

Sex	Pair bond status	Laying date			Clutch size			Egg volume (cc)			
		No. birds	Mean	SD	No. birds	Mean	SD	No. birds	No. Eggs	Mean	SD
Female	Retained	2468	7.7 Nov.	19.9	2553	2.02	0.51	1500	2923	37.23	5.76
	Changed	1073	12.4 Nov.	18.9	1103	1.96	0.55	639	1182	36.45	2.95
		$t = 6.54$, $df = 3539$, $P < 0.001$			$t = 3.19$, $df = 3654$, $P < 0.02$			$t = 4.43$, $df = 4103$, $P < 0.001$			
Male	Retained	2327	30.1 Oct.	12.9	2551	1.98	0.44	1985	3835	37.29	3.76
	Changed	798	3.0 Nov.	15.3	852	1.96	0.43	685	1279	36.63	2.99
		$t = 7.02$, $df = 3123$, $P < 0.001$			$t = 1.22$, $df = 3401$, NS			$t = 6.07$, $df = 4112$, $P < 0.001$			

Poor breeding years were excluded and only middle-aged birds included in the data set. Gulls that changed mates performed less well than those that retained mates.

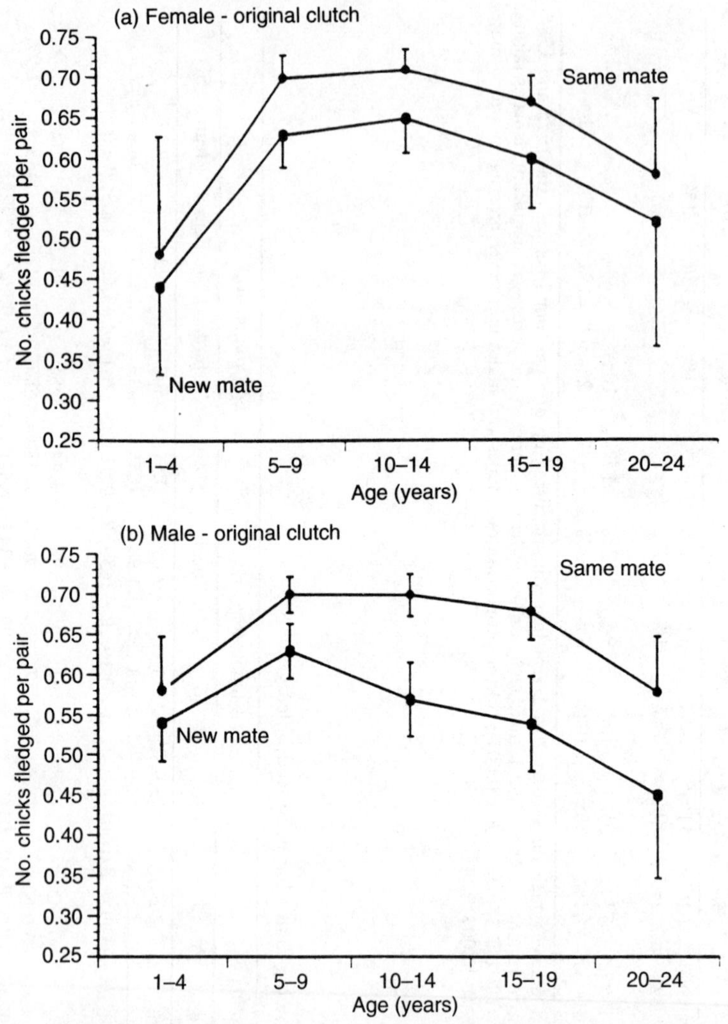

Fig. 16.2 The influence of mate change on the breeding biology of known aged Red-billed Gulls. Fledging success of (a) females and (b) males raising original clutches (first nesting attempt for season). Difference in breeding success between those retaining and changing for males 10–14 years of age ($t = 2.4$, $df = 1291$, $P < 0.02$).

male age^2), from the model, pair duration remained nonsignificant. The year effect remained significant.

Discussion

Comparison with other lariids and causes of mate change

Red-billed gulls like most gulls and terns are potentially long-lived and this coupled with a high degree of faithfulness to the locality of first breeding, creates the opportunity for long-term pair bonding. Mate change from one season to the next ranged from 75% for Caspian Terns

Table 16.4 Relationship between the type of mate change (death or divorce of mate) and the fate of clutches in the following year

Sex	Pair bond status	Original clutch: No. birds	Eggs laid	% Hatched	% Fledged	No. fledged per pair	SD	Replacement clutch: No. birds	Eggs laid	% Hatched	% Fledged	No. fledged per pair	SD
Females	Mate dead	104	197	59	30	0.58	0.71	32	60	28	2	0.03	0.18
	Divorced	247	450	63	36	0.66	0.81	55	96	42	14	0.24	0.47
	Significance of difference: $t = 0.87$, $df = 349$, NS								$t = 2.42$, $df = 85$, $P < 0.02$				
Males	Mate dead	80	153	62	30	0.58	0.77	29	49	43	20	0.34	0.72
	Divorced	320	589	60	30	0.55	0.73	90	164	41	16	0.29	0.59
	Significance of difference: $t = 0.32$, $df = 398$, NS								$t = 0.38$, $df = 117$, NS				

Poor breeding years were excluded and only middle-aged birds were included in the data set.

Table 16.5 Relationship between the type of mate change (death or divorce) and laying date, clutch size, and egg volume in the following year

Sex	Pair bond status	Laying date: No. birds	Mean	SD	Clutch size: No. birds	Mean	SD	Egg volume (cc): No. birds	No. eggs	Mean	SD
Females	Mate dead	101	7.8 Nov.	15.27	111	1.94	0.39	84	152	36.56	2.55
	Divorced	239	5.4 Nov.	14.84	255	1.90	0.41	205	368	36.09	2.73
	Significance of difference: $t = 1.35$, $df = 338$, NS				$t = 0.87$, $df = 364$, NS				$t = 1.35$, $df = 287$, NS		
Males	Mate dead	81	6.7 Nov.	13.23	91	1.98	0.42	67	126	36.13	3.10
	Divorced	306	2.8 Nov.	14.49	348	1.92	0.45	261	480	36.41	3.41
	Significance of difference: $t = 2.19$, $df = 385$, $P < 0.05$				$t = 1.15$, $df = 437$, NS				$t = 0.61$, $df = 226$, NS		

Poor breeding years were excluded and only middle-aged birds were included in the data set.

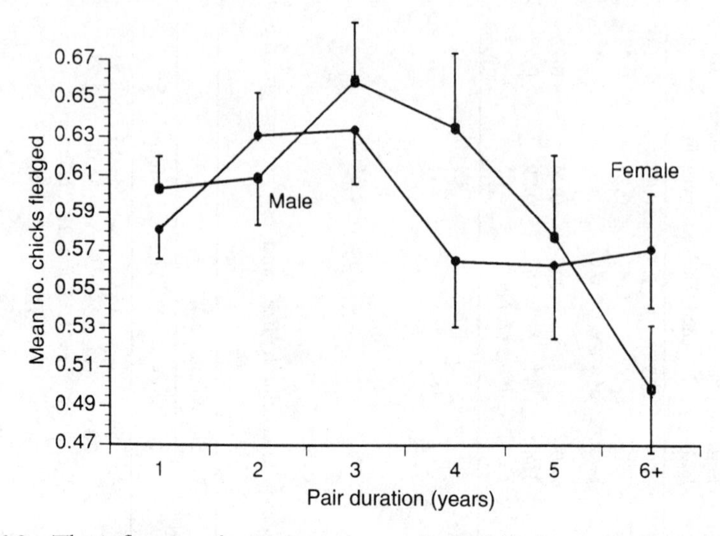

Fig. 16.3 The influence of pair duration on the fledging success of Red-billed Gulls, plotted prior to controlling for confounding variables (n = 10732 pairs). After controlling for year and age and age^2 (male n = 5006; female n = 6267), the only significant effect was year (t = –4.3, df = 2122, P < 0.001); female age approached significance (t = –1.5, df = 2122, P = 0.1–0.05). The data were analysed as a generalized linear model with a Poisson response and a log link function using the S-Plus 3.2 statistical package.

Sterna caspia to 13% for the Western Gull *Larus occidentalis*, and pair duration from 1 to 17 years (Table 16.6). Amongst the gull species, the reported divorce rates are remarkably similar. The reason for Caspian Terns having a high rate of mate change and a high divorce rate is probably because their nesting localities are much less stable (Cuthbert 1985; Johnston and Ryder 1987). Mate choice is also very similar in lariids with 66–76% of pairs having mates with an age or breeding experience difference of 1 year or less (Table 16.6).

Some aspects of mate fidelity in the Red-billed Gull are very similar to those reported for the Kittiwake Gull *Rissa tridactyla* (Coulson 1966; Coulson and Thomas 1980, 1983). In both species, over half of the pair bonds lasted only one breeding season and the divorce rate was highest for young individuals and birds breeding together for the first time. In the Kittiwake it was found that age and pair duration had a combined effect on reproductive success, as was originally reported in Red-billed Gulls (Mills 1973). However, in the current analysis the confounding effects of year and age were controlled for, and pair duration had no statistical effect on reproductive performance.

In the Red-billed Gull the highest incidence of divorce was found to be confined to unsuccessful breeders of young and newly established pairs, even after the effect of age was controlled for. Divorce was not apparent in unsuccessful pairs that had been together for 3 or more years. Besides

Table 16.6 Mate change and divorce in lariids

Species	% Changed mate	Divorce rate	%Pairs with partners with an age difference of 1 year or less (*n*)	Duration of pair bond (years)	Effect of previous reproductive success on: Successful: % Changed	Unsuccessful: % Changed	Source
Red-billed Gull	17 (9323)	10.5	68 (212)	1–17	14 (5714)	19 (2408)	Mills (1973), this study
Kittiwake Gull	36 (458)	26.1	72 (544)	1–14	17 (290)	52 (51)	Coulson (1966, 1972[1]); Coulson and Thomas (1980; 1983)
Herring Gull	21 (62)			1–6+			Drost *et al.* (1961); Tinbergen (1963)
Western Gull	13 (30)			1–3+			Pierotti (1981)
Glaucous-winged Gull	46 (13)	30.0		1–3+			Vermeer (1963)
Ring-billed Gull	38 (29)	28.0			2 (55)	43 (7)	Southern and Southern (1982); Fetterolf (1984)
Arctic Tern			66 (28)				Coulson and Horobin (1976)
Common Tern			76 (13)				Nisbet *et al.* (1984)
White-fronted Tern			72 (11)				Mills and Shaw (1980)
Caspian Tern	75 (24)	53.8		1–2+	50 (10)	50 (2)	Cuthbert (1985)
Fairy Tern	20 (15)						Dorward (1963)

Updated from Johnston and Ryder (1987).
[1]Data recalculated by Johnston and Ryder (1987).
Red-billed Gull *Larus novaehollandiae scopulinus*; Kittiwake Gull *Rissa tridactyla*; Herring Gull *Larus argentatus*; Western Gull *Larus occidentalis*; Glaucous-winged Gull *Larus glaucescens*; Ring-billed Gull *Larus delawarensis*; Arctic Tern *Sterna paradisaea*; Common Tern *Sterna hirundo*; White-fronted Tern *Sterna striata*; Caspian Tern *Sterna caspia*; Fairy Tern *Gygis alba*.

Kittiwakes, similar trends associated with previous reproductive success have been noted in other long-lived sea and shorebirds: Yellow-eyed Penguins *Megadyptes antipodes* (Richdale 1957), Little Penguins *Eudyptula minor* (Reilly and Cullen 1981), Gannets *Sula bassana* (Nelson 1978), Manx Shearwaters *Puffinus puffinus* (Brooke 1978), Fulmars *Fulmarus glacialis* (Ollason and Dunnet 1978), Blue Shags *Phalacrocorax atriceps* (Shaw 1986), Oystercatchers *Haematopus ostralegus* (Harris *et al.* 1987) and Short-tailed Shearwaters *Puffinus tenuirostris* (Bradley *et al.* 1990), although the effect of age was rarely considered in these studies.

Disadvantages of mate change

In the short term, there is a reduction in breeding performance in Red-billed Gulls after a mate change, as is the case for other seabird studies (e.g. Kittiwakes, Coulson 1966; Arctic Skua *Stercorarius parasiticus*, Davis 1976; Fulmars, Ollason and Dunnet 1978; and Short-tailed Shearwaters, Bradley *et al.* 1990), but not in Oystercatchers (Harris *et al.* 1987). Red-billed Gulls with new mates produced fewer chicks per pair, laid fewer and smaller eggs, and delayed clutch initiation longer than those that retained their partner. Amongst males where the cause of the change was the death of the former partner rather than divorce, fledging success was lower in pairs that laid replacement clutches. Similar significant differences in breeding success between birds that replaced dead and divorced mates has been reported for the Kittiwake (Coulson 1972). Mills (1973) suggested that delayed laying and lower clutch size of gulls changing mates may be due to the need to spend more time establishing a new bond which reduces the time available to forage for food. Coulson (1966) also argued that mate change in the Kittiwake results in less efficient breeding and hence new pairs expend greater energy to achieve the same success as pairs that retain their former partner. These suggestions have been supported from a behavioural study of the Kittiwake (Chardine 1987), which found that pairs that changed mates attended the nest site more frequently, and probably as a result of inefficiency, had less 'off-duty' time to forage. The reduction in breeding performance of individuals whose former partner dies probably results from the delay in the onset of laying due to the bird waiting in vain for the return of its former partner. Fetterolf (1984) studying pairing and pair dissolution in the Ring-billed Gull *Larus delawarensis* believed that establishing a new pair bond during the postbreeding period minimized disruption to breeding the next season (see also Williams Chapter 15 for empirical evidence in favour of a similar argument).

The greatest disadvantage of mate change was the hiatus in breeding following the change. This could be short or long term, and the effect was considerably greater in females than males because of the surplus of females in the population and the difficulty females had competing for mates. Overall, 60% of females and only 31% of males changing mates and surviving at least 1 year failed to breed the next season, whilst 16% of males and 32% of females never bred again. Lapses in breeding increased significantly with age, and females over 15 years of age found it extremely difficult to acquire another partner. Decreased breeding opportunity after a mate change obviously influences lifetime productivity.

A possible factor contributing to the break-up of pairs and the nonbreeding of some birds could be illness. However, a high proportion of individuals, particularly females, became long-term nonbreeders and this suggests that these individuals were of 'poor quality,' which makes them undesirable as partners to other birds as well (see WIMPS in Marzuff *et al.* Chapter 7).

Advantages of mate change

Whilst there are major disadvantages in changing partners, there are times when it is thought to be advantageous. In situations where there is incompatibility between members of the pair (Coulson 1966, 1972) and/or one is of poor quality which results in more strain being put on the other member, it would be an advantage for the pair to split up in the hope that they find a better partner. The higher proportion of divorces amongst unsuccessful newly established Red-billed Gull pairs suggests that there is incompatibility through errors in mate choice. Pairs that had been together for more than one season were less likely to change mates if they were unsuccessful. Evidence presented in this study indicates that errors in mate choice were more apparent in the younger individuals because mate changes amongst older individuals breeding together for the first time were reduced. Coulson (1972) has shown that incompatibility between members of Kittiwake pairs impacted on productivity. Successful Kittiwake pairs have a regular pattern of activity with both members sharing incubation duties equally (Coulson and Wooller 1984). In contrast, unsuccessful pairs characteristically do not co-ordinate their activities, have erratic incubation shifts, and on occasions do not relieve their partners for prolonged periods.

Courtship feeding in the Red-billed Gull also plays a stabilizing role in pair bonds, acting as an inducement for successful copulations and being nutritionally important to the female (Tasker and Mills 1981). Females that were well fed by their mate were able to spend most of the daylight hours in the nest territory whereas poorly provided females had to forage for themselves. In the following year, a greater proportion of the poorly provisioned females than the well provisioned females obtained a new mate.

The role of courtship feeding in mate change could have been direct or indirect. Indirectly, it can be a measure of the quality of the male because in Common Terns *Sterna hirundo* and Herring Gulls *Larus argentatus* the courtship feeding rate has been shown to be correlated with the time the males spend incubating and the frequency of chick feeding by the males (Nisbet 1973, Niebuhr 1981; Wiggins and Morris 1986). Thus males that are poor providers in courtship are generally incompetent in other aspects. Directly, inadequately provisioned females could have been undernourished at a time when they needed extra nutrition. Because courtship feeding occurs relatively late in the pairing process of Red-billed Gulls (Tasker and Mills 1981), assessment of the quality and suitability of the mate through courtship feeding is not a viable option within the season because divorce at that stage would mean that the female would have to delay breeding and this would be likely to lower the success in fledging young. Furthermore, the female would have difficulty finding a mate at this stage owing to the excess of females in the population. Within a season, the best option for the female, if she made a poor mate choice, would be to continue with that mate only to the end of that season and perhaps seek out extra-pair copulations in the hope of acquiring genetically superior offspring. In this study, the only female to solicit an extra-pair copulation during the potential fertile period was a female that was being inadequately fed by its mate. An individual may be able to compensate for a poor quality mate and fledge a chick, but the extra effort may be costly, jeopardizing its survival.

For a female Red-billed Gull there would be a trade-off between compensating for a poor mate and divorce, because if she changed mates and survived at least 1 year she would only have a 40% chance of breeding the next year and a 68% chance of ever breeding again. With advancing age the prospects of a female getting a new mate decrease significantly owing to a skewed sex ratio towards females. The only option some old females have to enable them to breed again after losing a mate is to form female–female partnerships and solicit extra-pair copulations or submit to a forced copulation.

Advantages of long-term monogamy

For those partners together for 5 or more years, breeding commenced on average 9 days earlier than those that were breeding together for the first time (Table 16.3). This may indicate that the members of the pair develop co-ordination and efficiency that enhance performance. It is a distinct advantage to breed as early as possible (Mills 1970), since early breeders have elevated and central sites which are less susceptible to flooding and predation. Early breeders also generally have nestlings present when food availability is increasing or at peak abundance rather than declining, and if they lose a clutch to predation, they still have time to lay a replacement which hatches near the peak in food abundance.

Over recent years, there has been a great deal of theorizing on the adaptive value of males maximizing their reproductive potential by copulating and fertilizing other paired females whose offspring they would not assist in raising (Trivers 1972; Birkhead and Møller 1992). We suspect that for a species like the Red-billed Gull which is long-lived and invests considerable time and energy in courtship feeding of the female and parental care during incubation and chick rearing, it is more advantageous for the male to strengthen the pair bond than to waste time philandering (Mills 1994). In a mating system such as this, both members have a lot to lose if the pairing is disrupted and unsuccessful. This study has shown that activities that weaken the bond, such as poor courtship feeding performance, jeopardize not only the current reproductive investment but also future reproduction. Males that did not feed their mates adequately had an increased probability of losing paternity, whereas those that were attentive prospered because the female resisted all extra-pair copulation attempts. In an earlier paper, Mills (1994) suggested the benefits of extra-pair copulations in Red-billed Gulls were more favourable to certain females than to males. If males obtained a successful extra-pair copulation it was more likely to be from a 'poor quality' female in female–female pairings which make up approximately 6% of pairings. These females have a lower probability of raising their young. In summary, males would maximize their productivity by strengthening the pair bond and in doing so would give themselves greater paternity protection, and greater likelihood of breeding the next season.

Summary

The number of lifetime partners of Red-billed Gulls varied from one to nine. An average of 54% of females and 51% of males had only one partner in their lives. Eighty-three per cent of pairs retained partners from one season to the next and 10.5% of pairs divorced. Mate fidelity was significantly related to the age of the pair member and how long they had been together. The probability of divorce decreased in relation to the number of chicks fledged the previous season. Pairs that replaced a divorced mate laid earlier, raised more chicks from replacement clutches, and were less likely to defer breeding than those that replaced a mate that had died. Females that received adequate courtship feeding resisted all extra-pair copulation attempts and were more likely to retain their partners next season. Poorly courtship fed females divorced the next season. Females were twice as likely as males never to breed again following the termination of a pair bond. Females paired with a new partner laid on average 4 days later, had significantly smaller clutches and eggs, and raised fewer chicks per pair than those that retained their partner. Controlling for year and age effects, reproductive success did not significantly change with pair duration. We hypothesize that in species like the Red-billed Gull, which have long-term pair bonds and invest considerable time and energy in courtship feeding and parental care, it is more advantageous to strengthen the pair bond than to philander to increase production.

Acknowledgements

Many people have assisted with field work over the past 30 years; in particular Jack Cowie, Rod Cossee, Bert Rebergen, Peter Shaw, Andy Garrick, Peter Moore, Chris Petyt, Ian Flux, and Jane Maxwell. The Zoology Department of Canterbury University kindly allowed use of the Edward Percival Field Station at Kaikoura during field work. The project over the past 5 years was funded by Earthwatch and its Research Corps and funding for computer analysis was provided by a New Zealand Lottery Board Science Grant. We are grateful to Dr Jeff Black for editorial and statistical help and Dr Allan Baker for constructive comments on the manuscript. Kathy and Joe Propersi assisted with data preparation and analyses and Dr Kelly Mara continued to provide statistical advice. In the present day scientific environment it is extremely difficult to obtain funding or institutional support for the continuation of long-term studies; therefore, we are extremely grateful to those who have actively advanced the value of the Red-billed Gull study. If it was not for Dr Ben Bell of the Biological Sciences Department, Victoria University, Wellington, for his assistance and help in providing access to a mainframe computer the study would not have been able to continue or be completed. We are also grateful to Drs M. J. Williams, M. C. Crawley, R. M. S. Sadleir, and Professor E. C. Young for moral support for the continuation of the study.

References

Birkhead, T. R. and Møller, A. P. (1992). *Sperm competition in birds. Evolutionary causes and consequences*. Academic Press, London.

Birkhead, T. R., Atkin, L., and Møller, A. P. (1987). Copulation behaviour of birds. *Behaviour*, **101**, 101–38.

Bradley, J. S., Wooller, R. D., Skira, I. J., and Serventy, D. L. (1990). The influence of mate retention and divorce upon reproductive success in short-tailed shearwaters *Puffinus tenuirostris*. *Journal of Animal Ecology*, **59**, 487–96.

Brooke, M. de L. (1978). Some factors affecting the laying date, incubation and breeding success of the Manx shearwater, *Puffinus puffinus*. *Journal of Animal Ecology*, **47**, 961–76.

Chardine, J. W. (1987). The influence of pair status on the breeding behaviour of the Kittiwake *Rissa tridactyla* before egg-laying. *Ibis*, **129**, 515–26.

Coulson, J. C. (1966). The influence of the pair bond and age on the breeding biology of the kittiwake gull *Rissa tridactyla*. *Journal of Animal Ecology*, **35**, 269–79.

Coulson, J. C. (1972). The significance of the pair bond in the Kittiwake. *Proceedings International Ornithological Congress*, **15**, 424–33.

Coulson, J. C. and Horobin, J. (1976). The influence of age on the breeding biology and survival of the Arctic Tern *Sterna paradisaea*. *Journal of Zoology*, **178**, 247–60.

Coulson, J. C. and Thomas, C. S. (1980). A study of the factors influencing the duration of the pair bond in the Kittiwake Gull *Rissa tridactyla*. *Proceedings International Ornithological Congress*, **17,** 823–33.

Coulson, J. C. and Thomas, C. S. (1983). Mate choice in the Kittiwake Gull. In *Mate Choice*. (ed. P. Bateson), pp. 361–76. Cambridge University Press, Cambridge.

Coulson, J. C. and Wooller, R. D. (1984). Incubation under natural conditions in the kittiwake gull, *Rissa tridactyla*. *Animal Behaviour*, **32,** 1204–15.

Cuthbert, F. J. (1985). Mate retention in Caspian Terns. *Condor*, **87,** 74–8.

Davis, J. W. F. (1976). Breeding success and experience in arctic skua (*Stercorarius parasiticus*). *Journal of Animal Ecology*, **45,** 531–7.

Dorward, D. F. (1963). The Fairy Tern *Gygis alba* on Ascension Island. *Ibis*, **103b,** 365–78.

Drost, R., Focke E., and Freytag, G. (1961). Entwicklung and Aufban einer Population der Libermoive, *Larus a. argentatus*. *Journal für Ornithologie*, **102,** 409–29.

Fetterolf, P. M. (1984). Pairing behavior and pair dissolution by Ring-billed Gulls during the post-breeding period. *Wilson Bulletin*, **96,** 711–14.

Harris, M. P., Safriel, U. N., Brooke, M. de L., and Britton, C. K. (1987). The pair bond and divorce among Oystercatchers *Haematopus ostralegus* on Skokholm Island, Wales *Ibis*, **129,** 45–57.

Johnston, V. H. and Ryder, J. P. (1987). Divorce in lariids: a review. *Colonial Waterbirds*, **10,** 16–26.

Mills, J. A. (1970). The population ecology of Red-billed Gulls *Larus novaehollandiae scopulinus* of known age. Unpublished Ph.D. thesis. University of Canterbury, Christchurch, New Zealand.

Mills, J. A. (1971). Sexing Red-billed Gulls from standard measurements. *New Zealand Journal of Marine and Freshwater Research*, **5,** 326–8.

Mills, J. A. (1973). The influence of age and pair bond on the breeding biology of the red-billed gull *Larus novaehollandiae scopulinus*. *Journal of Animal Ecology*, **42,** 147–62.

Mills, J. A. (1989). Red-billed Gull. In *Lifetime reproduction in birds* (ed. I. Newton), pp. 387–404. Academic Press, London.

Mills, J. A. (1991). Lifetime production in the Red-billed Gull. *Proceedings Ornithological Congress*, **20,** 1522–7.

Mills, J. A. (1994). Extra-pair copulations in the Red-billed Gull: females with high quality, attentive males resist. *Behaviour*, **128,** 41–64.

Mills, J. A. and Shaw, P. W. (1980).The influence of age on laying date, clutch size and egg size of the White-fronted Tern *Sterna striata*. *New Zealand Journal of Zoology*, **7,** 147–53.

Nelson, J. B. (1978). *The Sulidae*. Oxford University Press, Oxford.

Niebuhr, V. (1981). An investigation of courtship feeding in Herring Gulls. *Ibis*, **123,** 218–22.

Nisbet, I. C. T. (1973). Courtship-feeding, egg size and breeding success in Common Terns. *Nature*, **241,** 141–2.

Nisbet, I. C. T., Winchell, J. M., and Heise, A. E. (1984). Influence of age on the breeding biology of Common Terns. *Colonial Waterbirds*, **7,** 117–26.

Ollason, J. C. and Dunnet, G. M. (1978). Age, experience, and factors affecting the breeding success of the fulmar *Fulmarus glacialis* in Orkney. *Journal of Animal Ecology*, **47,** 961–76.

Pierotti, R. (1981). Male and female parental roles in the Western Gull under different environmental conditions. *Auk*, **98**, 532–49.

Reilly, P. N. and Cullen, J. M. (1981). The Little Penguin *Eudyptula minor* in Victoria. II. Breeding. *Emu*, **81**, 1–19.

Richdale, L. E. (1957). *A population study of penguins*. Clarendon Press, Oxford.

Shaw, P. (1986). Factors affecting the breeding performance of Antarctic Blue-eyed Shags *Phalacrocorax atriceps. Ornis Scandinavica*, **17**, 141–50.

Southern, L. K. and Southern, W. E. (1982). Mate fidelity in Ring-billed Gulls. *Journal of Field Ornithology*, **53**, 170–1.

Tasker, C. R. and Mills, J. A. (1981). A functional analysis of courtship feeding in the Red-billed Gull, *Larus novaehollandiae scopulinus. Behaviour*, **77**, 221–41.

Tinbergen, N. (1963). *The herring gull's world*. Collins, London.

Trivers, R. L. (1972). Parental investment and sexual selection. In *Sexual selection and the descent of man* (ed. B. Campbell), pp. 136–79. Aldine, Chicago.

Vermeer, K. (1963). The breeding ecology of the Glaucous-winged Gull (*Larus glaucescens*) on Mandarte Island. *Occasional Papers of the British Columbia Provincial Museum No. 13*.

Wiggins, D. A. and Morris, R. D. (1986). Criteria for female choice of mates: courtship feeding and parental care in the Common Tern. *American Naturalist*, **128**, 126–9.

17 *Dispersal, demography, and the persistence of partnerships in Indigo Buntings*

ROBERT B. PAYNE AND LAURA L. PAYNE

Introduction

In late spring and through the summer, the song of the Indigo Bunting *Passerina cyanea* (see drawing above) is a familiar sound in old fields and the edges of woodlands in eastern North America. The indigo blue males are conspicuous as they sing from the tops of small trees. The brown females nest near the ground in leafy shrubs and herbs. Buntings are long-distance migrants and spend the winter in flocks in Mexico and Central America. Pairs separate after the breeding season until the birds return in the following spring, from early May, the males about 10 days earlier than the females. Males usually settle on their breeding site of the previous year; females often settle on a new territory rather than reuniting with their old mate.

Although Indigo Buntings live in pairs in the breeding season, they often mate outside the pair bond. When they are nest building they call loudly, and males on neighbouring territories (1–2 ha) visit and attempt to copulate. Females also visit males on neighbouring territories and solicit copulation, then return to their social mate's territory to nest (Payne 1983*a*; Westneat *et al.* 1987). In addition, a male sometimes attracts more than one female to nest on his territory (Carey and Nolan 1979; Payne *et al.* 1988; Payne 1989, 1992).

In this chapter, we look at the causes and consequences of partners reuniting and the significance of persistent social pair bonds in relation to dispersal, time of arrival on territories, song themes, reproductive success, and demographics. We observed 221 pairs where both colour-banded partners returned to the study area. We also use a larger sample of pairs (n = 1184), where not all females were banded and not all partners returned, to describe the mating system and parental care in the buntings.

Background, study areas, and procedures

Buntings were observed from 1978 through 1994 in southern Michigan. One study area (1.4 km^2) was near Niles (41°55′N, 86°14′W) along roadsides and railways, in open woodland, in old fields, and in fields of maize and beans. The other (6 km^2) was at the E. S. George Reserve and neighbouring Pinckney State Recreation Area (42°27′N, 84°00′W) in woodland, old fields that had been cultivated before the 1930s, and shrubby swamps. At least six observers were active in the field each year through 1985, and in all we observed the birds for about 20 000 hours. The study areas were enlarged in the first years and maintained in later years to find dispersers.

Males were captured by playback of a song on their territory. Females were netted after they had a nest. Pairs were identified using the criteria of consistent resightings of birds and their colour bands, male singing behaviour, and female nest attendance within the male's territory. Unbanded adult males in areas where all males had been banded in previous years were considered to be new adult immigrants. In each year more than 90% of the breeding males and about half of the females were banded (Payne *et al.* 1988; Payne 1989). We restrict analysis of reproductive success to birds that were resident for at least 28 days, the time to nest and rear a brood to fledging; a pair shared a territory for as long as 100 days. Reproductive success of a pair was determined as the number of young buntings that fledged from all their nests in a season. We found no significant differences in mean success among years (Payne 1989).

Males advertise their territories by songs. They usually share their song themes in local neighbourhoods, which the first-year males learn and copy when they settle next to an established male. The variations in song may be important in territorial interactions between males and in attraction of a mate, but not in recognition of kin as the song themes do not follow kinship lines (Payne *et al.* 1987; Payne and Payne 1993*b*). Each male has a 'song theme' characterized by the occurrence and sequence of the song elements (Fig. 17.1). Males 'match' their songs when the same elements are delivered in the same sequence (Payne *et al.* 1981, 1988). The number of males with a song theme in a season averages three or four; some songs are unshared and others are matched by as many as 22 males. For returning females, we compared the songs of their mates between years.

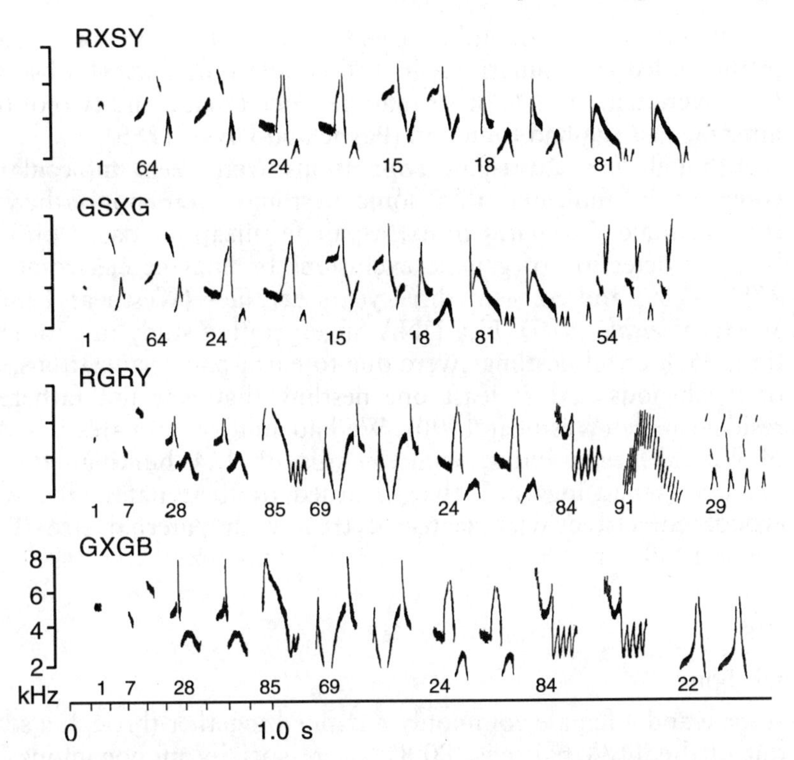

Fig. 17.1 Two song themes of Indigo Buntings at the E. S. George Reserve, Michigan. Letters indicate the colour bands of individual males. Numbers refer to the song elements in the catalogue of Thompson (1970).

'Breeding dispersal' was observed when a bird that bred in one year returned to a different territory in the next year. The study areas were large in relation to territory size, and a returning bunting was unlikely to be missed. We recognized a 'return' when a bird was seen on the study area in a later year, 'site fidelity' when it returned to the area where the male sang in the previous year, and 'breeding dispersal' when it settled elsewhere. Annual survival is 52–60% in males and is lower (33–47%) in females, and there is no change in survival, dispersal, or reproductive success in females with age (Payne 1989, 1992; Payne and Payne 1990, 1993a).

Extra-pair behaviour

Buntings sometimes copulate outside the pair bond so the estimates of male reproductive success depend on the accuracy of the behaviour observations. A few extra-pair copulations were observed (Payne 1983*a*; Westneat 1987*a*, Payne and Payne 1989). Intruding males often approached a female as she solicited copulation on her mate's territory, and the female also visited a male on a neighbouring territory and solicited there, mated, and returned to her nesting territory. Few visits by intruders led to copulations; in 156 copulations that we observed, three (2%) were with a male other than her social mate, and two of these were after the last egg had been laid (Payne and Payne 1989).

Although few extra-pair copulations were seen, molecular genetics comparisons indicated that some nestlings were not fathered by the resident male. Estimates of extra-pair fertilizations based on the probability of detection of genetic exclusions by enzyme electromorphs were 27%, 42%, and 32% in three years at Niles (Westneat 1986, 1987*b*; Westneat *et al.* 1987). In a DNA-fingerprinting study in another population, 35% of all nestlings were due to extra-pair fertilizations, and 48% of the broods had at least one nestling that was not fathered by the resident male (Westneat 1990). We had similar estimates based on morphological heritabilities; wing lengths of 124 buntings that we had banded as nestlings and that returned to their natal area were more closely correlated with maternal than with paternal size (Payne and Payne 1989).

Results

The pair bond

A male and a female commonly remained together through a season, and out of the 1184 pairings, 80.8% were socially monogamous. The proportion of territories in which more than one female bred was 25.8% at Niles (463 had one female, 139 had two, 20 had three, and two had four), and at the George Reserve it was 8.8% (494 had one female, 65 had two, and 1 had three). The frequency of polygyny differed between areas ($\chi^2 = 37.4$, $df = 1$, $P < 0.001$). Females that arrived late did not necessarily need to move to a new site, perhaps because female–female aggression was rare.

Although a male and female were often seen together, their social association was obvious mainly before her clutch was complete (the mate stays near and chases intruding males in the period of female fertility), and when the pair cared for the fledged brood.

Male and female buntings differ in their breeding behaviour. Females breed in their first and later seasons. They nest repeatedly though many clutches and broods are taken by predators and nests are lost in cool, rainy weather. The female incubates the three or four eggs, broods the young, and brings insects and spiders, and she calls and mobs a predator near the nest (Carey 1982; Payne 1992). Males usually do not feed the nestlings. Only 2% of the males with nestlings no older than 4 days fed them (a single feed by one male), and 10% of males with older nestlings, whereas 31% of males fed their fledglings (Westneat 1988*a*). Males also call and mob a predator near the nest. The young fledge in 8–9 days and are fed for another 2 weeks. After the nestlings fledge, the males often escort and feed them and the pair shares the care of the brood. Pairs often rear two broods in a season, and occasionally they rear three broods. When a pair renests, the time for a female to build a nest and lay a clutch is shorter when the male cares for the fledged brood than when he does not (Westneat 1988*a*, *b*; Payne 1989, 1992).

Males do not increase their care when the female is unable to feed the young. In the two cases when a female with nestlings was killed by a car, the male did not feed them and the nestlings starved. In the four cases when the male was killed while the female had nestlings, the brood survived, and in one a son returned and bred in the next year. When a male disappeared while his female was laying, his replacement showed no interest in the female's nest or brood. The female reared the brood (a bunting and a Brown-headed Cowbird *Molothrus ater*) by herself and her own young returned in the next year. Overall, the young are reared almost entirely by the female, in contrast to many monogamous songbirds in North America (Verner and Willson 1969).

Female change of mates within a breeding season

In both study areas the breeding season extended over several weeks, with a range of 100 days between the first egg in the earliest and latest nests at Niles, and 85 days at the George Reserve. The first nests were in late May in both areas. Mean date of the first egg for all nests at Niles was 16 June (SD 20.7 days, $n = 1464$ nests) and at the George Reserve it was 10 June (SD 18.0 days, $n = 996$). Mean nest dates varied among years in both areas (ANOVA, Niles, $F = 8.72$, $df = 7{,}1456$, $P < 0.0001$; George Reserve, $F = 7.58$, $df = 7{,}988$, $P < 0.0001$) and there were more late nests at Niles, but the extreme mean dates were no more than 10 days apart, and the variation among years accounted for only 4% of the total variance at Niles and 5% at the George Reserve. Because the season is long and the birds often breed more than once in a season, we compared reproductive success and mate fidelity within a season as well as between seasons.

Most females remained with their partner from one nest to the next, regardless of their success in the early nest. Other females changed mates and territories within a season. Both at Niles and at the George Reserve, 11% of the pairs separated after a nesting attempt, when the female moved to a new territory, and nested again with a new mate. Females were more likely to change mates when they were successful in the earlier attempt than when they were unsuccessful (Table 17.1). A female that paired with a new mate sometimes took her fledged young with her; the new male did not feed them. She often nested with the new mate after her young were independent.

Success was compared for females that remained with a mate and females that took a new mate. Females that remained were no more likely to be successful in a renest or second nest. Also, females that

Table 17.1 Fledging success of females in relation to whether they changed mates for their next nest within a breeding season

Area	Was earlier nest successful?	Did female change mates? No	Yes	χ^2	P
Niles	Yes	89	14	0.62	NS
	No	338	41		
George Reserve	Yes	43	10	4.41	<0.05
	No	115	10		
Overall	Yes	132	24	3.31	NS
	No	453	51		

Table 17.2 Consequence of mate change for females in early and later nests within a season

Area	Did female change mates?	Was earlier nest successful?	Did female change mates? No	Yes	χ^2	P
Niles	No	Yes	46	42	0.80	NS
		No	158	178		
	Yes	Yes	9	5	0.14	NS
		No	24	17		
George Reserve	No	Yes	29	14	1.07	NS
		No	66	47		
	Yes	Yes	5	5	0.20	NS
		No	6	4		

changed mates were no more likely to fledge young than females that remained with their old mate (Table 17.2).

Reproductive success of females in old and new pairs between seasons

Were females that reunited with the old partner more successful than those with a new mate? In 51% of the 221 pairs where a female and male from the previous year both returned, the female took a new mate. We compared:

(1) females that mated with an old mate,

(2) females whose old mate returned but who paired with a different male; and

(3) females whose former mates did not return (Table 17.3).

There were no significant differences in reproductive success among groups (ANOVA, $P > 0.05$). At the George Reserve the trend was for females to have higher success with a new male than with their old male, but the difference was not significant. Also, females were more likely to fledge young when they took a new mate (whether or not the old mate returned) than when they reunited with the old mate (Table 17.4). We

Table 17.3 Mean number of young fledged in the current season for females that reunited with an old mate, and females with a new mate

Area	Mate	Did old mate return?	*n*	Mean	SD	F^1	*P*
Niles	Old	Yes	73	2.26	2.00	0.88	NS
	New	Yes	75	2.17	1.86		
	New	No	59	2.08	2.18		
George Reserve	Old	Yes	36	2.00	2.01	1.82	NS
	New	Yes	37	2.78	4.12		
	New	No	32	3.72	1.93		

[1]F is calculated from one-way ANOVA.

Table 17.4 Proportion of females that fledged at least one young bunting after reuniting with an old mate, and when with a new mate

			Reproductive success			
Area	Mate	Did old mate return?	No	Yes	*G*	*P*
Niles	Old	Yes	25	48	1.13	NS
	New	Yes	24	51		
	New	No	24	35		
George Reserve	Old	Yes	15	21	6.10	<0.05
	New	Yes	7	30		
	New	No	6	26		

conclude that females generally were not more successful when they reunited with a previous mate.

Reproductive success in long-term pairs

Sixteen pairs bred together for 3 years or longer (13 for 3 years, two for 4 years, one for 5 years). We distinguished the success of each pair where a bird had more than one mate in a year, and we counted the fledglings separately for each pair.

The effect of mate familiarity and pair duration was tested in pairs that bred for 3 years or more. Their histories were examined to find whether mate fidelity was explained as a mutual attraction to each other or to the territory. In four of these 16 pairs the pair was together from the male's yearling season; in three of these they were together for a lifetime for each bird. However, in three of the 16 pairs, a female reunited with a previous mate after breeding with another for a year, and in pairs that reunited over several years, the pair was not always the first mating for either partner, so the females were not notably constant to the old mate, and the common theme in the long-term appeared to be that each member of the pair returned to the same territory.

Return to an old territory may give a pair the experience to nest earlier in the season, and opportunities to nest again and fledge more broods.

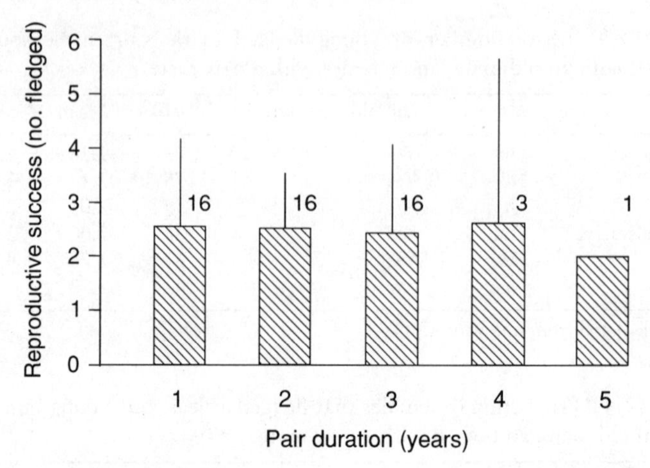

Fig. 17.2 Mean annual reproductive success (SD bars) in relation to pair duration (the number of years the male and female were together), in pairs that mated for at least 3 years.

To test the effect of pair duration on nest date, the date of the first egg of the season was compared, adjusted for mean date for all pairs in the area in that year. Mean nest date did not differ significantly in the 16 pairs that were together for 3 or more years, either across the years the pairs were together (ANOVA, $F = 2.45$, $df = 4{,}42$, $P = 0.06$), or across male age ($F = 2.02$, $df = 5{,}41$, $P = 0.10$). The tendency was for pairs to nest earlier when they were together for more years, but the correlation was not significant when allowance was made for male age ($r = -0.21$, $n = 47$, NS).

Mean annual reproductive success per pair was similar among pairs (ANOVA, $F = 1.49$, $df = 15{,}36$ NS), with age of the male ($F = 0.66$, $df = 5{,}46$, NS), and with number of years the pair had bred together ($F = 0.46$, $df = 4{,}47$ NS; Fig. 17.2). We conclude that there was neither an advantage nor a cost in reuniting with a previous partner.

Song themes

Indigo Bunting males behave differently towards playback of song of different individuals and recognize their neighbours by their songs (Emlen 1971). Females might recognize males by song and behave differently towards males that match the song theme of their mate and males with a different song theme (Payne 1983*a*). Within a season a female was no more likely to mate with another male that matched the song theme of her mate than with a male with another song theme (Payne 1983*b*). Between seasons, a returning female might associate the song of a male with her past success with him. If so, we predict that a female that was successful would seek out a male with the song theme of her old mate. In our observations, the females that had bred successfully in the previous

Table 17.5 Comparison of mate's song across years when old mate did not return but another male with the same song theme did return

Area	Did female fledge young in previous year?	Did mate in the next year have the same song theme as the previous mate?		
		Yes	No	P^1
Niles	No	0	11	0.039
	Yes	13	32	
George Reserve	No	1	4	0.409
	Yes	1	13	

[1]Fisher exact test

Table 17.6 Female breeding dispersal: effect of male song theme when the old mate did not return

Area	Did female disperse between years?	Did male on the old territory have the same song theme as her previous mate?			
		Yes	No	P^1	
Niles	Yes	30	23	0.287	NS
	No	5	3		
George Reserve	Yes	12	7	0.323	NS
	No	6	3		

[1]Fisher exact test

year and whose old mate did not return were more likely to pair with a male with their old mate's song theme than a male with a different song at Niles, but not at the George Reserve. In both areas most females paired with a male with a different song from their previous mates' song (Table 17.5). We also predict that females whose mate did not return are more likely to return to their old site when another male settles there with the old mate's song. However, females were no more likely to return to their old site when the new male had the same song than when he did not (Table 17.6). The results indicate that a female's attraction to a male did not depend on her previous association or success with a male with that song, and her return to an old territory was independent of the survival of the song theme of her old mate.

A demographic approach

In the absence of evidence that prior reproductive success, behavioural, or genetic traits determined the mate constancy of a pair, a null model based upon survival and dispersal (Payne and Payne 1990, 1993*b*) was used to describe the probability of whether a pair persists by chance from year to year. Male survival was 59% and female survival 47% at Niles, where most buntings were banded; the estimates of survival are similar on a continent-wide scale. Although most returning males (78%, combining ages) settled on the same territory from year to year, about half of the returning females whose old mate also returned to the study area settled on a new territory. The return of females to the same site or dispersal to a new site was independent of their reproductive success in the previous year, and of whether their old mate returned.

The probability P that a pair will remain together is:

$$(P \text{ male survival} \times P \text{ female survival} \times P \text{ male dispersal} \times P \text{ female dispersal})^{n},$$

where exponent n is the number of successive years together (total number of years -1). For a pair to breed together for 2 years ($n = 1$), the expected P is ($0.59 \times 0.47 \times 0.78 \times 0.5$), or 0.108, for 3 years ($n = 2$) P is 0.012, and for 4 years P is 0.001. With these estimates of survival and dispersal, it would be necessary to observe about 1000 banded pairs to find a banded pair that breeds together for as long as 4 years. This is about the number of pairs × years of buntings that we observed. The low survival gives a high probability that the mate will not return due to mortality, and decisions about whether or not to disperse would be costly in time and missed opportunity were a bunting to wait or search for its old mate.

Pair characteristics and reproductive success

We found no evidence of an effect of behavioural, morphological, or genetic similarity or complementarity upon the reproductive success of a pair as suggested by Halliday (1983). Females did not tend to mate with males with a song like their father's nor to avoid them; they mated at random with respect to song, when we accounted for the number of males with the song theme (Payne *et al.* 1987; Payne and Payne 1993*b*). Bunting pairs did no better when the male and female were relatively similar in body size than when they were dissimilar (Payne and Westneat 1988). The number of nestlings fledged was not related to the genetic similarity between mates, based on a sample of allozymes (Payne and Westneat 1988). There was no difference in success in the few pairs known from genealogies to be inbred ($r = 0.5$) and in the outbred pairs ($r < 0.25$) (Payne *et al.* 1987).

The number of females that bred with a male did not obviously affect the success of a pair. Where number of young fledged was known for the season, the success of females paired to a monogamous male did not differ significantly from the success of females paired to a polygynous male (Niles, monogamous, $n = 463$, mean 1.76, SD 1.79; polygynous, $n = 344$, mean 1.50, SD 3.48; $t = 1.95$, NS; George Reserve, monogamous, $n = 493$, mean 1.74, SD 1.76; polygynous, $n = 133$, mean 1.62, SD 3.33; $t = 0.72$, NS).

The effect of early experience on subsequent mate choice was tested by comparing local-born and immigrant buntings. Breeding populations of buntings are open to immigration and most breeders on the study area

Table 17.7 Reproductive success (1) of Indigo Buntings that bred on their natal area, and (2) of birds that immigrated as yearlings and adults

Reproductive	Area	(1) Local-born		(2) Immigrants			
		n	mean ± SD	*n*	Mean ± SD	*t*	*P*
Females							
Fleglings in lifetime	Niles	45	3.09 ± 3.64	228	3.86 ± 3.82	1.24	NS
Fledglings in year 1		40	1.78 ± 1.79	228	2.27 ± 1.73	1.67	NS
No. fledglings that returned		45	0.27 ± 0.49	228	0.29 ± 0.61	0.28	NS
Fledglings in lifetime	George Reserve	8	2.13 ± 1.73	186	3.17 ± 3.40	0.86	NS
Fledglings in year 1		6	2.00 ± 1.26	186	2.13 ± 1.80	0.18	NS
No. fledglings that returned		8	0.00	186	0.06 ± 0.08	—	—
Males							
Fledglings in lifetime	Niles	22	3.59 ± 3.74	209	3.96 ± 4.56	0.36	NS
Fledglings in year 1		19	0.74 ± 1.19	183	1.42 ± 1.81	1.61	NS
No. fledglings that returned		22	0.36 ± 0.65	209	0.24 ± 0.61	0.90	NS
Fledglings in lifetime	George Reserve	5	3.20 ± 5.63	277	2.88 ± 4.05	0.17	NS
Fledglings in year 1		2	0.00	222	1.20 ± 1.63	—	—
No. fledglings that returned		5	0.00	277	0.05 ± 0.23	—	—

were born elsewhere (Payne 1991). We found no tendency for buntings born there to seek out or avoid mating with other local-born birds, and pairs where both partners had returned to their natal area had about the same success as pairs where one or both were immigrants. In three pairs a local-born female returned for a second year when her local-born mate also returned; in only one of these pairs did she reunite with him. Reproductive success did not differ between the immigrants and local-born buntings in the number of young fledged in the first year, in their second year, or across their lifetimes, or the number of their young that returned in a later year (Table 17.7).

Discussion

Social monogamy and genetic uncertainty

Most pairs remain together within a season, and half of the surviving pairs reunite in the following year. Within a season, several factors influence the pair bond. Female Indigo Buntings alone provide nearly all the parental care in incubating, feeding, and brooding nestlings, and more nest failures are due to predation than to starvation or inclement weather (Payne 1992). Second, males are territorial and exclude other males from their territories, but the sexual defence is incomplete as some nestlings are fathered by a neighbouring male (Westneat 1987*b*, 1990). Aggression between females was rarely observed and does not appear to be responsible for their dispersion and monogamy. The Niles population with the higher rate of polygyny was the more densely settled, and we saw no aggression there between females that nested in a simultaneous polygynous relationship, and no increased starvation of nestlings or reduction in their reproductive success. Male parental care may contribute to within-season monogamy, as a female whose mate takes over the care of the fledglings begins a new nest in fewer days then females whose males do not care for the fledglings, and this display of male behaviour may encourage a female to stay on the territory and breed again with the same mate.

The low frequency of paternal care of the nestlings may be associated with the weak pair bond. There is little teamwork by a pair. Widowed females rear the young alone as do females whose mates provide no parental care; widowed males allow their young to starve. Males protect a mate from harassment by neighbouring males, but neighbours are attracted to a female only when she is sexually receptive and laying, and not when she has young. Males may also deter predators.

Why do males provide care in the context of genetic uncertainty? Variation in paternal care was not associated with whether the male was cuckolded, and males fed fledged buntings that genetically were not their own (Westneat 1988*a*, 1990). No male other than the social mate provided food to the nestlings. Whittingham *et al.* (1992) suggested that when there is a high probability that at least one young in a brood is a male's own and his care makes a difference in the number of young that fledge, he should feed them all, rather than adjusting his care to the degree of uncertainty, especially if the female does not increase her own care. Male buntings provide little individual care to the young that is not shared by the whole brood (e.g. nest defence), so the cost of caring for a brood of mixed paternity is no higher than if the brood were all their own. The benefits of paternal care to a mate's own young appear to outweigh the costs of caring for a brood of mixed paternity (Westneat and Sherman 1993). In contrast, the variation among species in paternal care may be associated with cuckoldry (Møller and Birkhead 1993). Males in species with a high frequency of extra-pair fertilizations provided little paternal care, and males in species with few extra-pair fertilizations usually fed the young. The behaviour of male buntings suggests that they do not discriminate their genetic offspring from offspring sired by other males, but care for their young according to average population conditions.

Variations in the within-season mate fidelity in Indigo Buntings can be explained as divergent behaviours of males and females each directed towards their own reproductive success. The high proportion of female buntings that changed a mate may be related to the long season and opportunities to find a better mate, and to whether the condition of their mates and habitats will change through the season. A male remains close to his mate during the period when she builds the nest and lays her eggs. Once she begins to incubate, he spends little time with her. He sings throughout the day (Thompson 1972) and may attract an additional female, and he intrudes into other territories where females are sexually receptive. A female centres her activity at the nest. She copulates with her social partner, but she may also copulate with a neighbouring male at her place or his (Payne 1983*a*; Westneat 1987*a*, 1990; Payne and Payne 1989) and she may copulate with as many as three males for a single brood (Quay 1988, 1989). Reproductive success for female buntings is largely independent of paternal care, as the males usually do not feed the young until after they fledge and the young can survive without them.

Social and ecological constraints on mate fidelity

The proportion of pairs that reunite for a second year may reflect population demography and the frequency of breeding dispersal, more than it reflects decisions made by individual birds to reunite or to mate with a new partner (see also Chapters 13 and 14). The low nest survival due to predation on eggs and nestlings, the uncertainty of paternity and associated low level of paternal care of the young, and the short life expectancy of adults are associated with the lack of a strong social bond between partners in the Indigo Bunting, and the breeding dispersal of females between seasons limits the likelihood that a pair will resume their old partnership.

The difference between males and females in breeding dispersal may be related to the annual changes in the vegetation of bunting habitats and to the differences of the sexes in social behaviour. Buntings breed in old fields and the edges of fields, swamps, and woods. These sites change from year to year as trees replace open habitats. In many territories that we watched over the years, the males returned to their site for as many years as they survived, but when the males did not return, the increasingly wooded sites were not re-occupied by other buntings, not even by yearling males. Also, bunting territories tend to occur in groups even when the habitat appears to be suitable over a wider area. This suggests that the males are attracted not only to a habitat, but also to the presence of other buntings, with opportunities for an additional mate and for extra-pair copulations. The advantages of social familiarity between males may in part determine the return of a male to his old territory. For most males a continuing factor from year to year was the return of one or more neighbours with whom they had established mutual territory boundaries in the previous year, with familiar and shared song themes (Payne and Payne 1993*a*, *b*). For the females there were no comparable social relationships. Although females may be familiar with a nest and feeding site, last year's information about the site may be obsolete because their habitat changes from year to year.

There was no obvious reproductive benefit or cost to buntings that reunited with their old mate over those that formed a new pair. The success of females in new pairs did not differ when the old male returned and when he did not. The comparisons between these new pairs, and pairs that mated together for a second season, show that females not in their old partnership were not disadvantaged, and widowed females were not disadvantaged over those with a choice of old and new mate. We do not know whether a female with a new male was courted by her old mate and then actively rejected him in 'divorce'. However, the frequency of breeding dispersal was the same in females whose mates died and in females that mated with a new partner when their old mate did return (Payne and Payne 1993*a*), and so females are free, not forced by circumstance or consequences, to seek a new partner.

Rowley (1983) suggested a lack of social selection for mate fidelity in species with high mortality. On a more proximate level, Ens *et al.* (1993) and Desrochers and Magrath (1993) suggested that high mortality leads to more opportunities for an individual to change to a new mate and form a new partnership. In the buntings there was little cost or gain of mate change, or of mate fidelity, and either interpretation is consistent with the low fidelity of pairs in these songbirds.

Summary

Although Indigo Buntings are usually socially monogamous within a breeding season, pairs often do not continue in permanent partnerships. In general the buntings were inconstant partners. Within a season, 11% of the nesting pairs separated after one nesting attempt and the female moved to nest with another male on another territory, even after a successful nesting. Between breeding seasons, when the old mate returned, only 49% of 221 pairs reunited. Only 16 pairs were constant for at least 3 years, and the longest partnership observed was for 5 years. Reproductive success in buntings that reunited with their old mates was generally the same as in new pairs. Success of pairs was independent of whether the male was monogamous or polygynous, of age, of morphological and genetic similarity among mates, and of whether the birds were immigrants or returns in their natal area. In pairs that resumed their partnerships for 3–5 years, success did not increase with the number of years the pair was together. When their old mate did not return, females were no more likely to mate with a male with the song theme of last-year's mate than with a male with a different song. The low survival of males and females and the breeding dispersal of females limits the proportion of survivors that continue as a pair from year to year. Female dispersal and mate change may be determined by the year to year change in habitat of their breeding sites, while male dispersal may be constrained by the benefits of social familiarity with their neighbours.

Acknowledgements

The University of Michigan gave access to the lands and facilities of the E. S. George Reserve. Amtrak and private owners allowed access to lands at Niles. Banding was carried out under permits from the State of Michigan and the U. S. Fish and Wildlife Service. We thank our field assistants, particularly K. D. Groschupf, C. A. Haas, R. E. Irwin, R. E. Jung, S. Doehlert Kielb, M. T. Stanback, D. F. Westneat, and J. L. Woods. K. E. Guire advised on statistical methods. For comments on the manuscript we thank J. M. Black, A. Craig, M. P. Lombardo, C. S. Parr, I. Newton, and J. L. Woods. The study was supported by the National Science Foundation.

References

Carey, M. (1982). An analysis of factors governing pair-bonding period and the onset of laying in Indigo Buntings. *Journal of Field Ornithology*, **53**, 240–8.

Carey, M. and Nolan, V., Jr. (1979). Population dynamics of Indigo Buntings and the evolution of avian polygyny. *Evolution*, **33**, 1180–92.

Desrochers, A. and Magrath, R. D. (1993). Environmental predictability and remating in European blackbirds. *Behavioral Ecology*, **6**, 271–5.

Ens, B. J., Safriel, U. N., and Harris, M. P. (1993). Divorce in the long-lived and monogamous oystercatcher, *Haematopus ostralegus*: incompatibility or choosing the better option? *Animal Behaviour*, **45**, 1199–217.

Emlen, S. T. (1971). The role of song in individual recognition in the Indigo Bunting. *Zeitschrift für Tierpsychologie*, **28**, 241–6.

Halliday, T. R. (1983). The study of mate choice. In *Mate choice* (ed. P. Bateson), pp. 3–32. Cambridge University Press, Cambridge.

Møller, A. P. and Birkhead, T. R. (1993). Certainty of paternity covaries with paternal care in birds. *Behavioral Ecology and Sociobiology*, **33**, 261–8.

Payne, R. B. (1983*a*). Bird songs, sexual selection, and female mating strategies. In *Social behavior of female vertebrates* (ed. S. K. Wasser), pp. 55–90. Academic Press, New York.

Payne, R. B. (1983*b*). The social context of song mimicry: song-matching dialects in indigo buntings (*Passerina cyanea*). *Animal Behaviour*, **31**, 788–805.

Payne, R. B. (1989). Indigo Bunting. In *Lifetime reproduction in birds* (ed. I. Newton), pp. 153–72. Academic Press, London.

Payne, R. B. (1991). Natal dispersal and population structure in a migratory songbird, the Indigo Bunting. *Evolution*, **45**, 49–62.

Payne, R. B. (1992). Indigo Bunting. In *The birds of North America*, no. 4 (ed. A. Poole, P. Stettenheim, and F. Gill), pp. 1–24. American Ornithologists' Union, Philadelphia.

Payne, R. B. and Payne, L. L. (1989). Heritability estimates and behaviour observations: extra-pair matings in indigo buntings. *Animal Behaviour*, **38**, 457–67.

Payne, R. B. and Payne, L. L. (1990). Survival estimates of Indigo Buntings: comparison of banding recoveries and local observations. *Condor*, **92**, 938–46.

Payne, R. B. and Payne, L. L. (1993*a*). Breeding dispersal in Indigo Buntings: circumstances and consequences for breeding success and population structure. *Condor*, **95**, 1–24.

Payne, R. B. and Payne, L. L. (1993*b*). Song copying and cultural transmission in indigo buntings. *Animal Behaviour*, **46**, 1045–65.

Payne, R. B. and Westneat, D. F. (1988). A genetic and behavioral analysis of mate choice and song neighborhoods in Indigo Buntings. *Evolution*, **42**, 935–47.

Payne, R. B., Thompson, W. L., Fiala, K. L., and Sweany, L. L. (1981). Local song traditions in Indigo Buntings: cultural transmission of behavior patterns across generations. *Behaviour*, **77**, 199–221.

Payne, R. B., Payne, L. L., and Doehlert, S. M. (1987). Song, mate choice and the question of kin recognition in a migratory songbird. *Animal Behaviour*, **35**, 35–47.

Payne, R. B., Payne, L. L., and Doehlert, S. M. (1988). Biological and cultural success of song memes in Indigo Buntings. *Ecology*, **69**, 104–17.

Quay, W. B. (1988). Marking of insemination encounters with cloacal microspheres. *North American Bird Bander*, **13**, 36–40.

Quay, W. B. (1989). Timing of sperm releases and inseminations in resident emberizids: a comparative study. *Condor*, 91, 941–61.

Rowley, I. (1983). Re-mating in birds. In *Mate choice* (ed. P. Bateson), pp. 331–60. Cambridge University Press, Cambridge.

Thompson, W. L. (1970). Song variation in a population of Indigo Buntings. *Auk*, **87**, 58–71.

Thompson, W. L. (1972). Singing behavior of the Indigo Bunting, *Passerina cyanea. Zeitschrift für Tierpsychologie*, **31**, 39–59.

Verner, J. and Willson, M. F. (1969). Mating systems, sexual dimorphism, and the role of male North American passerine birds in the nesting cycle. *Ornithological Monographs*, **9**, 1–76.

Westneat, D. F. (1986). Parental care and alternative mating tactics in the Indigo Bunting. Unpublished Ph.D. thesis. University of North Carolina, Raleigh.

Westneat, D. F. (1987*a*). Extra-pair copulations in a predominantly monogamous bird: observations of behaviour. *Animal Behaviour*, **35**, 865–76.

Westneat, D. F. (1987*b*). Extra-pair fertilizations in a predominantly monogamous bird: genetic evidence. *Animal Behaviour*, **35**, 877–86.

Westneat, D. F. (1988*a*). Male parental care and extra-pair copulations in the Indigo Bunting. *Auk*, **105**, 149–60.

Westneat, D. F. (1988*b*). The relationship among polygyny, male parental care, and female breeding success in the Indigo Bunting. *Auk*, **105**, 372–4.

Westneat, D. F. (1990). Genetic parentage in the indigo bunting: a study using DNA fingerprinting. *Behavioral Ecology and Sociobiology*, **27**, 67–76.

Westneat, D. F. and Sherman, P. W. (1993). Parentage and the evolution of parental behavior. *Behavioral Ecology*, **4**, 66–77.

Westneat, D. F., Frederick, P. C., and Wiley, R. H. (1987). The use of genetic markers to estimate the frequency of successful alternative reproductive tactics. *Behavioral Ecology and Sociobiology*, **21**, 35–45.

Whittingham, L. A., Taylor, P. D., and Robertson, R. J. (1992). Confidence of paternity and male parental care. *American Naturalist*, **139**, 1115–25.

Concluding perspectives

18 *Monogamy and sperm competition in birds*

T. R. BIRKHEAD AND A. P. MØLLER

Introduction

Monogamy in birds comprises two individuals of the opposite sex remaining together for the duration of one or more breeding events. Until recently monogamy was also assumed to imply an exclusive mating relationship between two individuals (Wittenberger and Wilson 1980), but recent behavioural and molecular studies (reviewed in Birkhead and Møller 1992) have shattered the illusion of sexual fidelity: in the majority of species extra-pair copulations and fertilizations outside the pair bond occur routinely. The consequences of this are far reaching, and include effects on fundamentally important life history traits, such as the extent of male parental care (Birkhead and Møller 1992; Møller and Birkhead 1993). The more immediate consequence in the present context is the need to distinguish between social and sexual monogamy (Gowaty 1985). In this book social monogamy is defined as partners remaining together and sexual monogamy as the absence of extra-pair paternity, but not necessarily the absence of extra-pair copulations. When a female is inseminated by two or more males during a single breeding episode the result is sperm competition (Parker 1970, 1984). This term now encompasses not only the processes occurring within the female reproductive tract, but all the morphology, physiology, and behaviour associated with multiple mating by females, from both a male and female perspective (Birkhead 1994).

Most emphasis in this book has been on the hypotheses for divorce or remaining together between breeding attempts. However, these cannot be separated from the evolutionary forces that have resulted in monogamy (*sensu* Wittenberger and Tilson 1980, and discussed in detail by Gowaty Chapter 2). Monogamy, mate fidelity, and divorce are all aspects of mate choice, a major component of sexual selection (Darwin 1871). Sperm competition is also an important part of mate choice (Møller 1991; Birkhead and Møller 1992) and our aim here is to review the relationships between sperm competition and mate fidelity within the framework of sexual selection.

Sexual selection

Sexual selection comprises two components: intrasexual selection (usually competition between males) and intersexual selection (usually female choice) (Darwin 1871). Because in species with an unbiased sex ratio each individual should be able to acquire a partner, it was once assumed that sexual selection would be minimal in monogamous animals. However, nothing could be further from the truth, and it is now clear that there are numerous opportunities for sexual selection in monogamous species (e.g. Fisher 1930; Kirkpatrick *et al.* 1990; Mock and Fujioka 1990; Andersson 1994; Møller 1994; Møller and Birkhead 1994). Male–male competition has never been controversial, but because the two components of sexual selection are often closely intertwined female choice is often difficult to distinguish from male–male competition. It is only in recent years that female choice has been demonstrated unequivocally (see Andersson 1994; Møller 1994). Sperm competition is an integral part of sexual selection, both in terms of male–male competition and female choice.

What do females gain from copulating with more than one male? The benefits females might obtain from copulating with a particular male can be divided into two categories: (1) direct and (2) indirect (genetic) (Westneat *et al.* 1990; Birkhead and Møller 1992). Direct benefits include resources such as food for themselves (i.e. through courtship feeding), or food for their offspring: Hoelzer's (1989) 'good parent' model (see also Heywood 1989). Females may also copulate with more than one male in order to replenish depleted sperm supplies, or as a hedge against one male having poor quality semen (see Sheldon 1994).

Indirect benefits are genetic benefits for the female's offspring. For example, females may copulate with more than one male to increase the genetic diversity of their offspring. The traditional view here is that in an unpredictable environment genetic diversity increases the chance of some offspring surviving (Williams 1975). While this seems an attractive possibility, there are theoretical reasons that make this an unlikely general explanation for females copulating with multiple males (see Birkhead and Møller 1992). However, later in this chapter we discuss the genetic diversity hypothesis further in the light of the number of lifetime partners that birds have (see Discussion). Another interpretation of the genetic diversity hypothesis has been proposed by Keller and Reeve (1994): they suggest that sperm may be sexually selected within the female reproductive tract and that females copulate with several males in order to ensure that their eggs are fertilized by males with high fertilization ability, and that their sons also inherit this trait, and their daughters inherit the tendency to copulate with multiple males. Other explanations for why females copulate with more than one male are the ideas that females choose males on the basis of either the attractiveness of their (genetically based) secondary sexual characteristics (Fisher 1930; Pomiankowski *et al.* 1991), or their genetic quality alone (Zahavi 1975; Hamilton and Zuk 1982; Iwasa *et al.* 1991). The idea that females chose males on the basis of their secondary sexual characteristics or genetic quality has been difficult to explain and a series of models (reviewed in Andersson 1994; Møller 1994) has been devised to account for how choice based on these criteria can occur. The links between the different models of female choice and monogamy are discussed below.

It is also clear that individuals vary considerably in quality (Clutton-Brock 1988; Newton 1989; Møller 1994), and in an ideal world each individual would find a partner with all the right attributes and remain with it for the rest of its breeding life. But it is not this simple and not all individuals obtain their 'preferred option' (Ens *et al.* 1993; Ens *et al.* Chapter 19). This is because mate choice is often constrained, resulting in a relatively poor or incompatible partnership (see Ens *et al.* Chapter 19). However, there are a number of ways that females can modify their choice of partner, which range from performing extra-pair copulations to changing partners. Females may remain with a social partner in order to retain a breeding opportunity but have a different sexual partner with whom they perform extra-pair copulations, as in the Splendid Fairy-wren *Malurus splendens* (Russell and Rowley Chapter 8) and Superb Fairy-wren *M. cyaneus* (Mulder *et al.* 1994). Females may also perform rapid mate switching, that is, pair and copulate with one male but then switch partners within a single fertile period. Because female birds store sperm, often for prolonged periods (Birkhead and Møller 1993*a*), rapid mate switching can also result in sperm competition and extra-pair paternity (see Birkhead *et al.* 1988). Finally, females may change breeding partners between breeding attempts within a season, or between seasons. Extra-pair copulation and divorce thus represent part of a continuum of female choice (Møller 1992). However, it is important to note that even where extra-pair copulation and divorce are absent, sexual selection can still occur. It does not cease with the acquisition of a social or sexual partner, but may continue throughout different stages of an individual's life, for example through differential parental investment (Møller 1994).

Extra-pair copulation and paternity

Extra-pair paternity is widespread among socially monogamous birds and occurs mainly through extra-pair copulation (Birkhead and Møller 1992), although there are some cases of extra-pair paternity resulting from rapid mate switching (e.g. Spotted Sandpiper *Actitis macularia*, Oring *et al.* 1992; European Starling *Sturnus vulgaris*, Pinxten *et al.* 1993; Shag *Phalacrocorax aristotelis*, Graves *et al.* 1993; Royal Penguin *Eudyptes schlegeli*, C. St. Clair personal communication; Adelie Penguin *Pygoscelis adeliae*, F. Hunter personal communication). Despite the prevalence of extra-pair paternity it is truly remarkable that so far, no obvious *ecological* correlates of extra-pair paternity between species have been detected (Birkhead and Møller 1992; Reyer 1994). In this book avian partnerships are divided into two categories, continuous and part-time. However, this particular ecological variable does not provide an explanation for levels of extra-pair copulation in birds (see Table 18.1).

Table 18.1 Relationship between selective pressures favouring social monogamy, number of lifetime social partners, category of pair type, and observed or predicted levels of extra-pair paternity

	No. social partners				Level of extra-pair paternity	
Category and species	Mean	Max	*n*	Pair type	Observed (%)	Predicted
Male care essential						
Bewick's Swan	1.1	7	(2501)	C	0	Low
Whooper Swan	1.2	4	(607)	C	0	Low
Mute Swan	1.3	4	(369)	C	?	Low
Cassin's Auklet	1.5	5	(594)	P	?	Low
Barnacle Goose	1.5	6	(2618)	C	0	Low
Blue Duck	1.6	4	(54)	C	0	Low
Sparrowhawk	1.9	5	(230)	P	Low	Low
Red-billed Gull	1.9	9	(1989)	P	?	Low
Short-tailed Shearwater	1.9	7	(328)	P	?	Low
Failed polygynists						
Great Tit	1.3	?	(1454)	P	16.2	High
Pinyon Jay	1.5	?	(141)	C/P	?	High
White-tailed Ptarmigan	1.5	5	(75)	P	?	High
Willow Ptarmigan	1.6	6	(348)	P	7.0	High
Florida Scrub Jay	1.6	?	(552)	C	0	High
Splendid Fairy-wren	1.7	9	(265)	C	65–100	High
Indigo Bunting	?	?		P	35.0	High
Enforced monogamy						
No species in book	—	—	—	P	—	High[1]

C = continuous partnership, P = part-time partnership (see Black Chapter 1). Values for extra-pair paternity are all percentage of extra-pair offspring, except for Willow Ptarmigan where the value is for percentage of broods. The value for Great Tit is the mean from several studies (from Møller and Birkhead 1995). Predicted levels (high or low) of extra-pair paternity are those predicted by the constrained female hypothesis.

[1]Predicted levels for enforced monogamy may be low because the opportunities are low and/or the cost of extra/pair copulations for females is high, but there may also be forced extra-pair copulations (see text).

Within the continuous partnership category, for example, both very low (zero) and very high levels of extra-pair paternity occur (e.g. swans and geese and fairy-wrens, respectively). Similarly, within the part-time category, low and high levels of extra-pair paternity are also likely to exist (e.g. low: shearwaters, gulls, and penguins; high: Indigo Buntings *Passerina cyanea*).

A number of other correlates have, however, been identified. For example, extra-pair paternity appears to be more common in passerines than in nonpasserines (see Møller and Birkhead 1994; Birkhead and Møller 1995). Controlling for phylogeny, high levels of extra-pair paternity are associated with:

(1) reduced paternal care (Møller and Birkhead 1993);
(2) relatively brightly coloured males (Møller and Birkhead 1994);
(3) males that sing variable/complex songs (A. P. Møller and T. R. Birkhead unpublished data); and
(4) males with relatively large testes (Møller and Briskie 1995).

These factors and the interrelationships between them are discussed below.

In Chapter 2 Gowaty proposes an ecological hypothesis to explain the occurrence of extra-pair copulations: the 'constrained female hypothesis'. This includes the idea that females seek extra-pair copulations when their choice of breeding partner is constrained and/or when the ecological circumstances are such that females can rear young with little or no help from the male partner, or without incurring the cost of reduced male care. Trivers (1972) first suggested that extra-pair copulations could be costly to females in terms of reduced paternal care (see also Birkhead and Møller 1992). More recently, Mulder *et al.* (1994) suggested that the degree of extra-pair paternity may 'reflect a . . . balance between female choice of male genetic quality and the importance of male assistance with parental care'. In other words in those ecological circumstances where females can rear young with little or no help from the male partner (1) because food is superabundant (e.g. Aquatic Warbler *Acrocephalus paludicola*; Schulze-Hagen *et al.* 1995), (2) because helpers take up some of the slack (e.g. Superb Fairy-wren; Mulder *et al.* 1994), or (3) because male partners do not adjust their investment in relation to the likelihood of paternity (e.g. Whittingham *et al.* 1993), then females can afford to engage in extra-pair copulations in order to improve the genetic quality (attractiveness or viability) of their offspring.

Our aim here is to evaluate the constrained female hypothesis. Before doing that, however, it is necessary first to re-consider the evolutionary forces favouring social monogamy, since this provides the framework for the constrained female hypothesis.

Monogamy

Social monogamy evolves as a consequence of the costs and benefits to each sex. Because female reproductive success is usually limited by time and energy constraints, it can be increased by assistance, for example from a male partner. On the other hand, males are less likely to benefit from social monogamy because their reproductive success is usually limited by the number of females they can inseminate. In our view (c.f. Gowaty Chapter 2), Wittenberger and Tilson (1980) identified three main hypotheses for the evolution of social monogamy:

(1) where male care is essential;

(2) failed polygamists; and

(3) enforced monogamy by males monopolizing females.

Gowaty (Chapter 2) has also discussed the selective pressures favouring monogamy, and has outlined three hypotheses from Wittenberger and Tilson's (1980) seminal paper:

(1) male care essential;

(2) polygamy threshold not reached; and

(3) female–female aggression.

In this chapter we have adopted a slightly different approach: our initial category is identical to Gowaty's, but we have combined her second and third categories into a single one, our 'failed polygamists'; this is because males, for example, can fail to be polygynous either because they do not attract additional females or because of female–female aggression (Slagsvold and Lifjeld 1994).

When male parental care (feeding and protection of young) is essential for female reproductive success, the choice of partner in terms of good parenting skills (Heywood 1989; Hoelzer 1989) may be paramount. Indeed, among species in which male care is essential both partners stand to benefit from monogamy and both will benefit from selecting a partner with good parenting abilities. This is the only category in which pair members may have largely similar interests, and comes close to what Mock *et al.* (Chapter 3) refer to as 'True Monogamy'. However, even within this system there is scope for sexual conflict and sexual selection (see below).

Although Mock *et al.* (Chapter 3) have suggested that monogamy and male care are intimately bound together, our failed polygamist category comprises species where male care is *not* essential for *some* female reproductive success. Under these circumstances female choice of social partner may be much less important than, say, choice of habitat in terms of female success (see e.g. Desrocher and Magrath Chapter 9). In this category monogamy occurs because males are failed polygynists and females are failed polyandrists (see Davies 1992). That is, individuals attempt to achieve their preferred option but fail to do so, either because of their own attributes or simply because the opportunity does not arise. However, treating failed polygamists (i.e. this category of monogamous species) and regularly polygamous species separately (e.g. Wittenberger and Tilson 1980; Ens *et al.* Chapter 19) may confound the thinking here and make it more difficult to detect patterns. In the present context it may be more profitable to consider polygamy as part of a continuum. Indeed, all mating systems may comprise a continuum, spanning polyandry, monogamy, and polygyny, as Davies (1992) has shown for the Dunnock *Prunella modularis*. Where females are not dependent upon males for some reproductive success, that is, where males are monogamous because they are unable to attract additional partners, females should settle in the territory that will provide them with the best option for rearing young without male help (polygyny threshold for females, Orians 1969; Davies 1989). Since there may not be a direct relationship between male quality and territory quality, females may attempt to get the best of both worlds, by settling on the territory of one male, but seeking copulations elsewhere (Møller 1991).

Enforced monogamy is a special case where female choice may be highly constrained. Wittenberger and Tilson (1980) referred to this as 'female defence monogamy', and identified dabbling ducks as an important example. However, as McKinney (1985, 1986) has pointed out, the term 'female defence monogamy' ignores the fact that females play an active role in their initial choice of partner. In many dabbling ducks *Anas* spp. both females and males apparently choose a partner in the autumn, but the male then drives off other males and escorts the female continuously until breeding starts several months later in the spring (McKinney *et al.* 1983; McKinney 1985, 1986; Sorenson 1994*a,b*). By escorting the female in this way the male effectively monopolizes the female and may restrict her ability to modify her choice (McKinney 1985). Males provide no paternal care and the pair bond breaks down as soon as the female starts incubating (McKinney 1985, 1986; see Williams and McKinney Chapter 4 for some examples of ducks with persistent partnerships). Competition between males is intense because females are in short supply. Forced extra-pair copulation, which is frequent in this category, may have evolved in response to this intense male–male competition (McKinney *et al.* 1983; McKinney 1985, 1986; Sorenson 1994*a,b*).

Uncertainty of paternity and paternal care

Ever since Trivers (1972) it has been thought that males should not invest in offspring to which they are not genetically related, and he proposed that cuckolded males should reduce their investment in one way or another. This association between male investment and paternity has received considerable theoretical attention (Maynard Smith 1977; Houston and Davies 1985; Whittingham *et al.* 1992; Westneat and Sherman 1993), but predictions arising from theoretical models have been diverse mainly because the assumptions of the models have differed. However, the results from empirical studies are also mixed. In some socially monogamous species, males *do not* adjust their parental investment in line with their likelihood of paternity (*sensu* Schwagmeyer and Mock 1993) (e.g. Dunnock, Davies 1992; Tree Swallow *Tachycineta bicolor*, Whittingham *et al.* 1993; Red-winged Blackbird *Agelaius phoeniceus*, Westneat 1995). In others, male investment is related to paternity. Dixon *et al.* (1994) conducted a pairwise comparison of the *same* male Reed Buntings *Emberiza schoeniclus* under different circumstances (e.g. high and low level of cuckoldry) and compared first and second broods within a season they found that male investment was negatively related to the extent of extra-pair paternity. Similarly, Lubjuhn *et al.* (1993) measured the anti-predator response of male and female Great Tits *Parus major* towards a stuffed owl placed near their nests and then conducted paternity analyses on their broods. They found that male Great Tits defended their nests more aggressively the greater the number of true genetic offspring they contained. In neither case is it known how males assess their likelihood of paternity.

In systems such as co-operative polyandry or co-operative breeding, males *do* adjust their level of care in line with the likelihood of paternity (Burke *et al.* 1989; Davies 1992; Davies *et al.* 1992; Mulder *et al.* 1994) and it may be that the presence of additional caring individuals offsets the disadvantage of their own reduced care. It has been argued that in a conventional system (i.e. without additional carers), because males cannot distinguish their own true offspring from extra-pair offspring, and hence are unable to direct care specifically at their true genetic offspring, any overall reduction in their level of care penalizes any of their true offspring and thus reduces their fitness (Hatchwell and Davies 1990; Davies *et al.* 1992).

A comparative study revealed that levels of male care during the nestling period were significantly and negatively associated with levels of extra-pair paternity (Møller and Birkhead 1993). In other words, in species with high levels of extra-pair paternity, male investment was lower than in other species. It therefore seems likely that male care is linked with 'confidence of paternity' in the sense that levels of male care are, to some extent, evolutionarily 'fixed' according to average levels of extra-pair paternity they are likely to experience. However, it is important to note that this does not imply a uniform response. This is because the costs of providing care will vary markedly between males of different quality, and the benefits of care in terms of reproductive success will also be highly variable, and will depend, for example, on local feeding conditions (see below).

The constrained female: social monogamy and male care

A subset of the 'constrained female hypothesis' (Gowaty Chapter 2: Fig. 2.2) states that with increasing female quality or environmental quality, a female's fitness increases up to an asymptote. Below a certain

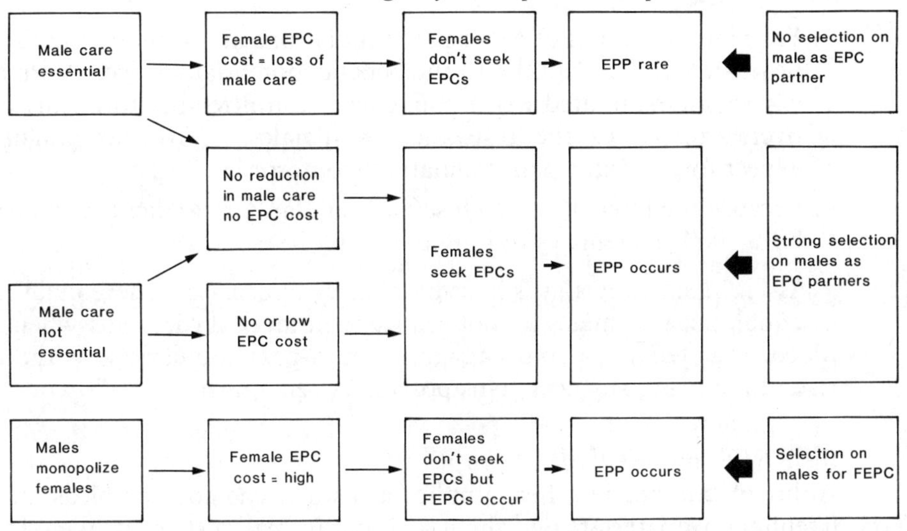

Fig. 18.1 Relationship between evolutionary forces favouring social monogamy (*three left hand boxes*) and the likelihood of extra-pair copulation (EPC), and selection pressures on males (*three right hand boxes*); EPP = extra-pair paternity, FEPC = forced extra-pair copulation. See text for further details.

level of female or environmental quality females depend upon male help (Gowaty's 'helpful coercion') and females are therefore less likely to engage in extra-pair copulations because they risk losing the male's help, which would result in an inevitable reduction in female fitness. At the other extreme, high quality females or those in good quality environments are not dependent on male assistance and are therefore free to engage in extra-pair copulations. The prediction is, therefore, that levels of extra-pair paternity will be higher in this situation. We discuss the predictions regarding levels of extra-pair paternity and divorce in three categories of monogamous birds (summarized in Fig. 18.1).

Where females are dependent on male assistance to breed successfully

In this category of birds extra-pair copulations may be too costly for females. If, as a consequence of a female engaging in extra-pair copulations, a male partner withholds some parental care (see above) and the cost of chick rearing for the female is increased, or withholds all care resulting in reproductive failure, it will pay females to remain faithful and retain their partner's full assistance. Several predictions arise from this hypothesis.

1. Females should rarely seek extra-pair copulations (unless it is linked with mate change, see below) and be reluctant to engage in extra-pair copulations initiated by males.
2. As a consequence levels of extra-pair paternity should be low.

3. For males the costs of extra-pair copulations are probably small (Birkhead and Møller 1992) and the benefits potentially large, so we might expect male-initiated extra-pair copulation attempts to occur. Moreover, because of the resistance of females (above) we might also expect forced extra-pair copulations to occur.
4. Because the interests of each sex are similar, we predict that males will be as likely as females to initiate divorce.

As Fig. 18.1 shows, it is possible that in some cases where male care is essential, because males do not reduce their level of care even when cuckolded, that females could engage in extra-pair copulations without the risk of incurring any cost. This possibility remains to be explored.

Male care not essential

Although the heading here implies a distinct category of birds, and for simplicity we have treated them as such, in reality there is likely to be a continuum between male care being essential and being nonessential. Where male care is not essential for some female reproductive success, then above a certain minimum, the quality of the male partner might be virtually irrelevant, especially if his investment other than defending the territory is small. Under such circumstances females can modify their choice of sexual partner through extra-pair copulation. The main predictions are therefore that because the cost of infidelity (in terms of reduced assistance from their partner) is relatively low because females can rear some young unaided, females should seek and engage in extra-pair copulations. However, the cost in terms of reduced paternal investment may still be important, and females will trade the loss of male care against the gain in genetic quality of their offspring. Extra-pair paternity should therefore be more frequent in this category than in those species in which male care is essential (see above). A major consequence of this is that once females start to seek better genes from extra-pair copulation partners there will be strong selection on males to advertise their quality, through either plumage or song, and to have the reproductive capability to perform extra-pair copulations (see below). In terms of divorce, where females choose territories, as may occur in species where males are potential or realized polygynists (Alatalo *et al.* 1986), then the 'better option' is likely to be a better territory, rather than a better male. Among those species where male care is not essential, we would expect females to be the most likely to initiate divorce. However, these predictions are likely to be confounded by a number of different factors, including the fact that males are more philopatric than females (Greenwood 1980). Indeed, if females are more likely to initiate divorce this may contribute to male philopatry!

Although the cost of extra-pair copulation might be relatively low for birds in this category, it is not negligible and females may still have to provide their social partner with some paternity in order to ensure his continued assistance. With no paternity at all males would be under strong selection to desert (see Whittingham *et al.* 1992). There is increasing evidence that females are able to control levels of extra-pair paternity (Lifjeld and Robertson 1992; Birkhead and Møller 1993*b*) and do so to ensure that their social partner fathers some offspring. This is illustrated by Mulder *et al.*'s (1994) study of Superb Fairy-Wrens: levels of extra-pair paternity are significantly higher in those females living in groups comprising numerous care-giving helpers than they are in socially monogamous pairs without helpers. In the latter situation the female depends much more on the care from her social partner, and hence appears to allow him a greater share of paternity to gain his assistance (Mulder *et al.* 1994).

Male care not essential—females monopolized by males

Where females are monopolized by males, female choice is obviously constrained. We might predict therefore that females would seek extra-pair copulations in order to exert some choice of which male fertilizes their eggs. However, enforced monogamy may severely constrain any opportunities females have for extra-pair copulations. Because competition for females and matings is intense, forced extra-pair copulation may be the only way some males achieve paternity. We might also predict that neither sex will benefit from retaining the same partner so divorce rates between seasons will be high.

Some preliminary tests of the predictions

In Table 18.1 we have categorized each of the case study species from the earlier chapters in terms of the selective forces favouring monogamy and the recorded levels of extra-pair paternity. To undertake a preliminary test of the constrained female hypothesis we have used data presented in Table 18.1 and combined these with other data from the literature (Birkhead and Møller 1992; Møller and Birkhead 1995; and unpublished data).

1. Where females are dependent on male assistance to breed successfully there is increasing evidence that females rarely seek extra-pair copulations, indicating that levels of extra-pair paternity should be low. Females in this category often refuse extra-pair copulations (e.g. Kittiwake *Rissa tridactyla*, Chardine 1986, 1987; Lesser Black-backed Gull *Larus fuscus*, MacRoberts 1973; Red-billed Gull *Larus novaehollandiae scopulinus*, (Mills 1994; Mills *et al.* Chapter 16). As a consequence, levels of extra-pair paternity are also expected to be low. We have compared levels of extra-pair paternity in species in which male care is either assumed to be essential (e.g. in seabirds and raptors; Wittenberger and Tilson 1980), or has been shown by observation or experiment to be essential, with all

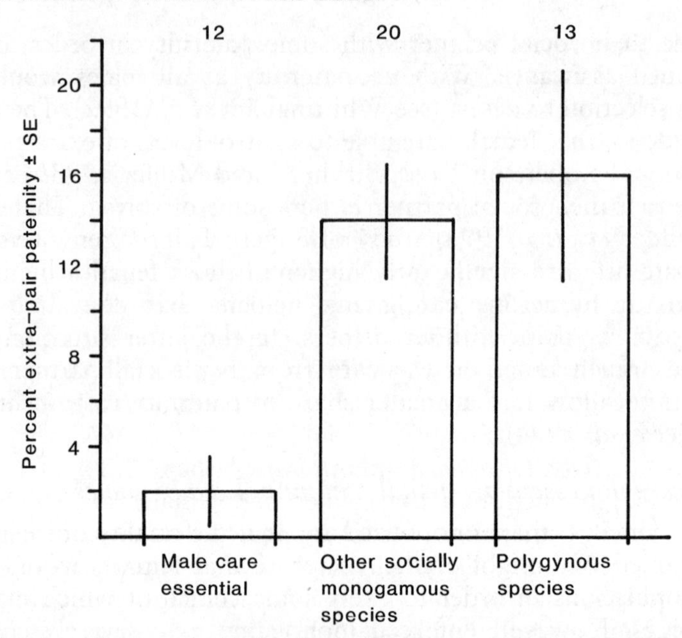

Fig. 18.2 A preliminary analysis of the effect of the necessity of male parental care on extra-pair paternity. In species in which experiments or observations have demonstrated that male care is vital for female reproductive success (*left hand column*) levels of extra-pair paternity are considerably lower than in either other socially monogamous species or polygynous species, that is species in which male care is not essential for some female reproductive success (see Wittenberger and Tilson 1980; also see text). Numbers above each column indicate number of species in the comparison (data from Table 18.1 and Birkhead and Møller 1992, 1994).

other species. This is a relatively crude comparison because the 'other' category may well contain species for which subsequent study may show male care to be vital. Also, in this analysis we have not controlled for phylogenetic effects. However, as Fig. 18.2 demonstrates, in species where male care is essential levels of extra-pair paternity are generally low (as in Table 18.1).

While the hypothesis predicts that in general female-initiated extra-pair copulations will be infrequent, some may occur as a mechanism to facilitate mate change. This seems to be the case in a number of species (for example, Common Guillemot *Uria aalge*, Birkhead *et al.* 1985; Oystercatchers *Haematopus ostralegus*, Heg *et al.* 1993; Spotted Sandpiper, Oring *et al.* 1992). This bet-hedging occurs early in the season in species such as Adelie Penguins (F. Hunter, cited in Williams Chapter 15) where partners may return to the breeding site at different times. It also appears to occur at the end of the season in Razorbills *Alca torda* (Wagner 1991) and Common Guillemots (T. R. Birkhead personal observation), when the female is left alone at the site once the male partner has departed with the chick. Similarly, female birds of prey in poor quality territories or with a poor quality male may visit other males and engage in extra-pair copulations (Newton and Wyllie Chapter 14; see also Poole 1985) as part of a mate change process.

A further prediction is that while it may be costly for females to engage in extra pair copulation it will not be costly for males and hence we would expect to see male-initiated extra-pair copulations. Indeed, in many species of seabird extra-pair copulations seem to be initiated predominantly by males, and because females are reluctant to engage in extra-pair copulations in many cases these are also forced (e.g. Common Guillemots, Birkhead *et al.* 1985; Hatchwell 1988; Emperor Penguin *Aptenodytes fosteri*, Prûvost 1953; Sooty Tern *Sterna fuscata* C. J. Feare personal communication, cited in Birkhead and Møller 1992: p. 109; Laysan Albatross *Diomedea immutabilis*, Fisher 1971).

The final prediction is that males and females are equally likely to divorce their partners in order to improve the quality of their partner (in terms of parenting ability). Although several authors in this book have indicated that females are more likely than males to initiate divorce, this again is confounded by male philopatry. Overall, there are probably relatively few unambiguous data on which sex initiates divorce. Divorce in birds where male care is essential should occur as a result of differences in compatibility or quality. This in turn may be reflected in reproductive success, with incompatible pairs being less likely to breed successfully. Although young Red-billed Gulls that fail to rear chicks are more likely to divorce (Mills *et al.* Chapter 16), the relationship between breeding failure and divorce is relatively weak overall (Ens *et al.* Chapter 19), possibly because one partner compensates for the other. However, there may be an interesting sexual conflict here: where a discrepancy in the quality of partners exists the poorer quality individual will want to retain the other (better quality individual) as a partner (Petrie and Hunter 1994) and may be prepared to invest differentially (Burley 1986, 1988) in order to do so. The better quality individual, on the other hand, despite its reduced effort, might do better to divorce and obtain a better quality partner.

2. Where male care is not essential we expect females to engage in extra-pair copulations because although they may incur a cost of reduced male care, the cost may be outweighed by the indirect (genetic) benefits to their offspring. Similarly, where males do not reduce their level of care in response to the loss of paternity, females will be free to seek extra-pair copulations. In both cases there will be strong selection on males to advertise their genetic quality. It is not surprising therefore that in socially monogamous species in which levels of extra-pair paternity are high, plumage sexual dimorphism is relatively extreme (Møller and Birkhead 1994) and song structure is relatively complex. Since relative sperm

numbers play a central role in determining the outcome of sperm competition (Martin *et al.* 1974; Colegrave *et al.* 1995) males will also be under strong selection to be able to perform effective extra-pair copulations when the opportunity arises. We therefore predict that males of these species will have relatively large testes (Møller 1991; Møller and Briskie 1995), and relatively large seminal glomera containing high numbers of spermatozoa (Birkhead *et al.* 1993*a*).

Unconstrained female choice for extra-pair copulation partners is similar in many respects to the situation in lekking birds (Møller 1992) and the result in both cases is that some males are disproportionately successful in obtaining copulations and paternity (e.g. Dixon *et al.* 1994; Mulder *et al.* 1994). It seems likely that female choice is responsible for the extreme secondary sexual characters in the males of these species (Darwin 1871; see also above). However, the situation in socially monogamous birds differs from that in lekking species in that male partners usually provide some parental care. But as Burley (1986, 1988; see also Burley *et al.* 1994) has shown, attractive males (i.e. the type of male chosen by females as extra-pair copulation partners) tend to invest relatively less in paternal care than unattractive males. Depending on their attractiveness males have two ways to increase their fitness. Attractive males engage in extra-pair copulations and invest relatively little in their offspring, (even though they are likely to be their genetic father), while unattractive males invest most in rearing offspring in their own nest—even if these are sometimes fathered by another male. The latter constitutes making the best of a bad job.

3. Where monogamy is enforced by males and female choice of partner is constrained we might also expect females to seek extra-pair copulations. However, contrary to this prediction, it appears that females in this category only rarely seek or initiate extra-pair copulations particularly *Anas* species living in open water habitat (McKinney *et al.* 1983; Sorenson 1994*a*,*b*). There are two reasons why this might be the case. First, it may be a consequence of close guarding by males, as in dabbing ducks (e.g. Goodburn 1984), which, facilitated by their open habitat, may prevent females from gaining uninterrupted access to other males. Second, extra-pair copulations may be extremely costly for females. In ducks, extra-pair copulation attempts often end up involving several males and are so vigorous that females are sometimes drowned (McKinney *et al.* 1983). In addition, following an extra-pair copulation, females are often subject to forced pair copulations by their partner. The presence of a phallus may be associated with forced copulation (both pair and extra-pair) in dabbling ducks. Given that females apparently rarely seek extra-pair copulations and refuse all extra-pair attempts by males, the occurrence of extra-pair paternity (Evarts and Williams 1987) indicates that forced extra-pair copulations result in the fertilization of

eggs and that males are able to control females. Females probably benefit from monopolization and intense mate guarding by males because it protects them from harassment by other males (e.g. Ashcroft 1976; McKinney 1985). On the other hand, the sons of females fertilized by forced extra-pair copulations may also be successful in using this strategy (see also Weatherhead and Robertson 1979; Harvey and Bennet 1985).

A second prediction relating to enforced monogamy is that the inter-seasonal divorce rates will be high and as McKinney (1985) has shown, this seems to be the case (see also Williams and McKinney Chapter 4).

Discussion

The results of our preliminary test of the constrained female hypothesis look promising, but before we can draw firm conclusions it will be necessary to obtain more data, and perform more rigorous tests, controlling for phylogeny (see Harvey and Pagel 1991) and potentially confounding effects. Wittenberger and Tilson's (1980) original hypotheses for social monogamy appear to have stood the test of time reasonably well, and provide a good starting point for making predictions about why individuals of each sex should modify their choice of partner, either through extra-pair copulations or divorce. The mixture of social and sexual monogamy that we observe in birds will depend upon the relative costs and benefits of particular strategies for each sex and the outcome of the conflict between the sexes. In those species in which both members of the pair need to co-operate to rear any offspring, the interests of each sex are broadly similar and the costs of extra-pair copulations for females outweigh any possible benefits. Social and sexual monogamy therefore coincide. In other species, however, there is a conflict of interests between the sexes and females appear to be able to balance the costs and genetic benefits by controlling the paternity of their offspring to produce a judicial mixture of pair- and extra-pair paternity (see Birkhead and Møller 1993*b*; Birkhead *et al.* 1993*a*,*b*; Lifjeld *et al.* 1994; Mulder *et al.* 1994). In a minority of species males appear to be able to monopolize and control females making extra-pair copulation too costly for females to initiate, but resulting in a situation in which forced extra-pair copulation by males is routine.

Gowaty's constrained female hypothesis proposes that observed levels of extra-pair paternity are the outcome of conflict and a coevolutionary arms race between males and females, with males selected to control females and females selected to resist male control. This concept is in tune with Parker's (1970) initial study of sperm competition and with other subsequent studies (e.g. Hrdy and Williams 1983; Briskie and Montgomerie 1992). Where Gowaty's theory breaks new ground is in the idea that it is *female* quality, or the quality of the environment in which females breed, that determines levels of extra-pair paternity. Our results (Fig. 18.2) provide support for this idea at the interspecific level, but Gowaty's hypothesis also provides a neat explanation for intraspecific variation in levels of extra-pair paternity (see above). It is difficult to test the intraspecific component of the hypothesis using existing data, simply because little is known about differences in female or environmental quality. However, as Gowaty (Chapter 2) points out, it should be possible to design field experiments to test this.

One result to emerge from this book is the fact that the mean number of social partners over a bird's lifetime is remarkably similar across species, regardless of either life span or the selective forces maintaining monogamy (Table 18.1). However, in terms of genetic partners, and the genetic diversity of offspring, there are likely to be marked differences between different categories of birds. In species where male care is essential, levels of extra pair copulation are low (Fig. 18.2) and it therefore follows that the lifetime number of genetic partners will also be low. In contrast, in the case of failed polygynists levels of extra-pair paternity are higher (Fig. 18.2) and hence the number of genetic partners will also be higher. If extra-pair behaviour was a way of increasing the genetic diversity of offspring we would have predicted that those species with a low number of lifetime social partners should have engaged more often in extra-pair copulations, but there is clearly no evidence for this. If genetic diversity were important, we might expect the total number of genetic partners to be similar across all categories of birds.

Conclusions and summary

Paternity data now exist for about 80 or so species of birds. At first sight this would appear to be a sufficiently large data set in order to conduct comparative analyses to test hypotheses for the evolution of monogamy. However, this is not the case. In terms of the hypothesis described here, there are still relatively few parentage studies of species in which male assistance is essential. Moreover, there are numerous confounding variables that need to be considered in such analyses, and this requires relatively large data sets. In addition, apart from seabirds and some raptors where it is known from observation that both partners are essential for successful reproduction, there have been few studies (but see Black *et al.* Chapter 5 and Hannon and Martin Chapter 10) that have tested directly for the effect of male care on female reproductive success, thus imposing a further constraint on testing this hypothesis.

The use of molecular techniques to assign paternity and the discovery of high levels of extra-pair paternity in birds has transformed avian social monogamy into an evolutionary enigma. The hypotheses discussed in this chapter provide a possible way to explain the apparent contradiction between social monogamy and sexual infidelity. The way forward is to continue to undertake long-term and comparative studies, but also to perform detailed, experimental studies of single species under field conditions to test specific predictions. By adopting an experimental approach we shall be able to overcome the numerous confounding variables that make the interpretation of monogamy such a challenge.

Acknowledgements

We are extremely grateful to Bruno Ens and Ben Hatchwell for critical and constructive comments on the manuscript. TRB's research on sperm competition is funded by BBSRC and the Royal Society and APM's research is funded by the Swedish National Research Council.

References

Alatalo, R. V., Lundberg, A., and Glynn, C. (1986). Female pied flycatchers choose territory quality and not male characteristics. *Nature*, **323,** 152–3.

Andersson, M. (1994). *Sexual selection*. Princeton University Press, Princeton.

Ashcroft, R. (1976). A function of the pairbond in the Common Eider. *Wildfowl*, **27,** 101–5.

Birkhead, T. R. (1994). Enduring sperm competition. *Journal of Avian Biology*, **25,** 167–70.

Birkhead, T. R. and Møller, A. P. (1992). *Sperm competition in birds: evolutionary causes and consequences*. Academic Press, London.

Birkhead, T. R. and Møller, A. P. (1993*a*). Sexual selection and the temporal separation of reproductive events: sperm storage data from reptiles, birds and mammals. *Biological Journal of the Linnean Society*, **50,** 295–311.

Birkhead, T. R. and Møller, A. P. (1993*b*). Why do male birds stop copulating while their partners are still fertile? *Animal Behaviour*, **45,** 105–18, 295–311.

Birkhead, T. R. and Møller, A. P. (1995). Extra-pair copulation and extra-pair paternity in birds. *Animal Behaviour*, **49,** 843–8.

Birkhead, T. R., Johnson, S. D., and Nettleship, D. N. (1985). Extra-pair matings and mate guarding in the common murre *Uria aalge*. *Animal Behaviour*, **33,** 608–19.

Birkhead, T. R., Pellatt, J. E., and Hunter, F. M. (1988). Extra-pair copulation and sperm competition in the zebra finch. *Nature*, **334,** 60–2.

Birkhead, T. R., Briskie, J. V., and Møller, A. P. (1993*a*). Male sperm reserves and copulation frequency in birds. *Behavioral Ecology and Sociobiology*, **32,** 85–93.

Briskie, J. V. and Montgomerie, R. (1992). Sperm size and sperm competition in birds. *Proceedings of the Royal Society, London B*, **247,** 89–95.

Burke, T., Davies, N. B., Bruford, M. W., and Hatchwell, B. J. (1989). Parental care and mating behavior of polyandrous dunnocks *Prunella modularis* related to paternity by DNA fingerprinting. *Nature*, **338,** 249–51.

Burley, N. (1986). Sexual selection for aesthetic traits in species with biparental care. *American Naturalist*, **127,** 415-45.

Burley, N. (1988). The differential allocation hypothesis: an experimental test. *American Naturalist*, **132,** 611–28.

Burley, N. T., Enstrom, D. A., and Chitwood, L. (1994). Extra-pair relations in zebra finches: differential male success results from female tactics. *Animal Behaviour*, **48,** 1031–41.

Chardine, J. W. (1986). Interference of copulation in a colony of marked Black-legged Kittiwakes. *Canadian Journal of Zoology*, **64**, 1416–21.

Chardine, J. W. (1987). The influence of pair-status on the breeding behaviour of the Kittiwake *Rissa tridactyla* before egg-laying. *Ibis*, **129**, 515–26.

Colegrave, N., Birkhead, T. R., and Lessells, C. M. (1995). Sperm precedence in zebra finches does not require special mechanisms of sperm competition. *Proceedings of the Royal Society, London B*, **259**, 223–8.

Clutton-Brock, T. H. (1988). *Reproductive success*. Chicago University Press, Chicago.

Darwin, C. (1871). *The descent of man, and selection in relation to sex*, John Murray, London.

Davies, N. B. (1989). Sexual conflict and the polygamy threshold. *Animal Bahaviour*, **38**, 226–34.

Davies, N. B. (1992). *Dunnock behaviour and social evolution*, Oxford University Press, Oxford.

Davies, N. B., Hatchwell, B. J., Robson, T., and Burke, T. (1992). Paternity and parental effort in dunnocks *Prunella modularis*: how good are male chick-feeding rules? *Animal Behaviour*, **43**, 729–46.

Dixon, A., Ross, D., O'Malley, S. L. C., and Burke, T. (1994). Paternal investment inversely related to degree of extra-pair paternity in the reed bunting (*Emberiza schoeniclus*). *Nature*, **371**, 698–700.

Ens, B. J., Safriel, U. N., and Harris, M. P. (1993). Divorce in the long-lived and monogamous oystercatcher *Haemotopus ostralegus*: incompatability or choosing a better option? *Animal Behaviour*, **45**, 1199–217.

Evarts, S. and Williams, C. J. (1987). Multiple paternity in a wild population of mallards. *Auk*, **104**, 597–602.

Fisher, H. I. (1971). The Laysan Albatross: its incubation, hatching, and associated behaviors. *Living Bird*, **10**, 19–78.

Fisher, R. A. (1930). *The genetical theory of natural selection*. Clarendon Press, Oxford.

Goodburn, S. F. (1984). Mate guarding in the Mallard *Anas platyrhynchos*. *Ornis Scandinavica*, **15**, 261–5.

Gowaty, P. A. (1985). Multiple parentage and apparent monogamy in birds. In *Avian monogamy* (ed. P. A. Gowaty and D. W. Mock) pp. 11–21. American Ornithologists' Union, Washington, DC.

Graves, J., Ortega-Ruano, J., and Slater, P. J. B. (1993). Extra-pair copulations and paternity in shags: do females choose better males? *Proceeding of the Royal Society B.*, **253**, 3–7.

Greenwood, P. J. (1980). Mating systems, philopatry and dispersal in birds and mammals. *Animal Behaviour*, **28**, 1140–62.

Hamilton, W. D. and Zuk, M. (1982). Heritable true fitness and bright birds: a role for parasites? *Science*, **218**, 384–7.

Harvey, P. H. and Bennet, P. M. (1985). Sexual dimorphism and reproductive strategies. In *Human sexual dimorphism*. (ed. J. Ghesquire, R. D. Martin, and F. Newcombe), pp. 43–59. Taylor and Francis.

Harvey, P. H. and Pagel, M. D. (1991). *The comparative method in evolutionary biology*. Oxford University Press, Oxford.

Hatchwell, B. J. (1988). Intraspecific variation in extra-pair copulation and mate defence in common guillemots *Uria aalge*. *Behaviour*, **107**, 157–85.

Hatchwell, B. J. and Davies, N. B. (1990). Provisioning of nestlings by dunnocks, *Prunella modularis*, in pairs and trios: compensation reactions by males and females. *Behavioural Ecology and Sociobiology*, **27**, 199–209.

Heg, D., Ens, B. J., Burke, T., Jenkins, L., and Kruijt, J. P. (1993). Why does the typically monogamous oystercatcher (*Haemotopus ostralegus*) engage in extra-pair copulations? *Behaviour*, **126**, 247–87.

Heywood, J. S. (1989). Sexual Selection by the Handicap Mechanism. *Evolution*, **43**, 1387–97.

Hoelzer, G. A. (1989). The good parent process of sexual selection. *Animal Behaviour*, **38**, 1067–78.

Houston, A. I. and Davies, N. B. (1985). The evolution of cooperation and life history in the dunnock, *Prunella modularis*. In *Behavioural ecology: ecological consequences of adaptive behaviour*. (ed. R. M. Sibly and R. H. Smith), pp. 471–87. Blackwell, Oxford.

Hrdy, S. B. and Williams, G. C. (1983). Behavioral biology and the double standard. In: *Social behavior of female vertebrates*. (ed. S. K. Wasser), pp. 3–35. Academic Press, New York.

Iwasa, Y., Pomiankowski, A., and Neee, S. (1991). The Evolution of Costly Mate Preferences II. The 'handicap' principle, *Evolution*, **45**, 1431–42.

Keller, L. and Reeve, H. K 1994. Why do females mate with multiple males? The sexually selected sperm hypothesis. *Advances in the Study of Behaviour*, **24**, 291–315.

Kirkpatrick, M., Price, T., and Arnold, S. J. (1990). The Darwin–Fisher theory of sexual selection in monogamous birds. *Evolution*, **44**, 180–193.

Lifjeld, J. T. and Robertson, R. J. (1992). Female control of extra-pair fertilization in tree swallows. *Behavioural Ecology and Sociobiology*, **31**, 89–96.

Lifjeld, J. T., Dunn, P. O., and Westneat, D. F. (1994). Sexual selection by sperm competition in birds: male–male competition or female choice? *Journal of Avian Biology*, **25**, 244–50.

Lubjuhn, T., Curio, E., Muth, S. C., Brun, J., and Epplen, J. T. (1993). Influence of extra-pair paternity on parental care in great tits (*Parus major*.) In *DNA fingerprinting: state of the science*, (ed. S. D. J. Pena, R. Chakraborty, J. T. Epplen, and A. J. Jeffreys), pp. 379–85. Birkhauser Verlag, Switzerland.

McKinney, F. (1985). Primary and secondary male reproductive strategies of dabbling ducks. In *Avian monogamy* (ed. P. A. Gowaty and D. W. Mock) pp. 68–82. American ornithologists' Union, Washington, D.C.

McKinney, F. (1986). Ecological factors influencing the social systems of migratory dabbling ducks. In *Ecological aspects of social evolution*. (ed. D. I. Rubenstein, and R. W. Wrangham) pp. 153–71 Princeton University Press, Princeton.

McKinney, G., Derrickson, S. R., and Mineau, P. (1983). Forced copulation in waterfowl. *Behaviour*, **86**, 250–94.

MacRoberts, M. H. (1973). Extramarital courting in lesser black-backed and herring gulls. *Zeitschrift für Tierpsychologie*, **32**, 62–74.

Martin, P. A., Reimers, T. J., Lodge, J. R., and Dziuk, P. J. (1974). The effect of ratios and numbers of spermatozoa mixed from two males on proportions of offspring. *Journal of Reproductive Fertility*, **39**, 251–8.

Maynard Smith, J. (1977). Parental investment: a prospective analysis. *Animal Behaviour*, **25**, 1–9.

Mills, J. A. (1994). Extra-pair copulations in the red-billed gull: females with high quality, attentive males resist. *Behaviour*, **128,** 41–64.

Mock, D. W. and Fujioka, M. (1990). Monogamy and long term pair-bonding in vertebrates. *Trends in Ecology and Evolution*, **5,** 39–43.

Møller, A. P. (1991). Sperm competition, sperm depletion, paternal care and relative testis size in birds. *American Naturalist*, **137,** 882–906.

Møller, A. P. (1992). Frequency of female copulations with multiple males and sexual selection. *American Naturalist*, **139,** 1089–101.

Møller, A. P. (1994). *Sexual selection and the barn swallow*. Oxford University Press, Oxford.

Møller, A. P. and Birkhead, T. R. (1993). Certainty of paternity covaries with paternal care in birds. *Behavioral Ecology and Sociobiology*, **33,** 261–8.

Møller, A. P. and Birkhead, T. R. (1994). The evolution of plumage brightness in birds is related to extra-pair paternity. *Evolution*, **48,** 1089–100.

Møller, A. P. and Briskie, J. V. (1995). Extra-pair paternity, sperm competition and the evolution of testis size in birds. *Behavioral Ecology and Sociobiology*, **36,** 357–65.

Mulder, R. A., Dunn, P. O., Cockburn, A., Lazenby-Cohen, K. A., and Howell, M. J. (1994). Helpers liberate female fairy-wrens from constraints on extra-pair mate choice. *Proceedings of The Royal Society of London B.*, **255,** 223–9.

Newton, I. (1989). *Lifetime reproduction in birds*. Academic Press, London.

Orians, G. H. (1969). On the evolution of mating systems in birds and mammals. *American Naturalist*, **103,** 589–603.

Oring, L. W., Fleischer, R. C., Reed, J. M., and Marsden, K. E. (1992). Cuckoldry through stored sperm in the sequentially polyandrous spotted sandpiper. *Nature*, **359,** 631–3.

Parker, G. A. (1970). Sperm competition and its evolutionary consequences in the insects. *Biological Reviews*, **45,** 525–67.

Parker, G. A. (1984). Sperm competition and the evolution of animal mating strategies. In: *Sperm competition and the evolution of animal mating systems* (ed. R. L. Smith), pp. 1–60. Academic Press, Orlando.

Petrie, M. and Hunter, F. M. (1994). Intraspecific variation in copulation frequency: An effect of mismatch in partner attractiveness? *Behaviour*, **12,** 265–77.

Pinxten, R., Hanotte, O., Eens, M., Verheyen, R. F., Dhondt, A. A., and Burke, T. (1993). Extra-pair paternity and intraspecific brood parasitism in the European starling, *Sturnus vulgaris*: evidence from DNA fingerprinting. *Animal Behaviour*, **45,** 795–809.

Pomiankowski, A., Iwasa, Y., and Nee, S. (1991). The evolution of costly mate preferences. I. Fisher and biased mutation. *Evolution*, **45,** 1422–30.

Poole, A. (1985). Courtship feeding and Osprey reproduction. *Auk*, **102,** 479–92.

Prûvost, J. (1953). Formation des couples, ponte et incubation chez le manchot empereur. *Alauda*, **21,** 141–56.

Reyer, U. (1994). 'Do-si-do your partner': Report on the Annual Conference of the BOU. *Ibis*, **136,** 110–12.

Schulze-Hagen, K., Leisler, B., Birkhead, T. R., and Dyrcz, A. (1995). Prolonged copulation, sperm reserves and sperm competition in the Aquatic Warbler *Acrocephalus paludicola*. *Ibis*, **137,** 85–91.

Schwagmeyer, P. L. and Mock, D. W. (1993). Shaken confidence of paternity. *Animal Behaviour*, **46**, 1020–22.

Sheldon, B. C. (1994). Male phenotype, fertility, and the pursuit of extra-pair copulations by female birds. *Proceedings of the Royal Society of London, B.*, **257**, 25–30.

Slagsvold, T. and Lifjeld, J. T. (1994). Polygyny in birds: the role of competition between females for male parental care. *American Naturalist*, **143**, 59–94.

Sorenson, L. G. (1994*a*). Forced extra-pair copulation and mate guarding in the white-cheeked pintail: timing and trade-offs in an asynchronously breeding duck. *Animal Behaviour*, **48**, 519–33.

Sorenson, L. G. (1994*b*). Forced extra-pair copulation in the White-cheeked Pintail: male tactics and female responses. *Condor*, **96**, 400–10.

Trivers, R. L. (1972). Parental investment and sexual selection. In: *Sexual selection and the descent of man, 1871–1971*. (ed B. Campbell) pp. 136–79, Chicago.

Wagner, R. H. (1991). The use of extra-pair copulations for mate appraisal by razorbills, *Alca torda*. *Behavioral Ecology*, **2**, 198–203.

Weatherhead, P. J. and Robertson, R. J. (1979). Offspring quality and the polygyny threshold: 'The sexy son hypothesis'. *American Naturalist*, **113**, 201–8.

Westneat, D. F. (1995). Paternity and paternal behavior in the red-winged blackbird *Agelaius phoeniceus*. *Animal Behaviour*, **49**, 21–35.

Westneat, D. F. and Sherman, P. W. (1993). Parentage and the evolution of parental behavior. *Behavioral Ecology*, **4**, 66–77.

Westneat, D. F., Sherman, P. W., and Morton, M. L. (1990). The ecology and evolution of extra-pair copulations in birds. *Current Ornithology*, **7**, 331–69.

Whittingham, L. A., Taylor, P. D., and Robertson, R. J. (1992). Confidence of paternity and male parental care. *American Naturalist*, **139**, 1115–25.

Whittingham, L. A., Dunn, P. O., and Robertson, R. J. (1993). Confidence of paternity and male parental care: an experimental study in tree swallows. *Animal Behaviour*, **46**, 139–47.

Williams, G. C. (1975). *Sex and evolution*. Princeton University Press, Princeton.

Wittenberger, J. L. and Tilson, R. L. (1980). The evolution of monogamy: hypotheses and evidence. *Annual Review of Ecology and Systematics*, **11**, 197–232.

Zahavi, A. (1975). Mate selection—a selection for a handicap. *Journal of Theoretical Biology*, **53**, 205–14.

19 *Mate fidelity and divorce in monogamous birds*

BRUNO J. ENS, SHARMILA CHOUDHURY, AND JEFFREY M. BLACK

Introduction

In contrast to most other phyla in the animal kingdom, the majority of bird species form a monogamous *pair bond* during the breeding season (Lack 1968). To be called a pair bond, the association between male and female should last considerably longer than the time needed for copulation, but the bond may be broken as soon as the female has finished laying eggs, as happens in several species of ducks. At the other extreme are some geese and swans, where male and female are close together every day and night of the year, even during migrations over thousands of kilometres. When we speak of monogamous, we mean *socially monogamous*, not *sexually monogamous*, since it has become clear that the first does not necessarily imply the latter (see Gowaty Chapter 2; Birkhead and Møller Chapter 18). Especially in songbirds, extra-pair paternity seems common (Birkhead and Møller 1992).

When both members of the pair survive the nonbreeding season they can either breed together again, or not. Pairs that breed together again are said to have *reunited*, while pairs that do not do so are said to have *divorced*. The pros and cons of this 'human' terminology are discussed in the introduction of this book. Convenience has led us to use a single measure for mate fidelity, the rate of divorce. An alternative would have been to use the rate of reunion in species that break the pair bond at the end of each breeding season and the rate of divorce in species with continuous partnerships. Two things should be clear though. First, in a given year the proportion of surviving pairs that divorce equals one minus the proportion of surviving pairs that reunite. Second, the costs of divorce must be the benefits of reuniting, and vice versa.

In species with a well-defined breeding season two types of divorce are possible. In a *between-season divorce* both members of a pair are alive but not breeding together in the subsequent breeding season. In a *within-season divorce* both members of a pair are alive but do not breed together in a subsequent breeding attempt within the same breeding season. Since this chapter deals exclusively with between-season divorce, we shall dispense with the prefix 'between-season'.

Although it is theoretically possible that surviving pair members either always reunite or always divorce, this hardly ever happens; usually, a variable proportion of pairs reunite. This leads to the central question of this review: why do monogamous birds sometimes remain faithful to their old mate and sometimes not? We cannot simply study the adaptiveness of the 'decision to divorce', since even in the case of a single pair, two individuals are involved. In fact, as we shall see, pairs may break up due to the action of a third individual, so that the resulting divorce must be understood as the outcome of the decisions made by at least three individuals (the action of the individual that triggered the divorce and the reactions of the other individuals). In this respect, the study of divorce resembles the study of mating systems in general, which, according to Davies (1992), should be viewed as 'outcomes of the decisions made by individuals, each selected to maximise its own success'. Thus, our primary aim will be to identify the relevant individual decisions underlying mate fidelity and divorce, and to assess how the costs and benefits vary with social context, that is, the social relationships of the focal individual.

In the first part of this review we describe the known causes of pair formation and mate change. Next, we discuss the logical status of the various hypotheses on divorce and their interrelationships. To give the reader a feeling for the diversity across species we then perform a cross-species comparison of divorce rates. However, the bulk of this chapter consists of reviewing the empirical evidence presented in the species chapters on the causes as well as the presumed costs and benefits of mate change and mate fidelity. Measurement of these costs and benefits poses many problems, both conceptual and practical. Conceptual, because, as discussed, the mate fidelity or divorce that we observe is not the outcome of a single decision made by a single individual. Practical, because in many species we can only measure the reproductive success of an individual as the joint achievement with its mate. With a different mate, reproductive success might have been different, especially in species where both parents have extended parental care. In the absence of experiments, statistics must be used to control for the attributes of the mate and other confounding factors. Owing to the complexity of the problem and the peculiarities of each data set, the necessary statistics may become quite involved and different authors have adopted different approaches. Thus this chapter also highlights the success of the various authors in passing through this statistical quagmire. Different data sets were adjusted in different ways and we can never be sure that new insights would not dictate other adjustments. We have therefore attempted as much as possible to list the unadjusted 'raw' values in the summary tables and graphs, followed by a discussion of possible confounding factors.

Mechanisms for pair formation and mate change

Timing of pair formation and divorces

To discuss the timing of the formation of old and new pairs, or their break-up, we distinguish between migratory and resident species and between species with continuous and part-time partnerships.

In *migratory species with part-time partnerships*, the first birds to return to the breeding grounds are very often the males. For instance in ptarmigan, migratory penguins, and Indigo Buntings *Passerina cyanea* the males arrive early to set up territories, nearly always on their breeding site of the previous year. Females arrive later and find a mate and nest. They are more likely than males to disperse to a different, but nearly always nearby, territory. Male ptarmigan display to the returning females and sometimes even pursue and coerce them back to the territory. Usually, old ptarmigan pairs re-form at the nest site that they occupied the previous season (Hannon and Martin Chapter 10). In Red-billed Gulls *Larus novaehollandiae scopulinus* the pair may then move to a new nest site (Mills *et al.* Chapter 16). While reunion occurs after the return to the breeding area in three shelduck species, reunion on the wintering grounds has been recorded in two migratory species of sea ducks. In several migratory duck species, formation of new pairs takes place in large flocks on the wintering grounds (Williams and McKinney Chapter 4). In contrast, in Red-billed Gulls new pairs are formed at nest sites defended by unattached males, or near the club, that is, the roost of non-breeding birds (Mills *et al.* Chapter 16). This happens after the old pairs have re-formed, since, as in many other species (e.g. Ainley *et al.* 1983, 1990; Coulson and Thomas 1983), both males and females tend to return earlier as they get older. Indeed, in the majority of migratory birds with part-time partnerships the late arrival of young birds probably contributes to the correlation between the ages of mates (see later). One might also suspect that divorce in such species is simply the result of old pairs not re-forming, but this is not necessarily the case. In Oystercatchers *Haematopus ostralegus*, old pairs can be seen together on their territory early in the season, before an ensuing divorce (see drawing on title page; Ens *et al.* 1993).

Even more interesting is the suggestion that for Ring-billed Gulls *Larus delawarensis* divorce occurs during the postbreeding period (Fetterolf 1984). Of course, all pair bonds dissolve after the breeding season in migratory species with part-time partnerships, but the issue is whether we can identify social behaviours during the postbreeding season that predict mate fidelity in the next year. In penguins, the postnuptial moult, prior to departure at the end of the season, apparently plays such a role. Usually, birds moult with their breeding partner of that year and remain faithful the next year. In the Macaroni Penguin *Eudyptes chrysolophus*, 26% of 27 birds that moulted with a different partner instead bred with this new mate the following year (Williams and Rodwell 1992). Finally, at the end of the breeding season, male Razorbills *Alca torda* escort the juveniles to sea, while the females delay departure. Apart from defending the nest site, the females may also consort with other males and even engage in extra-pair copulations, which can only have a social function at that time of year (Wagner 1991). Thus, there are many indications that pair formation and divorce not only occur immediately prior to breeding, but also during the postbreeding season.

Much of the above probably applies to *resident birds with part-time partnerships* if 'date that the individual enters breeding condition' is substituted for 'time of arrival'. In Great Tits *Parus major* for instance, members of a breeding pair behave as two solitary individuals in winter, even when they have the same home range. In late winter males leave the flock on warm days and start to sing. Female Great Tits that have entered breeding condition and occupy an overlapping home range start visiting such males and may pair with the first unmated male that they encounter, according to Dhondt *et al.* (Chapter 13). Similarly, male and female Sparrowhawks *Accipiter nisus* live in overlapping home ranges, and in spring females start visiting nest sites, where the males may offer them food. A change of home range necessarily leads to a change of potential mates, but there is little evidence on when home ranges are changed (Newton and Wyllie Chapter 14).

In *resident birds with continuous partnerships*, the timing of pair formation seems less closely linked to the breeding season. In Splendid Fairy-wrens *Malurus splendens* vacancies occur at any time of year and are filled immediately (Russell and Rowley Chapter 8). Similarly, in Florida Scrub Jays *Aphelocoma coerulescens* pair formation often occurs within the territory of a bird that loses its mate. However, it also occurs in the budding territory of a novice, that is, a male helper usurping parts of the territory of its parents and the neighbours. Some influence of the breeding season is still apparent; for instance pair formation in Florida Scrub Jays often takes many months, but in the breeding season may take only a few days (Fitzpatrick and Woolfenden Chapter 7). In Pinyon Jays *Gymnorhinus cyanocephalus* competition within a sex and choice of partners probably occur in the flock during the nonbreeding season

(Marzluff and Balda Chapter 7). There is little information on divorce, in part because it is so rare in these species.

In *migratory species with continuous partnerships*, like Barnacle Geese *Branta leucopsis*, courtship and pair formation take place on the wintering grounds and the majority of pairs then remain together until parted by death, or, occasionally, divorce. In the Bewick's Swan *Cygnus columbianus bewickii* only temporary liaisons have been observed on the wintering grounds and it is suspected that pairs are formed during spring migration (Rees *et al.* Chapter 6). Divorce is rare in these species, but in Barnacle Geese 65% of divorces happen during the 5 months that the birds are on the wintering grounds (Black *et al.* Chapter 5).

The causes of divorce

If we look at the sequence of events, divorce can come about in at least three different ways (Fig. 19.1), each of which will be discussed in more detail below:

(1) one partner may *desert* the other;

(2) one partner may be chased away by a *usurper*; and

(3) an old partner may find itself *pre-empted* by a new bird that arrived earlier.

Theoretically it is possible that both partners desert each other, both are chased out by usurpers or both are pre-empted, but to date there are no supporting observations. For instance, Oystercatchers can lose their nesting territory as a pair (Ens *et al.* 1993). However, although Oystercatchers without a nesting territory always divorce in the end, they remain paired for quite a while after losing their nesting territory. Thus, the actual divorce may be due to the female deserting a male without a nesting territory. Another theoretical possibility is that one of the partners actively *rejects* its former mate and chases it away. Again, we are not aware of any observations that suggest that such rejections of former mates ever occur in nature. Thus, to our knowledge, Fig. 19.1 represents all well-established causes of divorce, if it is acknowledged that males may also desert, be kicked out by a usurper, or be pre-empted.

This classification of social causes must be distinguished from a functional classification on the number of pair members that is expected to benefit from the divorce: both members, one member, or neither member. In some cases it is quite obvious which individual we expect to benefit (Fig. 19.1). Thus, once we know how divorces come about, we can focus our questions about the functional value of divorce. What benefit did a bird gain from deserting its mate, what caused a bird to become a usurper and why did the pre-empted bird give in to pre-emption? Since such questions require very detailed observations, which are generally lacking, we must be content, for the time being, with a qualitative review

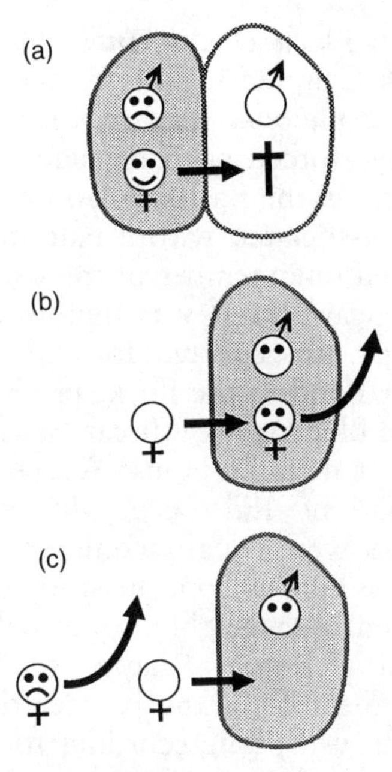

Fig. 19.1 Scheme depicting the different social causes of divorce: (a) desertion; (b) usurpation; (c) pre-emption. For convenience, we have illustrated all three mechanisms for a territorial species, where the male stays put in the territory and the female moves. The territory where the divorce occurs is shaded.

of the possibilities. This qualitative review will lead to some suggestions on species where behavioural observations were less detailed.

1. There is evidence for *desertion* in Blue Ducks *Hymenolaimus malacorhynchos* (Williams and McKinney Chapter 4), ptarmigan (Hannon and Martin Chapter 10), Great Tits (Dhondt *et al.* Chapter 13), penguins (Williams Chapter 15), Oystercatchers (Ens *et al.* 1993), Magpies *Pica pica* (Baeyens 1981), and House Wrens *Troglodytes aedon* (Freed 1987). All these species defend an all-purpose territory during the breeding season and usually females deserted males. The exception is the Magpie, where male desertions were more common than female desertions. Often, deserting individuals moved only a short distance to fill a vacancy in a neighbouring territory, as happened in Blue Ducks, Oystercatchers, and Magpies. Adelie Penguin *Pygoscelis adeliae* females pair with the nearest unattached male if their old mate is absent upon their return, but such females may desert their new mate if the old mate later returns. Deserted males that retain their territories can be left unmated for the rest of the season as in Macaroni Penguins, ptarmigans and Oystercatchers.

2. '*Forced divorce*', where one of the partners is chased out by a usurper, has been reported for territorial species like Mute Swans *Cygnus olor* (Minton 1968), Blue Ducks (Williams and McKinney

Chapter 4), African Black Ducks *Anas sparsa* (Ball *et al.* 1978), Great Tits, (Dhondt *et al.* Chapter 13), Oystercatchers (Ens *et al.* 1993), Song Sparrows *Melospiza melodia* (Arcese 1989), and House Wrens (Freed 1987). Remarkably, forced divorce may also occur in nonterritorial species. For instance, in the Barnacle Goose, in 11 out of 28 cases where an additional bird associated with a pair for some time, the additional bird finally displaced one member of the original pair, but the extent to which the divorce was forced was unknown (Black *et al.* Chapter 5). Similar observations are reported for the Jackdaw *Corvus monedula*, which forms a pair bond in the flock, prior to the occupation of a nest site (Röell 1978). In Blue Ducks, African Black Ducks, and Song Sparrows usurpers were always male. In House Wrens, usurpers were mostly male, but female usurpations did occur. In contrast, in Oystercatchers, usurpers of each sex were equally common. The best evidence for what triggered usurpations comes from the study on Blue Ducks: 11 out of 14 divorces were forced by a neighbouring male whose mate had recently died (Williams and McKinney Chapter 4). Finally, the most intriguing question concerns the role of the partner that remains in the territory and mates with the usurper. According to Freed (1987) 'the tactic of mate choice for each sex may simply be to accept the winner of a takeover attempt'. Alternatively, until more detailed observations become available, the possibilities that *bystanders*, as they are called by Ens (1992), sometimes 'invite' potential usurpers, or that they are just as much 'victims' as the partners that are forced out, cannot be excluded.

3. It could be argued that *pre-emption*, where a bird re-pairs before the old partner has returned, is a special case of usurpation. In 20 out of 30 divorces of Willow Ptarmigans *Lagopus lagopus* the female probably switched because another female was with her previous mate when she returned in spring (Hannon and Martin Chapter 10). Other evidence for pre-emption comes from penguins (Williams Chapter 15) and Great Tits (Dhondt *et al.* Chapter 13), but it probably occurs in many species. For instance, in her well known study of the Song Sparrow, Nice (1937) noted that she did 'not have any certain case of a female joining a new mate when the old one was available; either the former mate was dead or already mated, or, in one or two cases, returned later than she did'. Pre-emption not only occurs in species with part-time partnerships, but it may also occur even in species with continuous partnerships, for example Barnacle Geese, if the original pair is accidentally split up during migration (Owen *et al.* 1988).

If divorce is initiated by only one individual, as the observations suggest, we can draw three conclusions. First, if divorced males are territorial and unmated, they were probably deserted by their female. Otherwise, it is hard to see how the males could be unmated. Second, usurpation or pre-emption instead of desertion is a likely cause of nonbreeding among divorced birds, if the divorced individuals do not defend a territory, or if nonbreeding after divorce occurs in both sexes. This may well be the case in the Red-billed Gull where many males and females do not breed (Mills *et al.* Chapter 16). Although nonbreeding occurs in both sexes, Desrochers and Magrath (Chapter 9) believe that usurpations were an unlikely cause for divorce in European Blackbirds *Turdus merula*, because they do not see how this could lead to the correlation between divorce and poor success. The study on Blue Ducks solves their problem: new (and therefore poorly reproducing) pairs are most vulnerable to usurpations (Williams and McKinney Chapter 4). Third, the fact that divorced females often move, while divorced males often stay on the territory, suggests that females choose to desert, but does not prove it. The females may have been pre-empted or forced out by a usurper.

Towards a sound theoretical framework

Lack of detailed behavioural observations is not our only problem. We also lack a well organized theoretical framework. According to the latest count there now exist no fewer than 11 different hypotheses to explain mate fidelity and divorce in monogamous bird species (see Table 1.1 in Black Chapter 1). Choudhury (1995) grouped these hypotheses into three categories of functional explanations as to why divorce occurs:

(1) one or both pair members seek a new mate to improve reproductive success;
(2) one or both pair members seek a better territory, rather than a better mate; and
(3) neither pair member initiates divorce, but it is induced by some external event.

When Choudhury (1995) attempted to compare the hypotheses in terms of their predictions and their potential as viable alternative explanations she concluded that confusion still exists as to the underlying assumptions and predictions of the individual models. Some of the hypotheses only apply to particular situations, like migratory birds. There is also overlap between the hypotheses. For instance, several hypotheses have the same functional explanation that divorce results from individuals seeking to improve their breeding situation. Finally, some hypotheses that appear different may, in fact, describe different aspects of the same situation. For instance, the *better option hypothesis* stresses changes in the social environment that potentially allow a benefit to be made (e.g. the occurrence of a vacancy), while the *forced divorce hypothesis* stresses changes in the local social environment that require losses to be minimized (e.g. the arrival of a usurper). Thus, these hypotheses do not present a well-defined set of alternatives. Instead, they draw attention to particular aspects of the problem and point out possible mechanisms to which a more comprehensive theory of mate fidelity should pay attention.

The mathematical model of McNamara and Forslund (in press) is a first step towards such a comprehensive theory. It is not a general model of mate fidelity though, as it makes several rather specific assumptions. However, if we focus on the conceptual building blocks underlying the model we can interpret this particular model as one member of a more general family of models that explain divorce. In fact, McNamara and Forslund (in press) already vary some of the assumptions, most notably the assumption that only females choose. A short discussion of these building blocks, listing plausible alternative assumptions for each and followed by a brief description of the main conclusions of the model, seems a useful introduction to a more in-depth treatment of the issues later on.

1. *The criterion by which to judge the optimal 'divorce strategy'*. The choice of McNamara and Forslund (in press) for lifetime reproductive success seems logical and there are no practical alternatives. In fact, lack of data forces most empirical studies of divorce to focus on a component of lifetime fitness: annual reproductive success.

2. *The nature of the quality differences between individuals*. McNamara and Forslund (in press) assume that qualities are normally distributed within a sex and fixed for life. Reasonable alternatives include a skewed distribution with a few high quality individuals and many low quality individuals. Furthermore, quality may change with age and/or experience, instead of being fixed for life.

3. *The way male and female qualities combine*. If the male has 'quality' x and the female has 'quality' y, McNamara and Forslund (in press) assume that their annual reproductive success will amount to $x + y$, once the pair is experienced. Thus, possible effects of compatibility are not taken into account.

4. *The cause of divorce*. In the basic model McNamara and Forslund (in press) assume that only females choose to desert their mate. In an extension they investigate the consequences of allowing males to desert (or reject) their partner too. Thus, the model does not allow for usurpations and pre-emptions by either sex.

5. *The available options*. Since usurpations are not allowed, only unmated males are available as partners for females that desert their mate. If pre-emptions were allowed, all males whose mates had not yet arrived would be available as partners. Effectively all males would be potential partners, if usurpations were allowed to occur.

6. *The mechanism of partner assessment*. McNamara and Forslund (in press) assume that the female must breed with the male to fully assess his

quality. A reasonable alternative is that partners are able to fully assess each other during courtship on the basis of morphological and behavioural characters. It is also possible that paired individuals keep track of neighbours and assess their potential as future partners.

7. *The timing of the 'divorce decision'*. The mechanism of partner assessment that is assumed in the model necessarily leads to the assumption that individuals decide after breeding whether or not they desert their male. There is no such constraint on the timing of divorce when partners can assess each other without breeding together.

8. *The nature of the costs of (not) changing mate*. McNamara and Forslund (in press) investigate two costs of mate change and one cost of mate fidelity. These are respectively an initial inefficiency in reproducing with a new mate, an increased probability of death during mate searching, and reduced reproductive success due to waiting for a previous mate that has died overwinter. Alternative costs include a reduction in reproductive success with the current mate due to mate searching during breeding and an increased probability of remaining unmated.

In the model of McNamara and Forslund (in press) the female deserts if the expected future reproductive success of being faithful to the mate is less than the expected future reproductive success if the female deserts. This leads to a critical level of male quality such that the female should desert all males of a quality below this threshold and remain faithful to all males above this threshold. Although both the optimal threshold value and individual qualities are fixed for life, the probability that females are mated to low quality males decreases as they age and therefore the probability that females divorce will decrease with age too (prediction 1). What other insights do we gain from the model? To begin with, the amount of divorce depends in part on the degree to which repairing is assortative. If there is a low correlation in qualities in newly formed pairs very low quality individuals will always be deserted by their partners and will consequently change mates every year (prediction 2). Second, adding on to this the model tells us how the cost–benefit equation for divorce may be affected by the quality of the choosing individual. We expect that high quality individuals will be in great demand themselves and hence have a wider choice of potential mates (Burley 1977), thereby increasing the likelihood of improving on their current mate. As the quality of an individual increases, so does the range of partners that will not desert the individual, and it pays the individual to become more choosy. If the individual is above a certain quality level no individual will desert him/her. Choosiness then reaches a plateau. Since these high quality individuals are not deserted by their partners they typically have the lowest divorce rate in the population even though they are the most choosy (prediction 3). Finally, the model also tells us how the types of costs determine how divorce rate varies in long-lived and short-lived species.

1. The costs of reduced reproductive success following divorce lower divorce rates in short-lived species much more than in long-lived ones, since the former have few breeding attempts in life and cannot afford to sacrifice a season (prediction 4).
2. A mortality cost of divorce has the opposite effect of reducing divorce in long-lived species more (prediction 5); this is because long-lived species will lose a greater proportion of their lifetime success if they die prematurely.
3. The costs of remaining faithful to a mate that does not return increase divorce dramatically in short-lived species, but have very little effect on long-lived ones (prediction 6).

In conclusion, although the model of McNamara and Forslund (in press) does not allow for all possible scenarios of divorce, for example where birds seek a better territory instead of a better mate, or where divorce is forced on the pair through an external event, the model is a first step towards a set of mathematically deduced predictions which can now be tested by empirical studies.

Comparing divorce rates between species

Do species differ?

The studies on tits (Dhondt and Adriaensen 1994; Dhondt *et al.* Chapter 13) show that divorce rates may differ between populations of the same species and that this variation may be quite extreme. How should we compare species in such a situation? In Fig. 19.2 we have plotted all species for which at least two independent measurements were available. Despite considerable variability within some species there can be no doubt that divorce differs between species. This conclusion is supported by a closer examination of the two highly variable tit species. When Dhondt *et al.* (Chapter 13) compared Great Tits and Blue Tits *Parus caeruleus* in different plots, they found that in most plots divorce rates were higher among Blue Tits (see Fig. 13.1). Thus, the question why species differ is a valid one. In addition, we should ask why divorce rates are more variable in some species than in others.

Relating mate fidelity to survival and other species characteristics

Rowley (1983) hypothesized that there would be no selection on mate fidelity in short-lived species, because the chance that both members of a pair would survive until the next breeding season was so small. Ens *et al.* (1993) predicted a positive correlation between divorce rate and mortality rate for a different reason: vacancies would represent opportunities to

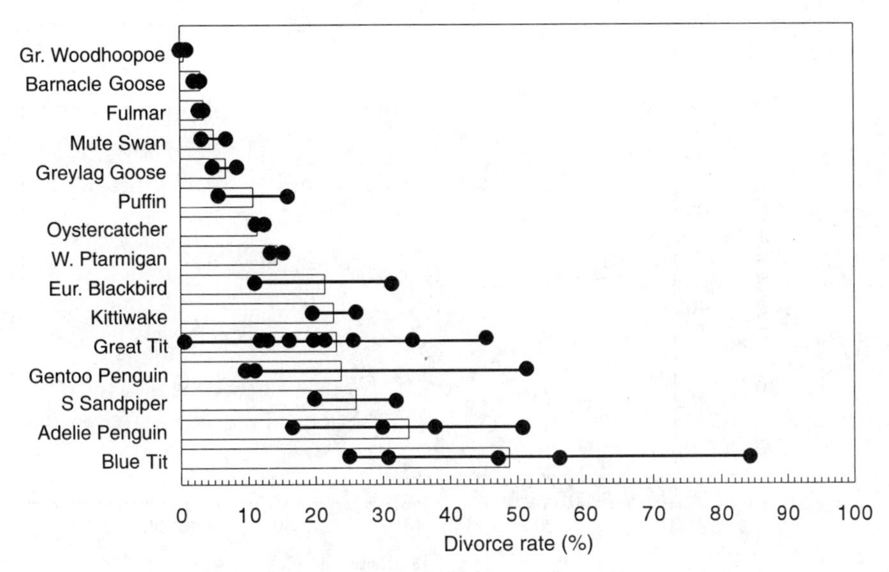

Fig. 19.2 Comparison of divorce rates in species where this parameter was assessed in two or more populations. The bar represents the average value for the species, while each dot represents a single measurement. All data come from Appendix 19.1.

change to a better mate at low cost and in species with high mortality there would be more vacancies. The model of McNamara and Forslund (in press) shows that both arguments are incomplete. According to the model there need not always be a positive correlation between divorce rate and mortality rate, though there often is. When reproduction with a new mate is initially very inefficient, a negative correlation is expected.

We cannot test the model of McNamara and Forslund (in press) in any detail, but the data in Appendix 19.1 on divorce rate, mortality rate, and social characteristics collected from over 100 different populations of 76 species of birds allow us to check if divorce rate increases with mortality rate and if there are additional variables that influence divorce rate. The table does not include species, or populations, with more than 5% polygamy. Otherwise, we included all studies which seemed to offer reliable data on mate fidelity (see legend to Appendix 19.1 for more details). Simply plotting each species' divorce rate against its mortality rate (Fig. 19.3) suggests that divorce rate increases with mortality rate, as expected, even though the scatter is high. A problem with this plot is that we cannot apply simple statistics, because data points are not independent, owing to phylogenetic relationships.

One solution to this problem is to apply the independent contrast method of Felsenstein (1985). This method concentrates on differences between species, that is, if species 1 and species 2 have a common ancestor (a node in the phylogenetic tree), the difference between these species in

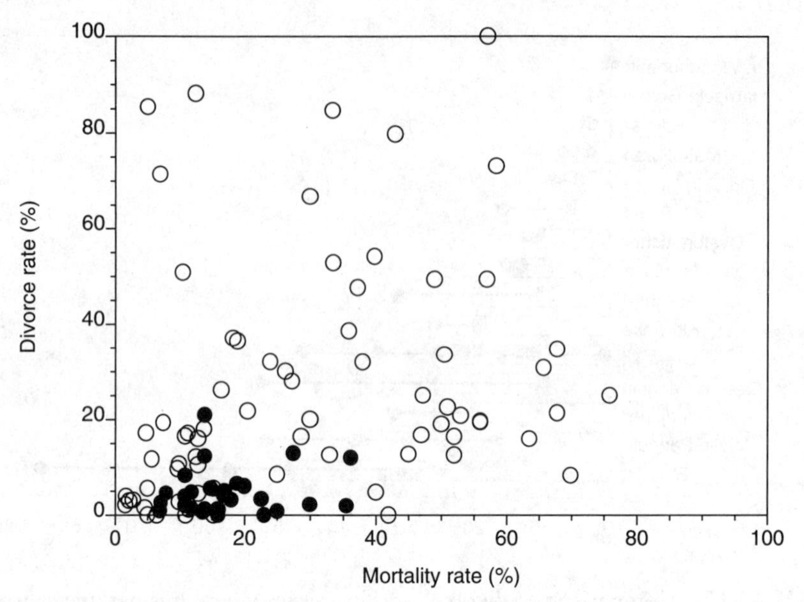

Fig. 19.3 Divorce rate as a function of mortality rate for different categories of species. Each dot is one measurement on one species. *Closed symbols* refer to species with continuous partnerships, while *open symbols* refer to species with part-time partnerships. Species with continuous partnerships are less likely to divorce for a given mortality rate than species that annually break up the pair bond (see text).

divorce rate (D_1 - D_2) is plotted against the difference in mortality rate ($M_1 - M_2$). Under the null hypothesis that evolutionary changes in divorce rate and mortality rate are unrelated, a positive difference in mortality rate should be associated with a positive difference in divorce rate no more often than with a negative difference. Clearly, the nodes in the phylogenetic tree are also themselves connected via common ancestors. These higher nodes are calculated as the average value of lower nodes. For statistical reasons we arcsine transformed the data and excluded studies with sample sizes of fewer than 20 pair-years from the analysis. Branch lengths were obtained by assuming that the ages of taxa were proportional to the number of species they contain (Grafen 1989). Above the family level we used Sibley and Ahlquist's (1990) molecular phylogeny, below the family level we used a branching pattern based on Sibley and Monroe's (1990) taxonomy, and multiple branching was assumed below the family level, that is, nodes in the phylogenetic tree were allowed to have more than two branches. Using the program written by Purvis and Rambout (1995), the 104 populations yielded 65 independent contrasts. The regression line was forced through the origin (Harvey and Pagel 1991).

When all contrasts were included in the sample, there was no relationship between mortality and divorce rate (r = 0.11, n = 65, P = 0.358).

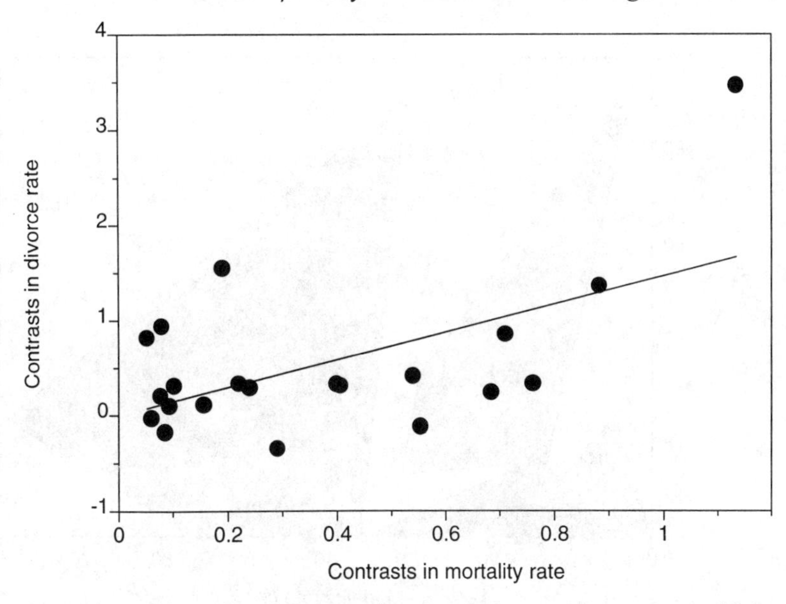

Fig. 19.4 The relationship between divorce and mortality rates at higher taxonomic levels. Data plotted are standardized linear contrasts (see text) and the regression line was forced through the origin ($r = 0.73$, $n = 22$, $P < 0.001$).

However, we know accurate estimates of both mortality and divorce rate are difficult to obtain. Since variables that are measured poorly will have the biggest effect at lower taxonomic levels and will tend to be averaged out at higher levels (Purvis and Harvey 1995), we looked for a relationship within the higher nodes only (one-third of the sample). Although there was a positive relationship between divorce and mortality rate at higher taxonomic levels (Fig. 19.4), suggesting that the evolution of the two traits is correlated, this relationship was dependent on one point on the graph. When this contrast (derived from the warblers, $n = 2$) was removed, the relationship was no longer significant. For families where we had sufficient data, we could study the contrasts within a family. We found a negative correlation for each family, but it was only significant for petrels (Fig. 19.5). Perhaps different combinations of traits were favoured in the early evolution of birds than during the subsequent evolution of this group. It is certainly striking that there is little overlap in the divorce rate—mortality rate combinations between different families (Fig. 19.5), suggesting strong phylogenetic constraints. What could these be? Neither the hypothesis of Rowley (1983), nor the hypothesis of Ens *et al.* (1993), has sufficient detail to provide an explanation. In contrast, the model of McNamara and Forslund (in press) potentially has, but we lack measurements of the appropriate variables. We can check though if some other easily measured life-history variables also affect divorce and mortality rates. We again employed the program of Purvis and Rambout

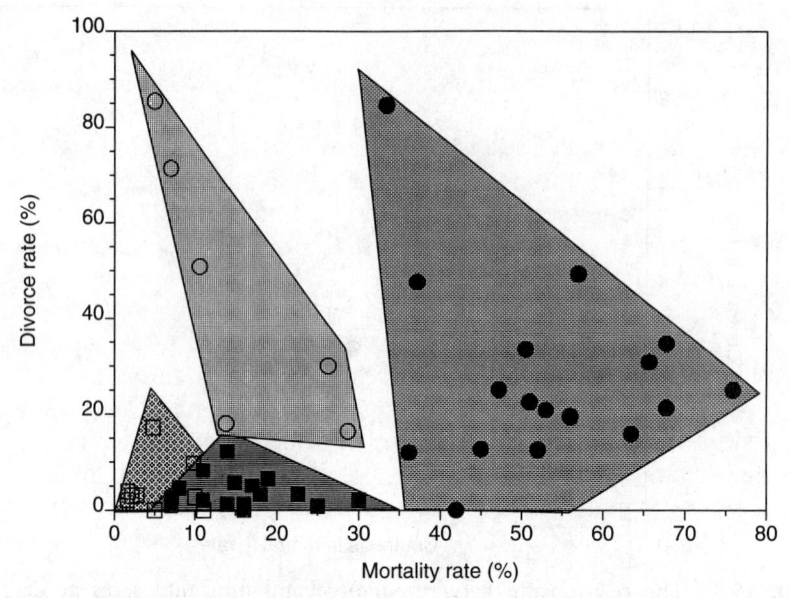

Fig. 19.5 The relationship between divorce rate and mortality rate within different taxonomic families, indicated with different symbols: petrels (*open squares*); penguins (*open circles*); wildfowl (*closed squares*); and tits (*closed circles*). Each symbol is one measurement on one species. The correlations for the standardized linean contrasts are negative for all families (petrels $r = -0.80$, $n = 6$, $P < 0.05$; penguins $r = -0.72$, $n = 3$, NS; wildfowl $r = -0.49$, $n = 10$, NS; tits $r = -0.64$, $n = 3$, NS).

(1995) to calculate independent contrasts and classified each species into the following categories (see Appendix 19.1):

(1) single-brooded or multiple-brooded;

(2) part-time or continuous partnerships;

(3) resident, semi-resident or migratory; and

(4) faithful or not faithful to the same nest site between years.

Species with high divorce rates tended to have part-time partnerships and be migratory more often than species with low divorce rates, which were more likely to have continuous partnerships (sign test on independent contrasts: $n = 12$, $P = 0.039$) and be resident ($n = 29$, $P < 0.001$). Multiple-brooded species tended to have higher mortality rates than single-brooded species ($n = 9$, $P = 0.039$). None of the other variables significantly affected divorce or mortality rate.

Summarizing, we are left with a surprising conclusion that there is no firm evidence for an increase in mortality rate with increasing divorce rate as was originally predicted. We must humbly admit that we have no good explanation.

What is a good mate?

If we want to understand divorce as part of a mate choice process one of our first questions should be: what is a good mate? In an attempt to find an answer, we concentrate on direct fitness benefits and ignore genetic fitness benefits since none of the studies in this book presents evidence on such indirect fitness benefits. Ultimately, time will tell if this is a serious deficiency, but in the discussion we shall present some arguments defending the approach.

Concentrating on direct benefits, we can distinguish three cases of mate preference. In the simplest case, mates are ranked similarly by all individuals of the opposite sex. For instance, males may always rear more young if mated to an old female than if mated to a young female. In a somewhat more complex case, the success of the pair may depend on the compatibility of the mates. For instance, young females may rear more young if mated to a young male than to an old male, while the converse applies to old females. Finally, in the most extreme case, the success of the individual may depend on the number of years it has been together with its particular mate. One important reason for distinguishing between these cases is that they affect the intensity of intrasexual competition for mates among experienced birds, that is, birds that have bred at least once. As shown by Fig. 19.6, competition among experienced

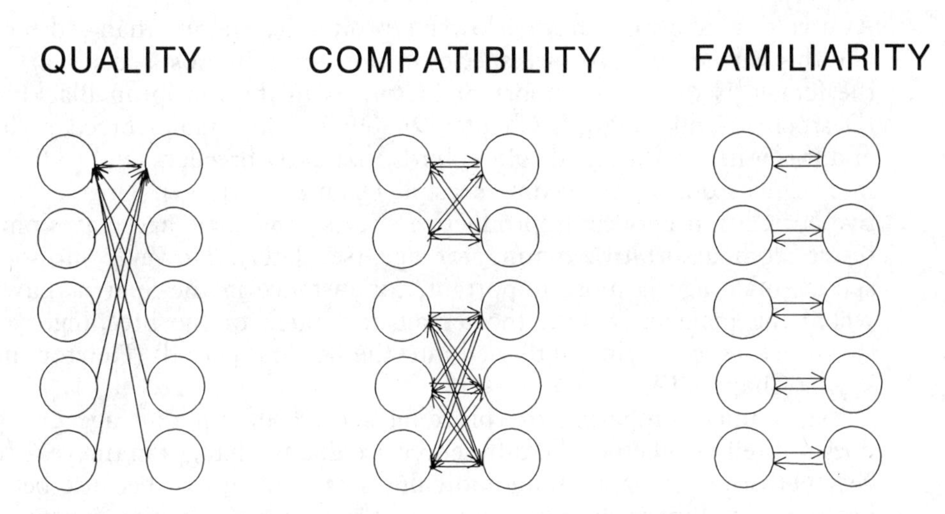

Fig. 19.6 The preference for a particular mate, and therefore the intensity of intrasexual competition for birds that bred at least once for (a) only quality differences between individuals, (b) only compatibility effects, (c) only strong effects of familiarity. *Circles* denote individual males or females and *arrows* show which mate each individual should prefer if it attempts to maximize its reproductive success. In the case of equal preferences more than one arrow is depicted.

breeders is most intense when all birds prefer only one individual of the opposite sex, especially when quality differences between individuals are large, while it is least intense if each individual prefers the mate with which it is most familiar. For competition to occur there need not be actual contests, only that a bird that mates with a high quality partner makes that partner less available to other birds.

Mate quality differences

The term quality is an easy way to describe the existence of differences in reproductive success between individuals. However, if we only measure reproductive success and subsequently conclude that the differences between individuals must be due to their quality, we have become trapped in a tautological statement. If we define the quality of an individual as those of its attributes that lead to high success, our task is to measure these attributes and seek to understand why they lead to high success. Below, we review the evidence for the bird species addressed in this book on derived measures of individual quality like age and breeding experience, and more fundamental measures of individual quality like foraging ability, morphology, social dominance, and territory quality. Some of these more fundamental measures are fixed for life, but others could underlie the change with age.

Age and breeding experience

As a rule, young birds have a lower reproductive success than old birds and this is true for most bird species in this book. In most species, age of the female is the more important factor, as in the European Blackbird (Desrochers and Magrath Chapter 9). Often, old females breed earlier and it is a nearly universal rule in birds that early breeders tend to fledge more chicks during a breeding season (Daan *et al.* 1989). Thus, a positive correlation between reproductive success and male age may simply result from assortative mating for age (see later). However, in some species male age is more important, for instance in the Sparrowhawk, where the male provisions the female for much of the breeding cycle (from before egg laying until well into the nestling period) (Newton and Wyllie Chapter 14).

Yet, simply comparing the breeding success of different age groups does not tell us whether the differences are due to changes in the year-to-year performance of the same individuals as they age, or because better breeders live longer. In the latter case the low reproductive success of young birds would be simply due to the selective inclusion of poorly reproducing short-lived individuals (Curio 1983; Nol and Smith 1987). The problem can also be viewed as one of the *selective death* of poorly reproducing individuals. On the basis of a thorough analysis Desrochers and Magrath (1993*a*) were able to exclude this possibility as an

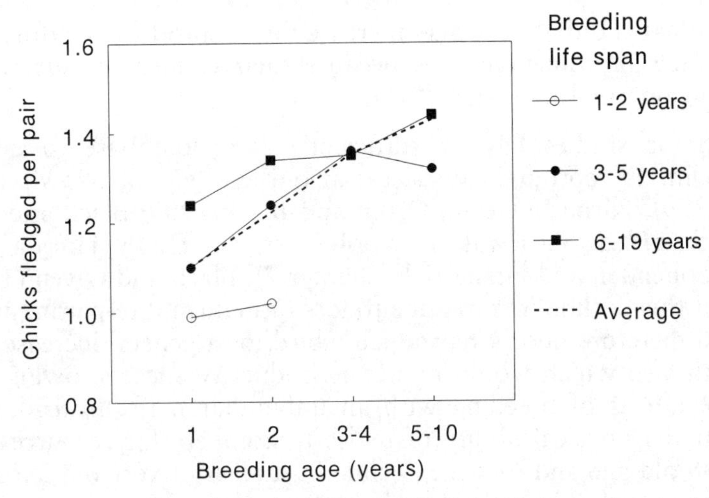

Fig. 19.7 An example of longer-lived individuals performing better throughout their life. Average number of chicks fledged per pair for female Kittiwakes of differing breeding experience according to their breeding life span (based on Table 16.1 in Thomas and Coulson 1988). The broken line indicates the average that would be found for each breeding age if all data were simply lumped.

explanation for the improved reproductive success of old female European Blackbirds. In contrast, for Short-tailed Shearwaters *Puffinus tenuirostris* it is known that long-lived individuals have higher success throughout their lives, but it is also true that for an individual reproductive success increases with age (Wooller *et al.* 1990). The same holds for the Kittiwake *Rissa tridactyla* (Thomas and Coulson 1988); the error that would result from lumping all ages is depicted in Fig. 19.7. Other examples of species where some individuals perform consistently better throughout their lives are reviewed by Clutton-Brock (1988), Newton (1989*a*) and Desrochers and Magrath (1993*a*). The majority of studies in this book did not check this possibility, which also bedevils the analysis of the relationship between success and pair duration.

Once selective death of poorly reproducing individuals is excluded, Nol and Smith (1987) list three possible explanations why, for an individual, reproductive success increases with age.

1. The effects of previous breeding experience may improve later breeding performance. Not surprisingly, age and breeding experience tend to be highly correlated making it difficult to identify whether the correlation between age and success is entirely due to the increase in success with breeding experience.
2. Older birds may be more productive because of an increase in general competence, that is, they improve in feeding skills, intraspecific competition, predator avoidance, or other skills.

3. Finally, older birds may increase their reproductive effort at the end of their reproductive lives because their chances of surviving to breed again are declining.

Some studies have a sufficiently large database to demonstrate a decline in reproductive success at old age (e.g. Sparrowhawks, Newton 1989*b*), Barnacle Geese, Owen and Black 1989; Black and Owen 1995; Short-tailed Shearwaters Wooller *et al.* 1990; Florida Scrub Jays, Woolfenden and Fitzpatrick Chapter 7). Black and Owen (1995) showed that this decline was not due to loss of mate and re-pairing with a young, and therefore poorly reproducing bird, or a general increase in re-pairing with age, which would reduce reproductive success, owing to the negative effects of breeding with an unfamiliar mate. Instead, they suggest that the physical ability of males to compete for resources deteriorates with old age and that the decline in success of very old females is due to their partnership with very old males.

Foraging ability

Depending on the rate of courtship feeding, female Red-billed Gulls are away from their nesting territory between 6 and 95% of the daylight hours (Mills *et al.* Chapter 16). Indeed, in gulls and terns in general, the quality of the male can be measured from the courtship feeding rate: (1) females that are inadequately provided during this phase can invest less in the eggs and the care of those eggs, while (2) males that are poor providers in courtship also tend to be 'incompetent' in incubation and chick feeding (Mills *et al.* Chapter 16). Thus, differences between individuals in foraging ability are an obvious candidate for quality differences and these differences could be fixed for life or change with age. Evidence for the latter is a food supplementation experiment that showed that poor reproductive success of young female European Blackbirds was indeed due to their poor foraging success (Desrochers 1992).

Morphology and social dominance

In birds, general morphology is usually fixed for life. Size may influence social dominance, as we would expect large birds to be better fighters. This may help in gaining access to the best resources, be they mates or territories. In Sparrowhawks lifetime reproductive success was positively correlated with body size in females, but not in males (Newton 1988). Other field evidence that large birds are more successful breeders comes from Barnacle Geese (large females pair at an earlier age and large birds of both sexes have higher reproductive success, Black *et al.* Chapter 5) and Bewick's Swans (large males have highest reproductive success, Rees *et al.* Chapter 6). Surprisingly, medium-sized males have highest reproductive success in Mute Swans and Whooper Swans *Cygnus cygnus* (Rees *et al.* Chapter 6). For Pinyon Jays laboratory observations suggest that social dominance increases with size in both sexes (Marzluff and Balda Chapter 7). Observations on Florida Scrub Jays suggest consistent differences in fighting ability between individuals in the absence of clear morphological differences (Woolfenden and Fitzpatrick Chapter 7). Sorensen (1994) reports that the most aggressive and dominant male White-cheeked Pintails *Anas bahamensis* guarded their mates most effectively against forced extra-pair copulations, but were also the most active in pursuing extra-pair copulations themselves.

In contrast to morphology, social dominance may change (usually increase) with age. In territorial birds this may come about through the effect of prior residence, an important topic with which we shall deal later. Especially complicated is the case of the Bewick's Swan, where dominance rank on the wintering ground increases with pair duration. However, pairs with cygnets dominate pairs without cygnets, possibly because males of large families invest more in aggression and defence (Black and Owen 1989). When the presence or absence of cygnets was taken into account, there was no effect of pair duration on dominance. Thus, the correlation between dominance and pair duration apparently results from the improved breeding success of old pairs (Rees *et al.* Chapter 6).

Territory quality

The majority of the species discussed in this book compete for breeding territories, which range from all-purpose territories to a simple nest site. Thus, consistent differences in reproductive success between individuals may be due not to intrinsic properties of the individuals, but to the quality of the territory they live in, especially if the tendency to breed on the same territory is high. For example, in European Blackbirds consistent variation in reproductive success between sites can be demonstrated (Desrochers and Magrath Chapter 9) and in Willow Ptarmigans more than half of the variability in date of laying and clutch size is due to differences between territories (Hannon and Martin Chapter 10), probably in both cases due to local differences in food supply. Local variation in food supply probably also explains the substantial and consistent variation in breeding performance of Sparrowhawks in different territories (Newton 1991). In Oystercatchers it is not the quality of the food supply, but the distance over which food has to be transported to the chicks that determines territory quality (Ens *et al.* 1992). Nest predation appears to be a major cause of nest loss in several species, including the Red-billed Gull, Indigo Bunting, Willow Ptarmigan, Florida Scrub Jay, and Splendid Fairy-wren, suggesting that a good breeding territory is one where nest predation is low. In species with helpers at the nest, helpers provide a work force that may increase reproductive success, for instance by reducing predation on the nest, or reducing the workload of the breeders and thereby enhancing their survival (see review by Emlen

1991). Thus, a good territory is one with a large work-force, although the helpers may not automatically go with the territory (see later).

Compatibility

The idea of compatibility implies that individuals are not intrinsically of poor quality, but when paired to an incompatible mate reproductive success may be lower than when paired to a compatible mate (see Black Chapter 1). One possible example is that animals may have to strike a balance between inbreeding and outbreeding (Bateson 1983). Another possible example would be if males and females specialize on different foods that yield high rewards at different times of the day, or different phases in the lunar cycle. If male and female share incubation they would do best by choosing a partner that specializes on a different food, so that their preferred feeding times do not overlap. Thus, the idea that mates can be compatible is obviously sensible. What is the evidence?

In penguins, the probability of having bred successfully the previous year was lower among birds that subsequently divorced, but only in species with more complex incubation routines and prolonged incubation shifts (Williams Chapter 15). There can be no doubt that if the partners in species with complex incubation routines do not alternate their behaviours properly, the eggs or chicks are at an increased risk from chilling, starving, or depredation during periods that both partners are at sea. However, such a state of affairs could also arise if one of the partners is a poor forager that needs much time to feed at sea. For instance, Hatch (1990) showed that failure of the male Fulmar *Fulmarus glacialis* to relieve the female soon after laying could result in breeding failures, while egg losses were associated with exceptionally long shifts throughout incubation. He also found that some birds tended to stay consistently longer at sea than others on feeding trips. Thus, the poor success of the pair may simply be due to one of the partners being of poor quality, not to them being incompatible. We know of no example in the literature of seabirds invoking compatibility that could not be explained by this simpler explanation. There may be cases where it takes the pair some time to synchronize their activities properly (Black Chapter 1), but that would be a case of familiarity, not compatibility. The only way to demonstrate compatibility conclusively would be to show that birds do better when paired to another partner who is not intrinsically of better quality.

Marzluff and Balda (Chapter 7) suggest that 'compatible Pinyon Jay partners may enhance each other's survival, whereas incompatible ones reduce their survival', because in 20.8% of cases, birds died the same autumn or winter as their mate, well above the 6.8% expected by chance. There is again no need to invoke the concept of compatibility, since the same result will follow if Pinyon Jays mate assortatively for viability.

Only the study of Barnacle Geese provides evidence that is hard to interpret in a different way (Black *et al.* Chapter 5). Large males and large or medium-sized females tended to have the highest reproductive success on average, but small individuals of either sex did better when mated to a small individual of the opposite sex.

We therefore conclude that the idea that mates can be compatible to a varying degree has great intuitive appeal, but there is remarkably little empirical evidence to date.

Familiarity with the mate (or site)

While the ideas of mate quality and compatibility imply that for each experienced individual there is a class of individuals of the opposite sex with which it would achieve highest reproductive success, the idea that familiarity is most important implies that there is only one such individual, namely the mate of the year(s) before (see Black Chapter 1). If familiarity is important, we expect reproductive success to increase, especially in the beginning, the longer the pair bond has endured. A simple plot of reproductive success against duration of the pair bond shows an increase, at least in the first few years, for many species in this book (Fig. 19.8). However, on its own this relationship is insufficient evidence for an effect of mate familiarity, since the increase might simply be due to the birds getting older, or more experienced in breeding in general, or selective exclusion of poor quality individuals from long pair durations (the problem of selective death in a new guise). These confounding effects must be excluded before we can accept an effect of familiarity with a particular mate. In European Blackbirds, no complicated statistics are required: the effect of pair duration already disappears if first-year females are excluded (Desrochers and Magrath Chapter 9). In Splendid Fairy-wrens the positive correlation between reproductive success and pair duration vanishes after controlling for the proportion of the breeding season that the female was present (some females enter a breeding group halfway through the breeding season), female age (older females reproduce better), and number of helpers (older groups have more helpers) (Russell and Rowley Chapter 8). After statistical correction, seven species (five with part-time partnerships) show no effect of pair duration, while eight species (six with continuous partnerships) remain, where there is still an effect of pair duration on reproductive success (Table 19.1). Indeed, the fact that species where familiarity seems to affect reproductive performance tend to have long-lasting pair bonds and few mates in a lifetime, often only one, supports the results (Fig. 19.9). However, the present analyses are clearly not the final word. In many species there were large differences in success between years, but not all studies controlled for this. Furthermore, although some studies checked whether the analyses suffered from pseudoreplication by repeating the analysis using a randomly selected value of each bird, no study claiming an effect of familiarity corrected for the possibility that individuals with long-lasting pair bonds are different from birds with short partnerships.

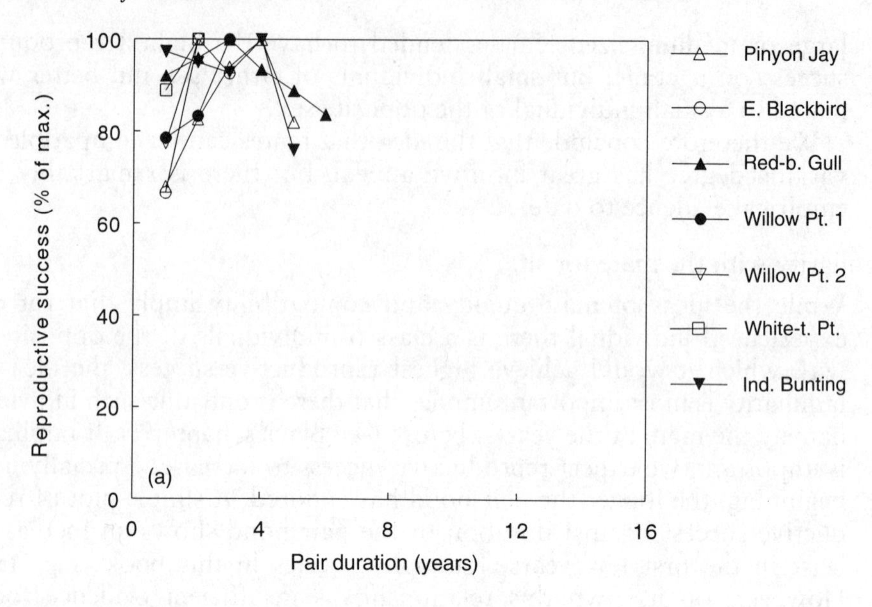

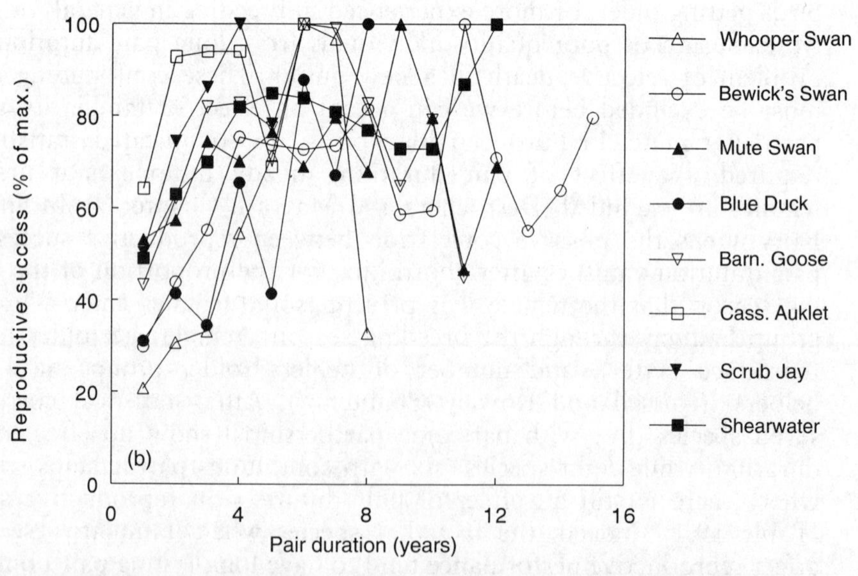

Fig. 19.8 Reproductive success (measured as fledglings produced per pair per year and represented as percentage of the maximum value) as a function of duration of the pair bond for different species in this book. Simple unadjusted averages for each pair duration are presented, so that sample sizes necessarily decrease with pair duration and the same pair is represented in several data points. In most studies this graph could be constructed from the data, if not already given. For the Indigo Bunting the graph applies to a selection of 16 pairs that had been together for at least 3 years. (a) Species without a clear effect of familiarity on success; (b) species where familiarity apparently improves reproductive success.

Table 19.1 The effect of pair duration and a change of mate on reproductive success (RS)

	RS increases with pair duration[1]	RS decreases following mate change[2]
Continuous partnerships		
Blue Duck	+	++
Barnacle Goose	+	++
Bewick's Swan	+	++
Whooper Swan	+	+
Mute Swan	+	+
Pinyon Jay	–	–
Florida Scrub Jay	+	+
Splendid Fairy-wren	–	–
Part-time partnerships		
European Blackbird	–	–
Willow Ptarmigan	–	–
White-tailed Ptarmigan	–	–
Cassin's Auklet	+	++
Short-tailed Shearwater	+	+
Great Tit		–
Red-billed Gull	–	++
Indigo Bunting	–	–

[1]+ = RS increases significantly with pair duratin, – = no significant change in RS with pair duration.
[2]++ = RS decreases significantly following a change of mate, + = RS decreases following a change of mate, bordering significance, – = no significant change in RS following a change of mate.

There are two further checks on the credibility of a mate familiarity (pair duration) effect. First, there must exist a plausible explanation why pairs improve reproduction the longer they have been together. Perhaps a fine tuning of pair members' co-operative, behavioural repertoire may explain the link between pair duration and reproductive success. There are numerous candidate behaviours in this regard that have been identified in the case study species (e.g. courtship feeding, dueting and other vocalization behaviour, nest protection against predators, change-over in shared incubation duties, co-ordinating parental care duties, alternating vigilance and feeding bouts in preparation for breeding: see Black Chapter 1), but only one of three case studies sought and found a correlation between a behaviour and pair duration. A whole range of behaviours in Pinyon Jays (Marzluff and Balda Chapter 7), and dominance in Bewick's Swans (Rees *et al.* Chapter 6) were not correlated with pair duration. In Barnacle Geese, longer-term pairs spent more time in the profitable positions at the edge of the feeding flock than younger pairs during winter. In line with this, the proportion of soft, contact calls declined, while the proportion of the loudest 'aggressive' calls increased as pairs grew older (Black *et al.* Chapter 5). This correlation suggests that Barnacle Geese may advance their social dominance if they remain together as a pair, but does not prove it. Second, in species where reproductive success

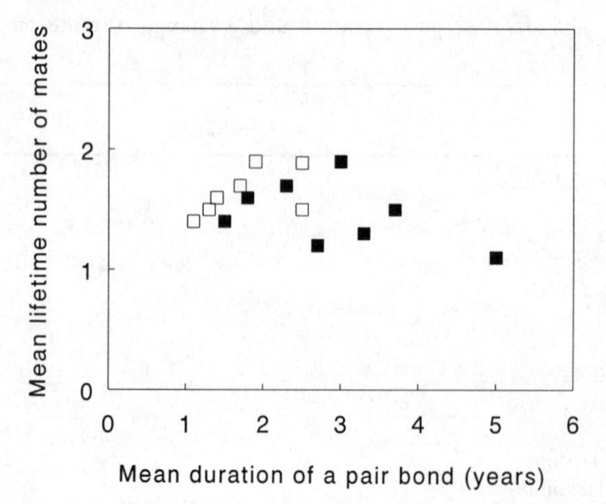

Fig. 19.9 Average number of mates in a lifetime and mean duration of the pair bond in species without an effect of familiarity on reproductive success (*open symbols*) and species where familiarity improves reproductive success (*closed symbols*).

improves with pair duration we expect a significant drop in reproductive success following a change of mate. This prediction is tested in the section below.

Possible costs and benefits of mate change

Should we expect the reproductive success of an individual to improve following a change of mate or not? It depends. If the change is forced upon the individual through the death of its former mate, there is no reason to expect an improvement. In fact, we particularly expect a negative effect under such circumstances for species where success increases with familiarity with the mate (see above). Divorce 'victims', that is, individuals that were deserted by their mate or forced out by usurper, are not expected to benefit either. Only individuals that 'chose to divorce', that is, who initiated the events that led to divorce, should benefit (Ens *et al.* 1993). However, even individuals that desert their mate may not always be able to avoid paying the cost of mate change.

At least four different costs of mate change can be envisaged.

1. *The costs of searching for a new mate*. The form of this cost depends on the timing of mate loss or desertion. If a new mate is sought before the bond with the old mate is broken, this may impair the current reproductive event. For instance, in a permanently resident species, shopping for new mates may impair communal defence of the territory and/or may induce the current mate to search for a new mate too. If a new mate is sought after the old mate has been deserted or died, birds may fail to

find a new mate in time so that they miss one or more breeding seasons and their reproductive success is impaired.

2. *The cost of fighting rivals.* The observation that usurpations occur proves that there are conditions under which individuals are willing to pay this price. At the same time the rarity of such reports may indicate that this cost is rarely worth paying, although at this stage it cannot be excluded that behavioural observations have generally not been sufficiently intensive to measure the frequency of usurpations. Many authors in this book feel that the cost of fighting rivals is generally too high, because they emphasize that only unpaired individuals are available as mates. The cost of a fight, most likely a survival risk, probably does not differ for individuals that have lost their mate or deserted, but there may be a difference in the frequency with which fights were necessary to obtain a new mate. Birds without a mate need a new mate anyway, so they may be more willing to pay this cost than paired individuals.

3. *The risk of ending up with a poorer mate.* Birds that start searching for a new mate have no guarantee that they will end up with a better mate than before. Indeed, the new mate may be of lower quality when only young, inexperienced birds are available. On theoretical grounds we expect a trade-off between the time spent searching for a new mate and the quality of the new mate. The cost will manifest itself as a drop in reproductive success for both birds that lost a mate through death and divorce.

4. *An initial inefficiency in reproduction with the new mate.* This cost should only occur in species where reproductive success increases with familiarity with the mate. Both birds that lost their mate and birds that divorced should suffer from it, but the magnitude of the decline in success may differ. There is no way in which deserting individuals can avoid paying this cost.

Several conclusions can be drawn from the above. First, a drop in success following a change of mate can have many different causes, whereas nonbreeding seems most likely to be due to the problem of finding a new mate. Second, it is important to distinguish between divorced birds and those whose mates died. In fact, for the divorced birds we should know whether they were the victim or chose to divorce. These distinctions are rarely made. Below we review what is known of the effects of a change of mate on the subsequent probability of (non)breeding and reproductive success.

Mate change and subsequent nonbreeding

In Barnacle Geese the time taken to re-pair after death of the mate or, more rarely, divorce is generally between 3 and 9 months, so some birds may miss a breeding season (Black *et al.* Chapter 5). Swans lose even more years if their mate dies: first-time breeding after mate loss (the time to re-pair + the time to breed for the first time after re-pairing), varies from 1.5 (0 + 1.5) to 3.4 (1.9 + 1.5) up to 4.9 (2.6 + 2.3) years for Mute, Whooper, and Bewick's Swans, respectively (Rees *et al.* Chapter 6). Another extreme example is the Red-billed Gull, where 32% of females and 16% of males never bred again following divorce or death of their partner, despite sometimes surviving for 10 or more years. Of gulls that bred again, 41% of females and 18% of males missed the first breeding season (Mills *et al.* Chapter 16). Clearly, finding a new mate can be very time-consuming in some species. In contrast, Cassin's Auklets *Ptychoramphus aleuticus* whose mates were removed generally obtained a replacement partner within a few days (Sydeman *et al.* Chapter 11). When male Oystercatchers were removed from the territory, the female either obtained a new mate quickly, or she lost the territory (Ens 1992).

Of 58 European Blackbird pairs that divorced, 21% of males and 10% of females failed to find a mate in the following year (Desrochers and Magrath Chapter 9). In Macaroni Penguins 39% of 18 divorced males remained unmated the year after mate change, and did not breed again for 2–4 years (even though they retained the old nest site), compared to only one female (6%). In Gentoo Penguins *Pygoscelis papua* all males bred in the year following divorce, but only 64% of females did (n = 14 pairs; Williams Chapter 15). Divorced Oystercatchers that lost breeding status were either deserted by their mate, or forced out of the territory by usurper (Ens *et al.* 1993) and the same might be true for the above species. In both male and female Red-billed Gulls that bred again, following a change of mate, divorced birds were more likely to miss the first year (Mills *et al.* Chapter 16). This would suggest that in this species many divorces come about through usurpation. Only for ptarmigan do we know with some certainty that the 20% of divorced males that remained unmated were actually deserted by their females (Hannon and Martin Chapter 10).

Mate change and reproductive success

One method to find out if mate change affects reproductive success is to compare the success of faithful pairs with the success of pairs that changed mates. There is a problem if the two types of pairs differ. For instance, old pairs may be more faithful (see later), or divorce may be more likely in pairs in poor territories. Another method is to compare the success of each individual in the last year with the old mate with the success in the first year with the new mate. This controls for many confounding factors on an individual basis, except for the increase in age and experience. The latter problems can be partly solved by comparing the change in success of birds that changed mate with the change in success of birds that remained faithful (Perrins and McCleery 1985; Desrochers and Magrath Chapter 9).

Table 19.1 shows that a change of mate had either little effect on reproductive success, or a negative effect. As predicted, this negative effect was most pronounced in species where reproductive success seemed to increase with the pair duration. However, in several cases the reduction in success was more likely to be due to the lower average age or breeding experience of the new partner than to the benefits of breeding experience with one particular partner. In Cassin's Auklet, for instance, mate change through divorce or mortality often led to pairing with an inexperienced bird (Sydeman *et al.* Chapter 11). All three species of swan had a lower reproductive success in the first year with a new mate, controlling for year and age, but not significantly so for Whooper Swans. The effect disappeared in Mute Swans if inexperienced new mates were excluded (Rees *et al.* Chapter 6). In Florida Scrub Jays newly established pairs always involved at least one individual not familiar with the site; new pairs also had fewer helpers—some were driven out, or did not follow (Fitzpatrick and Woolfenden Chapter 7).

What if mate change is narrowed down to cases of divorce only? As explained, we only expect choosers to improve. In the absence of detailed observations several authors have adopted the working assumption that the sex that moves, usually the female, is also the choosing sex. In European Blackbirds (Desrochers and Magrath Chapter 9), Oystercatchers (Harris *et al.* 1987), and Blue Tits (Dhondt and Adriaensen 1994), females fared better after divorce than males, as predicted, but the difference was significant only for the Blue Tit. In ptarmigan no change in success for 'deserting' females could be detected, but there was evidence that the birds were avoiding polygyny (Hannon and Martin Chapter 10). Although Desrochers and Magrath (Chapter 9) argue that in European Blackbirds females initiate divorce to obtain a better territory, they could not show that site quality improved for divorced females (nor did it improve for males). Perhaps the working assumption is wrong and not all female European Blackbirds that divorced actually chose to do so, a possibility that certainly applies to Oystercatchers. It has proven easier to demonstrate that birds that did not choose to divorce suffered a decline in success following divorce. Victimized Oystercatchers rarely regained breeding status in the same year (Ens *et al.* 1993). Similarly, deserted male ptarmigans clearly suffered from the change: they had a 49% chance of pairing with a yearling female and a 20% chance of remaining unmated (Hannon and Martin Chapter 10).

Preferred options and assortative mating

On the basis of the previous sections we should be able to define which mates should be preferred. In this section we shall seek to review how often birds succeed in mating with their preferred option. In species where the familiarity effect is of overriding importance in determining reproductive success pair members should prefer their old mate (Fig. 19.6). The only potential conflict is with first-time breeders and birds whose mate died. However, even in the species where familiarity seems important, there is also evidence for quality differences between individuals. Under a monogamous mating system quality differences imply that many birds will not be able to form a pair bond with their preferred option (Fig. 19.6). The question then is, which birds succeed and which birds fail and why?

The character for which we know at least part of the answer is age. In most cases, birds profit from an old, and therefore experienced, mate, although the magnitude of the effect varies between species and there may be a decline at old age. Who succeeds in this competition for experienced mates? The simpler answer is: the experienced mates of the opposite sex, since for nearly all species in this book the frequency with which mates are of similar age is higher than expected from randomly picking a male and a female from the population. According to Findlay (1987) we should not refer to this age correlation as 'assortative mating for age', even though population geneticists invented the term assortative mating and defined it as mated pairs being of the same phenotype more often than would occur by chance (Falconer 1981). In the view of Findlay 'it cannot be overemphasized that mating preferences can be inferred from population data only when the distribution of phenotypes (or genotypes) among mated pairs is a non-random sample of the distribution of available mates *when and where mating occurs*'. For a young bird entering the breeding population, many of the old birds may simply be 'unavailable', owing to high mate fidelity among experienced breeders. On the assumption that only birds whose mate died, divorced birds, and first-time breeders are 'available' to each other as mates, Reid (1988) calculated the extent of nonrandom choice required to obtain the age correlations observed in nine bird species with known life-history traits. In only one of the case study species did the continued association of pairs suffice to explain the observed age correlation. An alternative correction method is to ignore pairs that reunite and only look at the correlation between the ages of male and female at the time of first pair formation. Following such corrections, evidence for nonrandom choice persisted in Short-tailed Shearwaters, European Blackbirds, Barnacle Geese, Pinyon Jays, and (albeit weakly) Splendid Fairy-wrens, but disappeared in Florida Scrub Jays. Of course this still does not prove beyond doubt that the observed age correlation resulted from active choice.

Why would most old experienced birds be unavailable? In species with continuous partnerships the answer must be that the majority of old birds are not prepared to break their pair bond, or let it be broken. In species with part-time partnerships, the unavailability of old birds could be due to young birds arriving late at the breeding grounds (e.g. Ainley *et al.* 1983; Coulson and Thomas 1983), or, when the species is resident, because young birds come into breeding condition late (as suggested for the Great Tit). If this explains the unavailability of old experienced birds to the young birds, our next question is why it generally does not pay young birds to attempt usurpation, or to arrive earlier on the breeding grounds?

Variables that predict divorce

Breeding failure and/or age?

Traditionally, an increased probability of divorce following reproductive failure has been taken as evidence for the *incompatibility hypothesis*. It was implied that the experience of joint failure would be the cue mates use to detect that they were incompatible and stimulate them to divorce. There are three problems with this interpretation. First, reproductive failure need not be due to incompatibility, as already discussed. Second, the *better option*, the *errors of mate choice*, and the *habitat mediated hypotheses* also predict that unsuccessful pairs should be more likely to divorce than successful pairs. Third, mates may be able to assess each other in other ways than through breeding.

To find out if low reproductive success is indeed often followed by divorce we can make a cross-classification of reproductive success (0,1,2, etc. chicks fledged) against subsequent divorce (yes or no). This cross-classification has been analysed in two different ways. One method is to distinguish between failed pairs and successful pairs and compare the probability that the pair will subsequently divorce (Fig. 19.10). Alternatively, we may distinguish between pairs that divorced and pairs that did not, and compare their reproductive success in the previous year (Table 19.2). Either way, in many species pairs with a low reproductive success are more likely to divorce. In some species the difference is small and in the opposite direction (e.g. the Indigo Bunting), but there are no species where such a trend in the opposite direction is statistically significant. More important, even in species where failed pairs are significantly more likely to divorce than successful pairs, it is still the case that the vast majority of failed pairs do not divorce. Thus, even in these species reproductive failure is a poor predictor of divorce.

According to the model of McNamara and Forslund (in press) the propensity for a female to desert her mate will decrease with age. Since reproductive success tends to increase with age (see above), a correlation between divorce rate and reproductive success could therefore simply be due to young (and therefore unsuccessful) birds showing an increased tendency to divorce. After plotting the data on Kittiwakes presented by Coulson and Thomas (1983) it seems hard to escape the conclusion that female age, not previous reproductive success, is indeed the best predictor of divorce in this species (Fig. 19.11). In most species where the simple 'raw' relationship was investigated (Short-tailed Shearwaters,

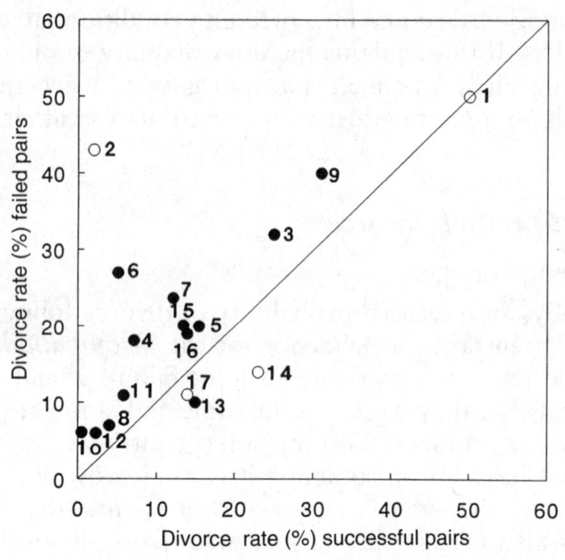

Fig. 19.10 Subsequent divorce rate of successful pairs plotted against subsequent divorce rate of failed pairs. If divorce rates were based on less than 50 pair-years *open symbols* were used, otherwise *closed symbols* were used. 1. Caspian Tern (Cuthbert 1985), 2. Ring-billed Gull (Southern and Southern 1982), 3. Kittiwake (Coulson and Thomas 1983), 4. Oystercatcher (Harris *et al.* 1987), 5. Little Penguin (Reilly and Cullen 1981), 6. Manx Shearwater (Brooke 1978), 7. Short-tailed Shearwater (Wooller and Bradley Chapter 12), 8. Fulmar (Ollason and Dunnet 1988), 9. European Blackbird (Desrochers and Magrath Chapter 9), 10. Florida Scrub Jay (Woolfenden and Fitzpatrick Chapter 7). 11. Cassin's Auklet (Sydeman *et al.* Chapter 11), 12. Whooper Swan (Rees *et al.* Chapter 6), 13. Indigo Bunting (Payne and Payne Chapter 17), 14. White-tailed Ptarmigan (Hannon and Martin Chapter 10), 15. Willow Ptarmigan (Hannon and Martin Chapter 10), 16. Red-billed Gull (Mills *et al.* Chapter 16), 17. Blue Duck (Williams and McKinney Chapter 4).

Red-billed Gulls, Willow Ptarmigans, White-tailed Ptarmigans *Lagopus leucurus*, Great Tits, Willow Tits *Parus montanus* Orell *et al.* 1994), the rate of divorce tended to decline with age. Only in the Barnacle Goose did divorce rate show no relationship to age, although it decreased with pair duration. In Cassin's Auklets and Willow Tits (Orell *et al.* 1994) reproductive success was not significantly related to subsequent divorce after controlling for age. In contrast, for European Blackbirds, Short-tailed Shearwaters, and Red-billed Gulls low success was related to an increased probability of divorce, independent of age.

Even after all confounding factors have been taken into account, a statistical correlation between low success and increased probability of divorce does not prove that the experience of low success caused the divorce. In both Florida Scrub Jay and Pinyon Jay, observations suggested that sudden poor performance of certain duties by a member of the pair, for example territory defence, triggered the rare cases of divorce

Table 19.2 Comparison of reproductive success (or a component thereof) in year A for pairs that either divorced or remained faithful in year B

	Do poor breeders divorce more in year B?[1]	Success in year A		Parameter of success[2] in year A	Source
		Divorced	Faithful		
Barnacle Goose	–	21%	21%	Recruit	Black *et al.* (Chapter 5)
Whooper Swan	–	25%	47%	Recruit	Rees *et al.* (Chapter 6)
Macaroni Penguin	+	26%	61%	Breed	Williams and Rodwell (1992)
Little Penguin	–	0.67	0.92	RS	Reilly and Cullen (1981)
Gentoo Penguin	–	74%	68%	Breed	Williams and Rodwell (1992)
Willow Ptarmigan	+	39%	62%	Fledge	S. Hannon (personal communication)
Blue Tit	+	0.03	0.28	Norm RS	Dhondt and Adriaensen (1994)
Great Tit	–	0.17	0.26	Norm RS	Dhondt *et al.* (Chapter 13)

[1]+ = success in year A significantly different for pairs that divorced in year B compared to pairs that did not divorce in year B; – = no significant difference in success.
2Recruit = probability of returning with one or more young from the breeding grounds; Breed = proportion of pairs that bred successfully; fledge = fledging success; RS = number of fledglings raised; norm RS = normalized reproductive success, i.e. number of fledglings minus the population average and dividing this difference by the population standard deviation.

in these species. Another possibility is that in pairs where one or both members continue to search for mates during the breeding season, reproductive success will be decreased due to parental neglect and the probability of divorce will be high. This effect will be especially marked if birds with low quality partners continue to search for mates. If we accept that in species with low rates of extra-pair paternity, extra-pair copulations may serve to locate alternative mates (e.g. Heg *et al.* 1993) the following observations support this scheme. The only female Red-billed Gull actively to solicit or engage in a successful extra-pair copulation during the fertile period was poorly provisioned by her mate (Mills *et al.* Chapter 16). Similarly, in Sparrowhawks it was those females that were least well fed by their mate in the prelaying period that most often visited other territories during this time (Newton and Wyllie Chapter 14). Thus, these birds started searching for a new mate before even having bred with their current mate.

The best way to test if low success leads to divorce is to perform an experiment. Orell *et al.* (1994) and Lindén (1991) manipulated brood sizes of Willow Tits and Great Tits, respectively, and found that, in all years, reduced broods produced fewer, and enlarged broods more, fledglings than control broods. In the Willow Tit, divorce rates for reduced, control, and enlarged broods were 9%, 12%, and 16%, respectively, suggesting that success increased the probability of divorce, although the differences were not statistically significant. In contrast, Lindén (1991) found a significant effect for Great Tits in the opposite direction: divorce rates were 50%, 33%, and 7%, for reduced, control,

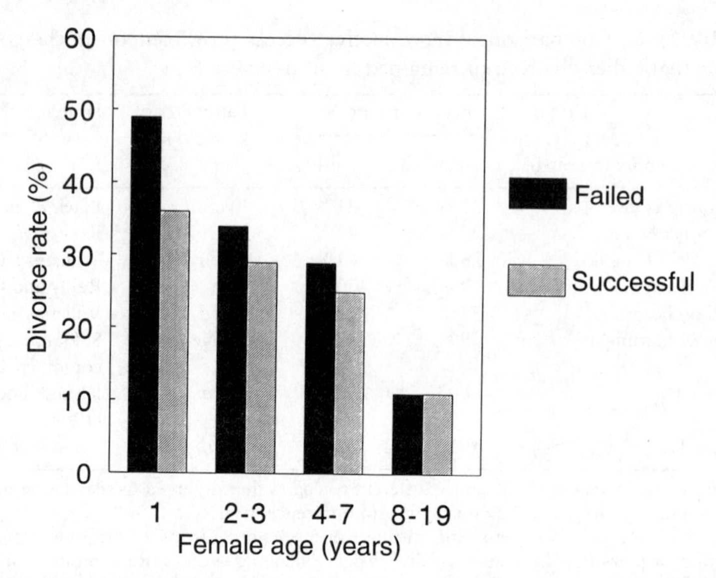

Fig. 19.11 The relationship between divorce rate and age of the female in Kittiwakes, separated for successful and unsuccessful pairs. Based on Table 16.5 in Coulson and Thomas (1983).

and enlarged broods, respectively. Since these results are contradictory, it is impossible to draw a firm conclusion.

Prospect of improvement

If divorce is brought about through choice by one or both partners, we would expect the probability of improvement for an individual to be the best predictor of divorce (Ens *et al.* 1993). As has become clear, the conceptual clarity of this view is marred by the difficulty of making it operational. For instance, birds that reproduce poorly will have a great scope for improvement, so we might expect most divorces among unsuccessful birds. However, if the poor success is due to both partners being of low quality, neither may stand much chance in securing a better mate, since poor quality individuals will be avoided by other birds. Thus, the challenge consists of specifying the options for each individual, including the cost of acquiring a new mate and the difference in quality with other birds. Below, we review the effect of vacancies, which may represent options to change mate at a low cost, and arrival time, which may indicate individual quality.

Availability of unpaired individuals

In species where familiarity with the mate does not improve reproductive success, mate fidelity may be a default situation: the limited opportunities to acquire a new mate at low cost prevent divorce (Freed 1987; Ens *et al.* 1993; Russell and Rowley Chapter 8; Hannon and Martin Chapter 10).

Table 19.3 Intraspecific correlations between divorce rate and mortality rate in a single population comparing different years (* indicates correlation coefficients that differ significantly from zero)

Species	r_s	n	Source
Barnacle Goose	+0.03	11	Black *et al.* (Chapter 5)
Oystercatcher[1]	+0.55*	11	Ens *et al.* (1993)
Oystercatcher[2]	+0.10	5	Ens *et al.* (1993)
Kittiwake	+0.45*	27	Coulson and Thomas (1985)
Willow Tit[3]	+0.14	6	Orell *et al.* (1994)
Willow Tit[4]	+0.26	6	Orell *et al.* (1994)

[1]Population on Skokholm, United Kingdom.
[2]Population on Schiermonnikoog, The Netherlands.
[3]Divorce rate against female survival rate.
[4]Divorce rate against male survival rate.

This implies that vacancies may trigger divorce and this has been observed in several species, including the Oystercatcher (Ens *et al.* 1993) and Magpie (Baeyens 1981). The nicest example is where experimental vacancies triggered divorce in neighbouring Oystercatcher pairs (Harris 1970). In Macaroni Penguins temporary removal experiments of males did not incite paired males to desert their mate. Instead, neighbouring, unpaired males moved into the vacated nest site, initiated sexual displays with the removed bird's female partner and defended the new territory or partner against other males (Williams Chapter 15).

On a more global level we predict for a given study population more divorces in years following high mortality in winter. Such correlations were positive for Kittiwakes, Oystercatchers, and Willow Tits, but significant only if sample sizes were large, that is, if the study had continued for 10 or more years (Table 19.3).

Rather than overall mortality rate, annual divorce rate for Barnacle Geese was positively correlated with the greater availability of older, single birds that had previously been paired, and negatively correlated with the number of young, unpaired birds in the population. Geese that have lost a mate are preferred as partners, since they are older and more experienced (Black *et al.* Chapter 5). It seems that some geese may make decisions based on 'better options' as they become available in the population, and avoid 'poor options'.

Relative arrival date

In some species arrival date has been shown to signal the quality of an individual and a change in arrival date between years could signal a change in quality. For instance, Møller (1994) showed, convincingly, that male Barn Swallows *Hirundo rustica* of high quality (i.e. with long tails) arrived earlier in the season than low quality males, especially in years when arrival was delayed due to bad weather. Whereas Møller found no effect of age on arrival date, it is well established that seabirds tend to arrive progressively earlier at the colony as they age (e.g. Ainley *et al.* 1983, 1990; Coulson and Thomas 1983; Pickering 1989). This difference is almost certainly related to the much greater longevity of the seabirds.

How long should a bird wait upon arrival on the breeding grounds for its partner to return, if early arrival signals high quality, both for itself and its mate? In short-lived species returning males should not waste precious time on waiting for last year's female, which will often never come (Rowley 1983). What is the answer for a long-lived species, assuming that quality increases with age and males generally arrive before females? If a female arrives before her old mate has returned, three things may have happened:

(1) he may have died;
(2) she may have improved much more in quality than he has; or
(3) the quality of the male may have declined.

The last two possibilities can be combined as one case, where the relative quality of the female now much exceeds the relative quality of the male. Either way, it seems that she has little to lose and much to gain by searching for a new mate immediately, perhaps keeping an eye on the nest site of the old mate, in case he returns before she has found a new mate. Similarly, if the female does not return within a few days of the male, this may be due to her death, or a relative decline in her quality. Now the male would do best to search for a new mate. Thus, unless familiarity strongly influences reproductive success, males and females of both short-lived and long-lived species should not waste time waiting for their partner. What do the birds do? According to Davis and Speirs (1990), male Adelie Penguins indiscriminately court all females that pass through their nesting territory. Similar indiscriminate behaviour of males is reported for ptarmigans (Hannon and Martin Chapter 10), Barn Swallows (Møller 1994), and Pied Flycatchers *Ficedula hypoleuca* (Dale and Slagsvold 1994). Positive evidence that asynchronous arrival leads to divorce comes from Kittiwakes (Coulson and Thomas 1983), Adelie Penguins (Ainley *et al.* 1983; Davis and Spiers 1990), and semi-resident populations of Great Tits (Dhondt *et al.* Chapter 13). In contrast, Williams (Chapter 15) reported no effect of asynchronous arrival on probability of divorce in Macaroni Penguins. Whereas male and female Adelie penguins tend to arrive on the same date, male Macaroni penguins tend to arrive 7 days before the females (T. D. Williams personal communication).

Perhaps we must distinguish between species where mean arrival dates do not differ much between the sexes and species where males arrive long before the females. In the latter species, returning females will always encounter their old mate upon their return, if he survived. However, if the female arrives especially early, she has the additional option of ignoring her old mate and pairing with attractive males whose mate has not arrived yet, that is, the early arriving females can pre-empt the late arriving females. This assumes that pre-emption does not carry high risk. Hannon and Martin (Chapter 10) argue that since the previous owners probably would have won a contest with a female that pre-empted them (due to their previous tenure on the territory), the original females probably moved elsewhere on their own accord. However, fights did sometimes occur between ptarmigan. Furthermore, late arriving female Adelie and Macaroni Penguin females do not always give in to the new female and may succeed in driving her out (Williams Chapter 15). Probably, the longer the new female has been in place, the more difficult it will be for the pre-empted female to dislodge her (see later).

Choosing a site or choosing a mate?

Some (many?) birds choose not mates, but breeding sites

Morse and Kress (1984) showed that experimental blocking of nest burrows caused Leach's Storm Petrel *Oceanodroma leucorhoa* to divorce and suggested that mate fidelity was a by-product of nest site fidelity. This view has several strong supporters in this book. Desrochers and Magrath (Chapter 9) argue that divorce in European Blackbirds is a side effect of females leaving low quality sites in search of better territories. Newton and Wyllie (Chapter 14) also believe that the main factor that keeps Sparrowhawk pairs together is site fidelity, and adds the twist that site fidelity is in turn contingent upon a consistent and reliable food supply. Finally, Payne and Payne (Chapter 17) complement this picture by suggesting that breeding dispersal of female Indigo Buntings is driven by year to year changes in habitat, while male dispersal is constrained by the benefits of social familiarity with their neighbours.

If mate fidelity is a simple by-product of site fidelity, we predict high divorce rates in species with low site fidelity. Comparing different penguin species, Williams (Chapter 15) concludes that there is no correlation between mate fidelity and nest site fidelity, despite the fact that most penguins have high mate fidelity and high site fidelity, while the Emperor Penguin *Aptenodytes forsteri* has neither. In the full-blown comparative analysis of the species listed in Appendix 19.1 we classified species as either site-faithful, when the majority of individuals returned to the same nest site (relative to its neighbours) as the year before, or not site-faithful. Of these, only five species were classified as not site-faithful and all five breed in colonies whose location may shift between years. In three of these, the Emperor Penguin (Williams Chapter 15), the Greater Flamingo *Phoenicopterus ruber roseus* (Cézilly and Johnson, in press), and the Caspian Tern *Sterna tschegrava* (Cuthbert 1985), pair members meet each other only during the breeding season and have very high divorce rates, supporting the suggestion that mate fidelity is a by-product of site fidelity. In contrast, the two remaining species with low site fidelity, the Pinyon Jay (Marzluff and Balda Chapter 7) and the White-fronted Bee-eater *Merops bullockoides* (Emlen 1990), have continuous partnerships and very low divorce rates. Even if we exclude the White-fronted Bee-eater, where members of a breeding group defend 'fixed' year-round feeding territories, we are still left with the Pinyon Jay that does not fit the scheme that mate fidelity is a by-product of site fidelity. Furthermore, migratory species with low divorce rates, like geese and swans, do not fit, because even though they show high nest fidelity, pair formation and mate change occurs on the wintering grounds. Thus, it is certainly not universally true that mate fidelity is a by-product of site fidelity.

We can also investigate the relationship between site fidelity and mate fidelity within a species. If the sexes disperse independently between seasons, we should be able to calculate the expected rate of mate change exactly from male and female survival and dispersal (Payne and Payne Chapter 17). To calculate the expected divorce rate we can ignore the survival rates as we only need to consider pairs where both members survived. Let d_f and d_m be the probabilities that a surviving female or male will disperse to a new site. Let us also assume that dispersing individuals can choose among n territories. If the sexes disperse independently, we expect a proportion of $(1 - d_m)(1 - d_f)$ pairs where both members reunite on the old breeding site, $d_m(1 - d_f) + (1 - d_m)d_f$ pairs which split because one member moves to a new site, $d_m d_f/n$ pairs where both members move to the same new site, and $d_m d_f(n - 1)/n$ pairs that split, because both move to a different site. Thus, the probability that the mates will reunite turns out to be $(1 - d_m)(1 - d_f) + d_m d_f/n$, which decreases with an increase in the number of available sites.

A more sophisticated analysis of Kittiwakes takes into account that the number of potential new sites increases with the distance from the old site (Fairweather and Coulson 1995). Mate retention of Kittiwakes that moved was significantly higher than expected by chance. For example, 44% of the Kittiwakes that moved more than 9 m retained their mate, whereas the chance expectation was only 7%. The conclusion that in Kittiwakes, mate fidelity is not a simple by-product of nest site fidelity, was further strengthened when the breeding colony (an old warehouse) was made unavailable in 1990. Although not a single individual could use its old nest in 1991, over 60% of individuals retained their mate of the year before (which is close to the long-term average of 74% published by Coulson and Thomas 1983), while breeding or attempting to breed in the surroundings (Fairweather and Coulson 1995).

Site fidelity and the competitive advantage of prior residence

Even if high mate fidelity is not necessarily a by-product of high site fidelity, it seems prudent to recognize that the two issues are closely connected in many species. For instance, Dhondt *et al.* (Chapter 13) argue that for a Great Tit the decision whether to stay put, move to a new home range, or become a winter-emigrant is more linked to the mechanism of settlement than to the pure decision of divorce. Thus, we must enquire into the costs and benefits of site fidelity. Below we shall argue that the competitive benefits that individuals accrue from prior residence explains the high site fidelity of many species.

Keeping in mind that Appendix 19.1 is probably biased against species with low site fidelity, we begin by noting that nearly all species listed in that Appendix defend some sort of nesting space, ranging from part of a cliff ledge, or a hole in a tree trunk, to vast tracts of habitat in which the birds permanently reside. Inevitably, this habitat will differ in quality from the birds' point of view. As a consequence, we may expect competition for the sites of highest quality, for example sites where the adults and/or eggs are safest from predators and/or parasites, and/or territories where food for the chicks is most plentiful. Some birds will succeed in excluding others; the ensuing 'despotic distribution' was first modelled by Fretwell (1972). These despots are not necessarily individuals of superior fighting ability, but may also benefit from the effect of prior residence. In fact, in Oystercatchers asymmetries due to differences in prior residence may be more important than asymmetries due to differences in fighting ability. According to Ens *et al.* (1995) this explains why nonbreeding Oystercatchers 'queue' for territories of high quality. Nonbreeding Oystercatchers can settle immediately in a territory of poor quality, or wait many years to settle in a territory of high quality. To benefit from prior residence, both within and between seasons, nonbreeders must commit themselves to a small area, even though ranging over a wider area would increase the likelihood of spotting a vacancy (see also Zack and Stutchbury 1992). In the course of its life the queuing individual builds up a high local, that is, site-dependent, dominance. Site dominance will not drop off steeply when the bird leaves its territory, but will gradually decline with distance from the territory. Thus, observations that divorced individuals have rarely moved more than one territory, as is the case for the majority of individuals in the majority of species, would fit within this framework.

Pre-emption: prior residence within and between seasons

The easiest way to profit from prior residence between seasons is to continue residence throughout the year. In some species only a segment of the population is resident throughout the year, and this segment will profit most from the competitive advantage of prior residence. Winter residency varies dramatically in different populations of Great and Blue Tits. Mate fidelity is highest in tit populations with a high incidence of winter residency (Dhondt *et al.* Chapter 13). According to Newton and Wyllie (Chapter 14) a similar situation applies to Sparrowhawks.

Yet, as the Oystercatcher example suggests, permanent residence is not necessary for an effect of prior residence between seasons to occur. What is necessary is that former neighbouring owners still recognize each other upon return from the wintering grounds. Although this is probably often the case, it is experimentally proven in only a few species (e.g. Godard 1991). The fact that pre-emption occurs shows that this advantage of between-season residency may be overruled by within-season residency. The removal experiments of Krebs (1982) reveal how the advantage of prior residence can be overcome. When Krebs removed male Great Tits from their territory, a new bird would quickly settle. If the old male was released after a short time period, he would regain his territory, but if release was delayed, the old owner invariably lost to the new bird. Initially, new birds engaged in a lot of aggression with the neighbours, but this declined when territory boundaries were settled. This cost of settlement, that each new owner needs to pay, could be the basis of the asymmetry. Thus, if the old owner waits too long to return from the wintering grounds, the new owner has paid the cost of settlement and the pay-off asymmetry now favours the new owner. The species chapters yielded several examples of pre-emptions that failed, but we know of no data that relate the success rate of pre-emption to the length of delay in the return of the old owner. Such data would be revealing.

Conclusions and prospects

Some of the studies reported in this book were begun half a century ago. In contrast, many of the specific ideas reported in this and other chapters are less than a decade old. Thus, it is not surprising that the long-term data are not always of the most suitable form to test these ideas. In this final section we want to investigate whether it is possible to make some specific suggestions to alleviate this situation. At the outset we want to stress though that these suggestions should not be read as alternatives for the long-term studies. To the contrary: long-term studies on populations of marked individuals should be continued for as long as the financial and motivational resources of the individual investigator allow. There is a special need for such studies on long-term resident passerines from the southern hemisphere, as this review is heavily biased towards long-lived, site-faithful nonpasserines breeding in the northern hemisphere.

What progress has been made?

We begin by asking what progress has been made since Rowley (1983) wrote his influential review. Five points are worth discussing.

1. First of all, there is now much better evidence that there are indeed species where reproductive success increases with familiarity with the mate. These species are generally characterized by the long duration of their pair bonds.

2. Rowley was baffled by the large variation in divorce rates between long-lived species. Since mate fidelity may simply result from the high costs of changing mate and/or territory there may be species with long-term pair bonds where these bonds confer no special advantage. In species with low mate fidelity there are probably neither large reproductive benefits from familiarity with the mate, nor large fitness costs from a change of mate and/or site.

3. Despite Rowley's warning that we 'should beware of the stud-breeders' simple outlook—that male plus female equals offspring' not much convincing evidence for compatibility has since accumulated.

4. Rowley extensively documented the improvement in reproduction with age. Although the basic picture has not changed we now know that we have to walk through a statistical minefield for a proper assessment of the effects of age, experience, and familiarity. Quality may also consistently differ between individuals throughout their lives. Such consistent differences open up the possibility of selective death of poor quality individuals, impeding studies of the effect of age, experience, and familiarity on reproductive success. The number of casualties this problem has made, that is, conclusions that need to be retracted, is at present unknown.

5. Both Rowley's and this review suffer from the need to perform the cost–benefit analysis of mate fidelity on the basis of indirect arguments and indirect data. Perhaps the most important progress is that we are beginning to learn the various causes of divorce. The description of these causes enables us to identify decisions that allow a more direct analysis of the costs and benefits. When to desert your mate? When to give in to a usurper? When to give in to pre-emption and when not? They also allow us to enquire into the frequency of the social conditions that favour desertion, usurpation, and pre-emption.

Measuring costs and benefits

Whichever divorce decision we study, the costs and benefits must be gauged in terms of fitness gains or losses. Most of the arguments reviewed in this chapter are based on reproductive success in a single year, but this is an incomplete measure of fitness for three reasons. First, it ignores effects on other fitness components, like survival. Second, it ignores the possibility that the offspring are not genetically related to one or both parents, owing to extra-pair copulations, intraspecific brood parasitism, and adoption of chicks. Third, it ignores the possibility that even lifetime production of offspring is not a good measure of fitness.

The first problem seems solvable. Many of the studies have a sufficient database to estimate all fitness components and this information can then be used to calculate the expected future reproductive success, that is, the total number of chicks the individual may expect to fledge over the rest of its lifetime (Grafen 1988). To calculate the expected future reproductive success for birds of different mating status, the matrix equations presented by Ens *et al.* (1995) may prove useful. With regard to the second problem we note that (1) techniques, most notably DNA-fingerprinting, are available to take the appropriate measurements, and (2) current evidence suggests that extra-pair paternity is low in species where both partners are necessary to raise the brood (Birkhead and Møller Chapter 18). Since in most, if not all, species with long-lasting pair bonds both partners are necessary to raise the brood, extra-pair paternity is probably low and the conclusions on the mate familiarity effect are not ruined. In fact, it may well be that extra-pair paternity is only common in songbirds. According to the review by Birkhead and Møller (1992) the only nonpasserine with high levels of extra-pair paternity known to date is the Shag *Phalacrocorax aristotelis* (Graves *et al.* 1993). The third problem is the toughest. However, at present, the empirical evidence for indirect genetic benefits seems weak, despite the theoretical interest. For instance, Sheldon (1994) concludes that the positive associations between male phenotype and success at obtaining extra-pair copulations or fertilizations are not necessarily due to females trying to obtain indirect genetic fitness for their offspring. Instead, females may pursue extra-pair copulations as insurance against the functional infertility of their mate.

Theoretical challenges

The proliferation of hypotheses on mate fidelity indicates that the problem is obviously very complex, involving many variables (Choudhury 1995; Black Chapter 1). In such cases mathematics offers the best hope of not getting lost in the logic. The model of McNamara and Forslund (in press) is an important first step. It would be nice to know though how the predictions of their model are affected by different assumptions on the mechanisms of mate assessment and mate change. McNamara and Forslund (in press) identify as their primary goal the development of a full population model that finds the evolutionarily stable divorce strategies for both males and females.

Another important theoretical challenge is to link theories on despotic distributions to ideas on mate choice and mate change. It is unsatisfactory to discuss site fidelity as a factor confounding the study of mate fidelity, as we and many authors before us have done. Once it is realized that the decision on when to settle (i.e. at what age) cannot be separated from the decision on where to settle (i.e. in what type of territory) it becomes possible to build a model of the recruitment decision that provides a partial explanation for both the despotic distribution and the problem of deferred maturity (Ens *et al.*, 1995). Similarly, the decision on where to settle cannot be separated from the decision with whom to mate (now or in the future). For example, nonbreeding Oystercatchers often settle as a pair, but single nonbreeders may fill a vacancy, or act as usurpers and cause a divorce in an established pair. Once an individual has settled, only neighbouring individuals are potential new mates. Somewhere in the course of this review, the ecological conditions that 'set the stage on which individuals play out their behavioural strategies' (Davies 1992) faded away. Developing a theory that successfully unifies competition for mates with competition for limiting resources in monogamous birds may bring the ecological conditions back to centre stage.

Empirical challenges

We do not have to wait for the theoreticians to do their work before collecting new measurements. Apart from the improved fitness measurements that we discussed above, more detailed behavioural observations are needed as these offer the best insights into the precise causes of divorce and mate fidelity. Anecdotal observations of forced divorce are surprisingly common. In some species usurpations may well be the most common cause of divorce and there are few species where we can be certain that it never happens: forced divorce is easily missed, since the actual take-over fight may take only a few hours. Even if the fighting continued over several days, few studies are intensive enough to be sure that they will pick up the event.

Detailed observations on the process of pair formation and divorce events should include detailed studies of courtship behaviour, as this may be the time when the partners attempt to assess each other's quality. (Though studies of courtship flourished in the heyday of ethology, they were usually performed on unmarked individuals and with different questions in mind). The extensive studies on the Pied Flycatcher may serve as an example of the detail that is required. Detailed experiments in the field and laboratory prove that females prefer brightly coloured males (Sætre *et al.* 1994). Female competition may explain why females, none the less, often fail to realize their preferred option and mate with an inferior male. In contrast, Dale and Slagsvold (1994) concluded that male Pied Flycatchers do not choose mates, because they do not vary the intensity of their courtship behaviour with the extent to which females have been experimentally handicapped. The authors suggest that discriminating males would reduce their changes of becoming polygynous and might even run the risk of remaining unmated.

Finally, this chapter makes abundantly clear that removal experiments, as well as other experimental manipulations, are absolutely vital to further our understanding. Apart from the observation of Harris *et al.* (1987) that 'it is impractical to study divorce in a species that rarely divorces', we can never be sure of having safely traversed the statistical quagmire without experiments.

Summary

This synthesis addresses the question why for a given species some pairs reunite, while other pairs divorce, within the constraint of a monogamous relationship. Although within-season divorce occurs, attention here focuses on mate fidelity between seasons. In most studies divorce is inferred from both birds of a pair being alive yet not breeding together, instead of detailed behavioural observations of the pair bond. As a result we are poorly informed on the causes of divorce, which include pre-emption (a new bird takes up the territory before the previous owner had returned), usurpation (a third bird evicts one of the partners), and desertion (one bird leaves its partner to form a pair bond with a new mate).

Although mate fidelity is generally considered a topic in sexual selection, hypotheses to explain divorce rates are not derived from sexual selection theory, but from post-hoc interpretations of empirical data. The recent proliferation of these intuitively argued hypotheses testifies to the surprising complexity of the problem. In their present form they do not constitute viable alternative hypotheses, but each offers insights on one or more aspects of the problem and points out possible mechanisms and conditions to which a more comprehensive theory of mate fidelity should pay attention. The mathematical model of McNamara and Forslund (in press) is a first step in this direction.

Comparing species, there is some indication that divorce rate increases with mortality rate at a high taxonomic level, but within taxonomic families the relationship is, rather surprisingly, more often in the reverse direction. Phylogeny apparently constrains the range of values that mortality and divorce rate can take, but it is unclear why.

Differences in reproductive success between pairs may be due to differences between individuals in quality, compatibility, or familiarity. Under a monogamous mating system, the existence of quality differences between individuals implies competition for mates of high quality, which will be especially intense if quality differences are large. There is abundant evidence for quality differences between individuals, either fixed for life or changing with age. It is generally not known how much of this variation is due to individuals changing with age, or to individuals of high quality surviving longer. In spite of its popularity as an idea, there is little empirical evidence for effects of compatibility on reproductive success, except in Barnacle Geese, where small individuals achieve highest success with small partners and large individuals with large partners. Finally, in several species with long-term pair bonds there was suggestive evidence for an improvement in reproductive success with pair duration. Further evidence that familiarity affected reproductive success in these species came from a notable decrease in reproductive success following a change of mate. What we do not fully understand though is how the mate familiarity effect operates.

In species that had high mate fidelity, yet in which long-term pair bonds seemed to carry no special advantages, divorce may have been constrained by high costs of mate change other than losing access to a familiar mate. These include the costs of searching for a mate (sometimes birds were unmated for several years), the costs of fighting rivals (usurpations typically involved physical fights), and the risk of ending up with a poor mate (often, new mates were young and inexperienced). Whether and how (through mortality, nonbreeding, or reduced reproductive success) these costs are paid depends on the precise mechanism of mate change.

Although it is often found that divorce rate is higher for pairs that failed to fledge chicks than for pairs that succeeded, the evidence that low success actually triggers divorce is not convincing. Usually, old partners reunite, even if they failed to rear chicks in the year before. Furthermore, in several species the correlation between low success and subsequent divorce is due to young birds being both unsuccessful and more likely to divorce. Finally, birds may be able to assess each other's quality without breeding together. Thus, the relationship between success and subsequent probability of divorce is not as informative as it once seemed to be and it is argued that the best predictor of desertion, which is only one route to divorce, should be the prospect of improvement. Vacancies may represent opportunities to change to a better mate at low cost, while relative changes in arrival time may signal relative changes in quality of the mate.

In some species, mate fidelity seems a by-product of fidelity to the nest site, but this is certainly not the case in species with high fidelity to the mate and low fidelity to the nest site and in species that form and break their pair bonds on the wintering grounds. It is argued that the competitive advantage of prior residence underlies high site fidelity. In migratory species within-season residency of a new bird may conflict with between-season residency of the old owner. Perhaps an old owner will regain the territory if she/he returns before the new owner has been present for a sufficiently long time period, while the new owner will successfully preempt the old owner if the latter fails to return before this critical period has ended.

Future progress in the field depends on the following.

1. More detailed behavioural observations, so that we learn the frequency with which different causes of divorce operate in different species and under different conditions. Whether birds have a pair bond should be inferred from their association behaviour, not simply from the act of breeding. Detailed studies of courtship behaviour may teach us how mates assess each other.

2. A methodology must be devised that allows us to assess individual bird quality before breeding, independently of the actual success achieved later on.

3. Such a methodology is vital in much needed field experiments that should include mate removal (permanent or temporary), and manipulation of mate and territory quality.

4. The value of these experiments is increased if the results can be compared with quantitative predictions. To cope with intrasexual competition and intersexual conflicts a game theory approach seems unavoidable.

Acknowledgements

We are grateful to the 28 researchers that provided unpublished details of mate fidelity and divorce in their study species (Appendix 19.1). Andy Purvis and Ian Owens gave us advice on the comparative analyses and helped us interpret the results. We are also grateful to Andy for supplying us with a CAIC computer program. We thank Murray Williams, Frank McKinney, Eileen Rees, John Marzluff, Glen Woolfenden, Eleanor Russell, Ian Rowley, André Desrochers, Susan Hannon, Ron Wooller, Stuart Bradley, André Dhondt, Ian Newton, Tony Williams, and Bob Payne for useful comments and criticisms. Kate Lessells didn't spare the rod, which was both painful and helpful.

References

Ainley, D. G., LeResche, R. E., and Sladen, W. J. L. (1983). *Breeding biology of the Adélie penguin*. University of California Press, Berkeley.

Ainley, D. G., Ribic, C. A., and Wood, R. C. (1990). A demographic study of the south polar skua *Catharacta maccormicki* at Cape Crozier. *Journal of Animal Ecology*, **59**, 1–20.

Arcese, P. (1989). Territory acquisition and loss in male song sparrows. *Animal Behaviour*, **37**, 45–55.

Ashcroft, R. E. (1976). Breeding biology and survival of Puffins. Unpublished D. Phil. thesis. University of Oxford.

Baeyens, G. (1981). Functional aspects of serial monogamy: the Magpie pair-bond in relation to its territorial system. *Ardea*, **69**, 145–66.

Ball, I. J., Frost, P. G. H., Siegfried, W. R., and McKinney, F. (1978). Territories and local movements of African Black Ducks. *Wildfowl*, **29**, 61–79.

Barrat, A. (1976). Quelques aspects de la biologie et de l'ecologie du Manchot Royal *Aptenodytes patagonicus* des Iles Crozet. *Comité nacional Francaise recherche Antarctique*, **40**, 9–51.

Bateson, P. (1983). Optimal outbreeding. In *Mate choice* (ed. P. Bateson), pp. 257–77. Cambridge University Press, Cambridge.

Birkhead, T. R., and Møller, A. P. (1992). *Sperm competition in birds: evolutionary causes and consequences*. Academic Press, London.

Black, J. M. and Owen, M. (1989). Agonistic behaviour in goose flocks: assessment, investment and reproductive success. *Animal Behaviour*, **36**, 199–209.

Black, J. M. and Owen, M. (1995). Reproductive performance and assortative pairing in relation to age in barnacle geese. *Journal of Animal Ecology*, **64**, 234–44.

Boekelheide, R. J. and Ainley, D. G. (1989). Age, resource availability, and breeding effort in Brandt's Cormorant. *Auk*, **106**, 389–401.

Bost, C. A. and Jouventin, P. (1990). Laying asynchrony in Gentoo Penguins on Crozet Island: causes and consequences. *Ornis Scandinavica*, **21**, 63–70.

Bradley, J. S., Wooller, R. D., Skira, I. J., and Serventy, D. L. (1990). The influence of mate retention and divorce upon reproductive success in short-tailed shearwaters *Puffinus tenuirostris*. *Journal of Animal Ecology*, **59**, 487–96.

Brooke, M. de L. (1978). Some factors affecting the laying date, incubation and breeding success of the Manx shearwater, *Puffinus puffinus*. *Journal of Animal Ecology*, **47**, 477–95.

Bryant, D. M. (1979). Reproductive costs in the house martin *Delichon urbica*. *Journal of Animal Ecology*, **48**, 655–75.

Burley, N. (1977). Parental investment, mate choice, and mate quality. *Proceedings of the National Academy of Sciences, U.S.A.*, **74**, 3476–9.

Cézilly, F. and Johnson, A. R. Re-mating between and within breeding seasons in the Greater Flamingo *Phoenicopterus ruber roseus*. *Ibis*. (In press.)

Choudhury, S. (1995). Divorce in birds: a review of the hypotheses. *Animal Behaviour*, **50**, 413–29.

Clutton-Brock, T. H. (1988). *Reproductive success: studies of individual variation in contrasting breeding systems*. University of Chicago Press, Chicago.

Cooke, F., Bousfield, M. A., and Sadura, A. (1981). Mate change and reproductive success in the lesser snow goose. *Condor*, **83**, 322–7.

Coulson, J. C. and Thomas, C. S. (1983). Mate choice in the kittiwake gull. In *Mate choice* (ed. P. Bateson), pp. 361–76. Cambridge University Press, Cambridge.

Coulson, J. C. and Thomas, C. S. (1985). Differences in the breeding performance of individual Kittiwake Gulls *Risa tridactyla* (L). In *Behavioural ecology: ecological consequences of adaptive behaviour* (ed. R. M. Sibly and R. H. Smith), pp. 489–503. Blackwell, London.

Curio, E. (1983). Why do young birds reproduce less well? *Ibis*, **125**, 400–4.

Cuthbert, F. J. (1985). Mate retention in the Caspian Tern. *Condor*, **87**, 74–8.

Daan, S., Dijkstra, C., Drent, R. H., and Meijer, T. (1989). Food supply and the annual timing of avian reproduction. *International Ornithology Congress Proceedings*, **19**, 392–407.

Dale, S. and Slagsvold, T. (1994). Male pied flycatchers do not choose mates. *Animal Behaviour*, **47**, 1197–205.

Davies, N. B. (1992). *Dunnock behaviour and social evolution*. Oxford University Press, Oxford.

Davis, L. S. and Speirs, E. A. H. (1990). Mate choice in penguins. In *Penguin biology* (ed. L. S. Davies and J. T. Darby), pp. 377–97. Academic Press, San Diego.

Delnicki, D. (1983). Mate changes by Black-bellied Whistling Ducks. *Auk*, **100**, 728–9.

Delius, J. d. (1965). A population study of Skylarks *Alauda arvensis*. *Ibis*, **107**, 466–92.

Desrochers, A. (1992). Age-related differences in reproduction by European Blackbirds: restraint or constraint? *Ecology*, **73**, 1128–31.

Desrochers, A. and Magrath, R. D. (1993*a*). Age-specific fecundity in European Blackbirds (*Turdus merula*): individual and population trends. *Auk*, **110**, 255–63.

Desrochers, A. and Magrath, R. D. (1993*b*). Environmental predictability and re-mating in European blackbirds. *Behavioral Ecology*, **4**, 271–5.

Dhondt, A. A. and Adriaensen, F. (1994). Causes and effects of divorce in the blue tit *Parus caeruleus*. *Journal of Animal Ecology*, **63**, 979–87.

Emlen, S. T. (1990). White-fronted Bee-eaters: helping in a colonially nesting species. In *Cooperative breeding in birds: long-term studies of ecology and behaviour* (ed. P. B. Stacey and W. D. Koenig), pp. 489–526. Cambridge University Press, Cambridge.

Emlen, S. T. (1991). Evolution of cooperative breeding in birds and mammals. In *Behavioural ecology: an evolutionary approach*, (3rd edn), (ed. J. R. Krebs and N. B. Davies), pp. 301–37. Blackwell, London.

Ens, B. J. (1992). The social prisoner: Causes of natural variation in reproductive success of the Oystercatcher. Unpublished Ph.D. thesis. University of Groningen.

Ens, B. J., Kersten, M., Brenninkmeijer, A., and Hulscher, J. B. (1992). Territory quality, parental effort and reproductive success of oystercatchers (*Haematopus ostralegus*). *Journal of Animal Ecology*, **61**, 703–15.

Ens, B. J., Safriel, U. N., and Harris, M. P. (1993). Divorce in the long-lived and monogamous oystercatcher *Haematopus ostralegus*: incompatibility or choosing the better option? *Animal Behaviour*, **45**, 1199–217.

Ens, B. J., Weissing, F. J., and Drent, R. H. (1995). The despotic distribution and deferred maturity: two sides of the same coin. *American Naturalist*, **146**, 625–50.

Evans Ogden, L. J. and Stutchbury, B. J. (In press). Hooded Warbler, In *Birds of North America* (ed. F. Gill, A. Poole, and P. Stettenheim). American Ornithologist's Union.

Fairweather, J. A. and Coulson, J. C. (1995). Mate retention in the kittiwake *Rissa tridactyla* and the significance of nest site tenacity. *Animal Behaviour*, **50**, 455–64.

Falconer, D. S. (1981). *Introduction to quantitative genetics*, (2nd edn). Longman, Essex.

Farkas, T. (1969). Notes on the biology and ethology of the Natal Robin *Cossypha natalensis*. *Ibis*, 281–92.

Felsenstein, J. (1985). Phylogenies and the comparative method. *American Naturalist*, **125**, 1–15.

Fetterolf, P. M. (1984). Pairing behaviour and pair dissolution by Ring-billed Gulls during the post-breeding period. *Wilson Bulletin*, **96**, 711–14.

Findlay, C. S. (1987). Non-random mating: a theoretical and empirical overview with special reference to birds. In *Avian genetics: a population and ecological approach* (ed. F. Cooke and P. A. Buckely), pp. 289–319. Academic Press, London.

Forslund, P. and Larsson, K. (1991). The effect of mate change and new partner's age on reproductive success in the barnacle goose *Branta leucopsis*. *Behavioral Ecology*, **2**, 116–22.

Freed, L. A. (1987). The long-term pair bond of tropical house wrens: advantage or constraint? *American Naturalist*, **130**, 507–25.

Fretwell, S. D. (1972). *Populations in a seasonal environment*. Princeton University Press, Princeton.

Godard, R. (1991). Long-term memory of individual neighbours in a migratory songbird. *Nature*, **350**, 228–9.

Grafen, A. (1988). On the uses of data on lifetime reproductive success. In *Reproductive success: studies of individual variation in contrasting breeding systems* (ed. T. H. Clutton-Brock), pp. 454–71. University of Chicago Press, Chicago.

Grafen, A. (1989). The phylogenetic regression. *Philosophical Transactions of the Royal Society of London*, **326**, 119–57.

Grant, B. R. and Grant, P. R. (1989). *Evolutionary dynamics of a natural population: the large cactus finch of the Galapagos*. University of Chicago Press, Chicago.

Gratto-Trevor, C. I. (1991). Parental care in Semipalmated Sandpipers *Calidris pusilla*: brood desertion by females. *Ibis*, **133**, 394–9.

Graves, J., Ortega-Ruano, J., and Slater, P. J. B. (1993). Extra-pair copulations and paternity in shags: do females choose better males? *Proceedings of the Royal Society of London*, **253**, 3–7.

Haig, S. M. and Oring, L. W. (1988). Mate, site, and territory fidelity in Piping Plovers. *Auk*, **105**, 268–77.

Harris, M. P. (1970). Territory limiting the size of the breeding population of the oystercatcher (*Haematopus ostralegus*)—a removal experiment. *Journal of Animal Ecology*, **39**, 707–13.

Harris, M. P. (1973). The biology of the Waved Albatross *Diomedea irrorata* of Hood Island, Galapagos. *Ibis*, **115**, 483–510.

Harris, M. P. (1979). Population dynamics of the Flightless Cormorant *Nannopterum harrisi*. *Ibis*, **121**, 135–46.

Harris, M. P., Safriel, U. N., Brooke, M. de L., and Britton, C. K. (1987). The pair bond and divorce among Oystercatchers *Haematopus ostralegus* on Skokholm Island, Wales. *Ibis*, **129**, 45–57.

Harvey, P. H. and Pagel, M. D. (1991). *The comparative method in evolutionary biology*. Oxford University Press, Oxford.

Hatch, S. A. (1990). Incubation rhythm in the Fulmar *Fulmarus glacialis*: annual variation and sex roles. *Ibis*, **132**, 515–24.

Hatch, S. A., Robert, B. D., and Fadely, B. S. (1993). Adult survival of Black-legged Kittiwakes *Risa tridactyla* in a Pacific colony. *Ibis*, **135**, 247–54.

Heg, D., Ens, B. J., Burke, T., Jenkins, L., and Kruijt, J. P. (1993). Why does the typically monogamous oystercatcher (*Haematopus ostralegus*) engage in extra-pair copulations? *Behaviour*, **126**, 247–87.

Holmes, R. T. (1971). Density, habitat and the mating system of the Western Sandpiper *Calidris mauri*. *Oecologia*, **7**, 191–208.

Isenman, P. (1971). Contribution a l'ethologie et a l'ecologie du Manchot Empereur *Aptenodytes forsteri* a la colonie de Pointe Geologie (Terre Adelie). *L'Oiseau et Recherche Francais Ornithologie*, **41**, 9–64.

Jehl, J. R. (1973). Breeding biology and systematic relationships of the Stilt Sandpiper. *Wilson Bulletin*, **85**, 115–47.

Jones, I. L. and Montgomerie, R. (1991). Mating and remating of least auklets (*Aethia pusilla*) relative to ornamental traits. *Behavioral Ecology*, **2**, 249–57.

Kingsford, R. T. (1990). Flock structure and pair bonds of the Australian Wood Duck *Chenonetta jubata*. *Wildfowl*, **41**, 75–82.

Kluijver, H. N. (1951). The population ecology of the Great Tit *Parus major*. *Ardea*, **39**, 1–135.

Krebs, J. R. (1982). Territorial defence in the great tit (*Parus major): do residents always win?* Behavioral Ecology and Sociobiology, **11**, 185–94.

Lack, D. (1968). *Ecological adaptations for breeding in birds*. Methuen, London.

LaCock, G. D., Duffy, D. C., and Cooper, J. (1987). Population dynamics of the African penguin *Spheniscus demersus* at Marcus Island in the Benguela upwelling ecosystem. *Biological Conservation*, **40**, 117–26.

Lindén, M. (1991). Divorce in great tits: chance or choice? An experimental approach. *American Naturalist*, **138**, 1039–48.

Macdonald, M. A. (1977). An analysis of the recovery of British-ringed Fulmars. *Bird Study* **24**, 208–14.

McNamara, J. M. and Forslund, P. Divorce rates in birds; predictions from an optimization model. *American Naturalist*. (In press.)

Marchant, S. and Higgins, P. (1990). *Handbook of Australian, New Zealand and Antarctic birds*. Oxford University Press, Oxford.

Minton, C. D. T. (1968). Pairing and breeding of Mute Swans. *Wildfowl*, **19**, 41–60.

Mjelstad, H. and Saetersdal, M. (1990). Reforming of resident Mallard pairs *Anas platyrhynchos*, rule rather than exception? *Wildfowl*, **41**, 150–1.

Møller, A. P. (1994). *Sexual selection and the barn swallow*. Oxford University Press, Oxford.

Morse, D. H. and Kress, S. W. (1984). The effect of burrow loss on mate choice in the Leach's Storm Petrel. *Auk*, **101**, 158–60.

Moulton, D. W. and Weller, M. W. (1984). Biology and conservation of the Laysan Duck *Anas laysanensis*. *Condor*, **86**, 105–17.

Nagata, H. (1986). Female choice in Middendorff's Grasshopper-warbler (*Locustella ochotensis*). *Auk*, **103**, 694–700.

Nelson, J. B. (1978). *The Sulidae: gannets and boobies*. Oxford University Press, Oxford.

Newton, I. (1988). Individual performance in Sparrowhawks: the ecology of two sexes. *Proceedings International Ornithological Congress*, **19**, 125–54.

Newton, I. (1989*a*). Synthesis. In *Lifetime reproduction in birds* (ed. I. Newton), pp. 441–69. Academic Press, London.

Newton, I. (1989*b*). Sparrowhawk. In *Lifetime reproduction in birds* (ed. I. Newton), pp. 279–96. Academic press, London.

Newton, I. (1991). Habitat variation and population regulations in Sparrowhawks. *Ibis*, **133**, (suppl.) 46–88.

Nice, M. M. (1937). Studies in the Life History of the Song Sparrow. I. A. population study of the Song Sparrow. *Transactions of the Linnaean Society of New York*, **4**, 1–247.

Nol, E. and Smith, J. N. M. (1987). Effects of age and breeding experience on seasonal reproductive success in the song sparrow. *Journal of Animal Ecology*, **56**, 301–13.

O'Donald, P. O. (1983). *The Arctic Skua: a study of the ecology and evolution of a seabird*. Cambridge University press, Cambridge.

Ollason, J. C. and Dunnet, G. M. (1978). Age, experience and other factors affecting the breeding success of the fulmar, *Fulmarus glacialis*, in Orkney. *Journal of Animal Ecology*, **47**, 961–76.

Ollason, J. C. and Dunnet, G. M. (1988). Variation in breeding success in Fulmars. In *Reproductive success* (ed. T. H. Clutton-Brock), pp. 263–78. University of Chicago Press, Chicago.

Orell, M., Rytkönen, S., and Koivula, K. (1994). Causes of divorce in the monogamous willow tit, *Parus montanus*, and consequences for reproductive success. *Animal Behaviour*, **48**, 1143–54.

Owen, M. and Black, J. M. (1989). Barnacle Geese. In *Lifetime reproduction in birds* (ed. I. Newton), pp. 349–62. Academic Press. London.

Owen, M., Black, J. M., and Liber, H. (1988). Pair bond duration and timing of its formation in Barnacle Geese (*Branta leucopsis*). In *Waterfowl in winter* (ed. M. Weller), pp. 23–38. University of Minnesota Press, Minneapolis.

Penney, R. L. (1968). Territory and social behaviour in the Adelie Penguin. *Antarctic Research Series*, **12**, 83–131.

Perrins, C. M. and McCleery, R. H. (1985). The effect of age and pair bond on the breeding success of Great Tits *Parus major, Ibis*, **127**, 306–15.

Petersen, Æ. (1981). Breeding biology and feeding ecology of Black Guillemots. Unpublished D.Phil. thesis. University of Oxford.

Pickering, S. P. C. (1989). Attendance patterns and behaviour in relation to experience and pair-bond formation in the Wandering Albatross *Diomedea exulans* at South Georgia. *Ibis*, **131**, 183–95.

Prevett, J. P. and MacInnes C. D. (1980). Family and other social groups in Snow Geese. *Wildlife Monographs*, **71**, 1–46.

Purvis, A. and Harvey, P. H. (1995). Mammal life history evolution: a comparative test of Charnov's model. *Journal of Zoology, London*, **237**, 259–83.

Purvis, A. and Rambout, A. (1995). Comparative analysis by independent contrasts (CAIC): an Apple Macintosh application for analysing comparative data. *Computer Applications for Biosciences*, **11**, 247–51.

Raveling, D. G. (1988). Mate retention in Giant Canada Geese. *Journal of Canadian Zoology*, **66**, 2766–8.

Reid, W. V. (1988). Age correlations within pairs of breeding birds. *Auk*, **105**, 278–85.

Reilly, P. N. and Cullen, J. M. (1981). The Little Penguin *Eudyptula minor* in Victoria, II. Breeding. *Emu*, **81**, 1–19.

Rice, D. W. and Kenyon, K. W. (1962). Breeding cycle and behaviour of Laysan and Black-Footed Albatrosses. *Auk*, **79**, 517–67.

Richdale, L. E. (1957). *A population study of penguins*. Oxford University Press, Oxford.

Richdale, L. E. and Warham, J. (1973). Survival, pair bond retention and nest-site tenacity in Buller's Mollymawk. *Ibis*, **115**, 257–63.

Robinson, D. (1990). The social organization of the Scarlet Robin *Petroica multicolor* and Flame Robin *P. phoenicea* in Southeastern Australia: a comparison between sedentary and migratory flycatchers. *Ibis*, **132**, 78–94.

Röell, A. (1978). Social behaviour of the jackdaw, *Corvus monedula*, in relation to its niche. *Behaviour*, **64**, 1–124.

Rowley, I. (1965). The life-history of the Superb Blue Wren *Malurus cyaneus*. *Emu*, **64**, 251–97.

Rowley, I. (1973). The comparative ecology of Australian corvids II. Social organisation and behaviour. *CSIRO Wildlife Research*, **18,** 25–65.

Rowley, I. (1983). Re-mating in birds. In *Mate choice* (ed. P. Bateson), pp. 331–60. Cambridge University Press, Cambridge.

Rowley, I. (1990). *Behavioral ecology of the galah Eolophus roseicapillus in the wheatbelt of western Australia.* Surrey Beatty, Chipping Norton.

Rowley, I. and G. Chapman (1991). The breeding biology, food, social organisation, demography and conservation of the Major Mitchell or Pink Cockatoo, *Cacatua leadbeateri*, on the Margin of the Western Australian Wheatbelt. *Australian Journal of Zoology*, **39,** 211–61.

Ryabitsev, V. K. (1993). *Territorial relations and community dynamics of birds in the subarctic* [in Russian]. Ekaterinburg, Nauka.

Savard, J. L. (1985). Evidence of long-term pair bonds in Barrow's Goldeneye (*Bucephala islandica*). *Auk*, **102,** 389–91.

Sheldon, B. C. (1994). Male phenotype, fertility, and the pursuit of extra-pair copulations by female birds. *Proceedings of the Royal Society of London*, **257,** 25–30.

Shields, W. M. (1984). Factors affecting nest and site fidelity in Adirondack Barn Swallows (*Hirundo rustica*). *Auk*, **101,** 780–9.

Sibley, G. C. and Ahlquist, J. E. (1990). *Phylogeny and classification of birds: a study in molecular biology*. Yale University Press, New Haven.

Sibley, C. G. and Monroe, B. L. (1990). *Distribution and taxonomy of birds of the world.* Yale University Press, New Haven.

Snow, D. W. (1958). *A study of blackbirds.* Allen and Unwin, London.

Soikkeli, M. (1967). Breeding cycle and population dynamics in the Dunlin (*Calidris alpina*). *Annales zoologici*, **4,** 158–198.

Sorenson, L. G. (1992). Variable mating system of a sedentary tropical duck: the White-cheeked Pintail (*Anas bahamensis bahamensis*). *Auk*, **109,** 277–92.

Sorenson, L. G. (1994). Forced extra-pair copulation and mate guarding in the white-cheeked pintail: timing and trade-offs in an asynchronously breeding duck. *Animal Behaviour*, **48,** 519–33.

Southern, L. K. and Southern, W. E. (1982). Mate fidelity in Ring-billed Gulls. *Journal of Field Ornithology*, **53,** 170–1.

Summers, R. W. (1983). The life cycle of the Upland Goose *Chloephaga picta* in the Falkland Islands. *Ibis*, **125,** 524–44.

Sætre, G.-P., Dale, S., and Slagsvold, T. (1994). Female pied flycatchers prefer brightly coloured males. *Animal Behaviour*, **48,** 1407–16.

Thomas, C. S. and Coulson, J. C. (1988). Reproductive success of Kittiwake Gulls, *Rissa tridactyla*. In *Reproductive success: studies of individual variation in contrasting breeding systems (*ed. T. H. Clutton-Brock), pp. 251–62. University of Chicago Press, Chicago.

Trivelpiece, W. Z. and Trivelpeice, S. G. (1990). Courtship period of Adelie, Gentoo and Chinstrap Penguins. In *Penguin biology* (ed. L. S. Davis and J. T. Darby), pp. 113–27. Academic Press, San Diego.

Vermeer, K. (1963). The breeding ecology of the Glaucous-winged Gull *Larus glaucescens* on Mandarte Island. *Occasional Papers of the British Columbia Provincial Museum*, No. 13.

Wagner, R. H. (1991). The use of extrapair copulations for mate appraisal by razorbills. *Behavioral Ecology*, 2, 198–203.

Walters, J. R. (1990). Red-cockaded Woodpeckers: a primitive cooperative breeder. In *Cooperative breeding in birds: long-term studies of ecology and behaviour* (ed. P. B. Stacey and W. D. Koenig), pp. 69–101. Cambridge University Press, Cambridge.

Warkentin, I. G., James, P. C., and Oliphant, L. W. (1991). Influence of site fidelity on mate switching in urban-breeding Merlins (*Falco columbarius*). *Auk*, **108**, 294–302.

Weimerskirch, H. (1990). The influence of age and experience on breeding performance of the antarctic fulmar Fulmarus *glacialoides. Journal of Animal Ecology*, **59**, 867–75.

Williams, M. (1973). Dispersionary behaviour and breeding of Shelduck *Tadorna tadorna* on the River Ythan Estuary. Unpublished Ph.D. thesis. Aberdeen University.

Williams, T. D. and Rodwell, S. (1992). Annual variation in return rate, mate and nest-site fidelity in breeding Gentoo and Macaroni Penguins. *Condor*, **94**, 636–45.

Wood, R. C. (1971). Population dynamics of breeding South Polar Skuas of unknown age. *Auk*, **88**, 805–14.

Wooller, R. D., Bradley, J. S., Skira, I. J., and Serventy, D. L. (1990). Reproductive success of short-tailed shearwaters *Puffinus tenuirostris* in relation to their age and breeding experience. *Journal of Animal Ecology*, **59**, 161–70.

Zack, S. and Stutchbury, B. J. (1992). Delayed breeding in avian social systems: territory quality and 'floater' tactics. *Behaviour*, **123**, 194–219.

Appendix 19.1 Species characteristics and divorce rates in monogamous birds (populations with more than 5% polygamy are excluded). The following variables are recorded. Brood: S = single-brooded, M = mutilple-brooded (more than 20% of successful pairs attempt a second brood). Pair bond: B = partnership habitually broken after the breeding season, Y = partnership maintained year round (some sort of association behaviour must be maintained throughout the year). Residency: R = resident year-round on territory (seabirds are classified as resident, if they visit their specific nest site throughout the year), S = semi-resident (no close attachment to breeding site during the nonbreeding season, but no migration over large distances either), M = migratory (migration spans distances of at least a few hundred kilometres). Fidelity to previous breeding site: F = strong fidelity (the majority of individuals must return to the 'exact' nesting site of the previous year; what matters is the spatial arrangement with respect to other individuals), N = no strong fidelity (this would include seabirds that return to the same colony, but can nest anywhere within that colony). Annual divorce rate (estimated from the number of pairs that divorced divided by the total number of pairs that survived to the next breeding season) and the number of pair-years on which the estimate is based. Annual mortality rate (when mortality rates between males and females differed, the average value was taken). All data listed by Rowley (1983) are also listed in this table, unless divorce rates seemed inflated because many birds that moved as a pair were missed in subsequent years. Studies with fewer than 5 pair-years were also excluded.

Species	Latin name	Broods	Pair bond	Residency	Site fidelity	Divorce rate %	Pair years	Mortality rate %	Source
Sphenisciformes									
King Penguin	*Aptenodytes patagonicus*	S	B	S	F	71.4	21	7.0	Barrat (1976)
Emperor Penguin	*Aptenodytes forsteri*	S	B	M	N	85.4	41	5.0	Isenman (1971)
Gentoo Penguin	*Pygoscelis papua*	S	B	S	F	51.5	68	—	Bost and Jouventin (1990)
Gentoo Penguin	*Pygoscelis papua*	S	B	S	F	10.0	376	—	Trivelpiece and Trivelpiece (1990)
Gentoo Penguin	*Pygoscelis papua*	S	B	R	F	10.0	245	—	Williams and Rodwell (1992)
Adelie Penguin	*Pygoscelis adeliae*	S	B	M	F	16.4	165	28.7	Penney (1968)
Adelie Penguin	*Pygoscelis adeliae*	S	B	M	F	30.0	50	26.3	Davis and Speirs (1990)
Adelie Penguin	*Pygoscelis adeliae*	S	B	M	F	38.0	458	—	Trivelpiece and Trivelpiece (1990)
Chinstrap Penguin	*Pygoscelis antarctica*	S	B	M	F	17.9	402	—	Trievelpice and Trivelpiece (1990)
Rockhopper Penguin	*Eudyptes chrysocome*	S	B	M	F	20.9	67	—	Moors and Cunningham[1]
Macaroni Penguin	*Eudyptes chrysolophus*	S	B	M	F	9.2	152	—	Williams and Rodwell (1992)
Yellow-eyed Penguin	*Megadyptes antipodes*	S	B	R	F	18.0	539	13.9	Richdale (1957)
Little Penguin	*Eudyptula minor*	M	Y	R	F	3.0	115	14.0	Reilly and Cullen (1981)
African Penguin	*Spheniscus demersus*	S	B	R	F	13.8	116	—	LaCock *et al.* (1987)
Procellariiformes									
Waved Albatross	*Diomedea irrorata*	S	B	M	F	0.0	310	5.0	Harris (1973)
Laysan Albatross	*Diomedea immutabilis*	S	B	M	F	2.1	330	1.6	Rice and Kenyon (1962)

Procellariiformes (continued)									
Buller's Mollymawk	*Diomedea bulleri*	S	B	M	F	0.0	400	11.1	Richdale and Warham (1973)
Fulmar	*Fulmarus glacialis*	S	B	M	F	3.9	51	1.7	Macdonald (1977)
Fulmar	*Fulmarus glacialis*	S	B	M	F	3.1	456	2.8	Ollason and Dunnet (1978)
Antarctic Fulmar	*Fulmarus glacialoides*	S	B	M	F	3.0	492	2.2	Weimerskirch (1990)
Cape Petrel	*Daption capense australe*	S	B	S	F	2.7	73	10.0	P. Sagar (personal communication)
Short-tailed Shearwater	*Puffinus tenuirostris*	S	B	M	F	17.2	1911	4.8	Bradley *et al.* (1990)
Manx Shearwater	*Puffinus puffinus*	S	B	M	F	9.7	175	9.8	Brooke (1978)
Pelecaniformes									
Gannet	*Sula bassana*	S	B	M	F	16.5	115	11.0	Nelson (1978)
Masked Booby	*Sula dactylatra*	S	Y	R	F	45.2	42	—	Kepler (1969)
Brandt's Cormorant	*Phalacrocorax penicillatus*	S	B	M	F	62.5	8	20.0	Boekelheide and Ainley (1989)
Flightless Cormorant	*Nannopterum harrisi*	M	B	—	—	88.1	151	12.6	Harris (1979)
Ciconiiformes									
Greater Flamingo	*Phoenicopterus ruber*	S	B	S	N	100.0	63	6.5	Cézilly and Johnson (in press)
Anseriformes									
Mute Swan	*Cygnus olor*	S	Y	S	F	10.6	492	22.6	Minton (1968)
Mute Swan	*Cygnus olor*	S	Y	S	F	0.7	603	18.9	Rees *et al.* (Chapter 6)
Whooper Swan	*Cygnus cygnus*	S	Y	M	F	5.6	305	14.9	Rees *et al.* (Chapter 6)
Bewick's Swan	*Cygnus bewickii*	S	Y	M	F	0.0	2220	16.0	Rees *et al.* (Chapter 6)
Pink-footed Goose	*Anser brachyrhynchus*	S	Y	M	F	4.6	109	8.0	J. Madsen (personal communication)[2]
Greenland White-fronted Goose	*Anser albifrons*	S	Y	M	F	1.2	173	14.0	A. D. Fox. (personal communication)[2]
Greylag Goose	*Anser anser*	S	Y	M	F	5.1	292	17.0	L. Nilsson (personal communication)[2]
Greylag Goose	*Anser anser*	S	Y	S	F	8.1	566	11.0	K. Kortrschal (personal communication)[2]
Lesser Snow Goose	*Anser caerulescens*	S	Y	M	F	0.0	238	25.0	Prevett and MacInnes (1980)
Lesser Snow Goose	*Anser caerulescens*	S	Y	M	F	0.8	604	25.0	Cooke *et al.* (1981)
Hawaiian Goose	*Branta sandvicensis*	S	Y	S	F	0.9	327	7.0	J. M. Black *et al.* (unpublished data)[2]
Giant Canada Goose	*Branta canadensis maxima*	S	Y	M	F	2.2	183	30.0	Raveling (1988)
Barnacle Goose	*Branta leucopsis*	S	Y	M	F	2.6	729	10.0	Forslund and Larsson (1991)
Barnacle Goose	*Branta leucopsis*	S	Y	M	F	2.0	5974	10.0	Black *et al.* (Chapter 5)
Barnacle Goose	*Branta leucopsis*	S	Y	M	F	4.6	393	10.0	B. Ebbinge (personal communication)[2]
Brent Goose	*Branta bernicla*	S	Y	M	F	4.5	1274	16.2	B. Ebbinge (personal communication)[2]
Magellan Goose	*Chloephaga picta*	S	Y	S	F	15.4	13	18.0	Summers (1983)

Appendix 19.1 Continued

Species	Latin name	Broods	Pair bond	Residency	Site fidelity	Divorce rate %	Pair years	Mortality rate %	Source
Black-bellied Whistling Duck	*Dendrocygna autumnalis*	—	—	—	—	10.5	19	—	Delnicki (1983)
Maned Duck	*Chenonetta jubata*	—	—	—	—	0	6	—	Kingsford (1990)
African Black Duck	*Anas sparsa*	—	Y	R	F	10.0	10		Ball *et al.* (1978)
Blue Duck	*Hymenolaimus malacorynchos*	S	Y	R	F	11.3	71	14.0	Williams and McKinney (Chapter 4)
Shelduck	*Tadorna tadorna*	—	—	—	—	19.5	41	—	Williams (1973)
White-cheeked Pintail	*Anas bahamensis*	S	B	S	F	34.1	44	39.8	Sorenson (1992)[3]
Laysan Duck	*Anas laysanensis*	S	B	S	—	27.3	11		Moulton and Weller (1984)
Mallard	*Anas platyrhynchos*	—	—	—	—	9.1	11	—	Mjelstad and Saetersdal (1990)
Barrow's Goldeneye	*Bucephala islandica*	S	B	M	F	16.7	6	32.3	Savard (1985)
Speckled Teal	*Anas flavirostris*	—	—	—	—	8.5	47	—	J. Port (personal communication)[3]
Falconiformes									
Sparrowhawk	*Accipiter nisus*	S	B/Y	R	F	11.3	230	35.0	Newton and Wyllie (Chapter 14)
Merlin	*Falco columbarius*	S	B	M	F	68.4	19	30.0	Warkentin *et al.* (1991)
Galliformes									
Willow Ptarmigan	*Lagopus lagopus*	P	B	S	F	16.9	183	49.0	Hannon and Martin (Chapter 10)
Willow Ptarmigan	*Lagopus lagopus*	S	B	S	F	13.9	36	52.0	Hannon and Martin (Chapter 10)
White-tailed Ptarmigan	*Lagopus leucurus*	S	B	S	F	19.5	41	47.0	Hannon and Martin (Chapter 10)
Gruiformes									
Florida Sandhill Crane	*Grus canadensis*	S	Y	S	F	3.9	257	11.0	S. Nesbitt and T. C. Tacha (personal communication)
Coot	*Fulica atra*	S	B	S	F	38.5	52	53.9	M. Brinkhof (personal communication)
Charadriiformes									
Oystercatcher	*Haematopus ostralegus*	S	B	M	F	10.7	577	10.0	Harris *et al.* (1987)
Pied Oystercatcher	*Haematopus ostralegus finschi*	S	B	M	F	12.2	139	12.7	P. Sagar (personal communication)
Grey Plover	*Pluvialis squatarola*	S	B	M	F	4.8	21	40.5	V. K. Ryabitsev (1993) and (personal communication)

Piping Plover	*Charadrius melodus*	S	B	M	F	66.7	30	30.0	Haig and Oring (1988)
Dunlin	*Calidris alpina*	S	B	M	F	27.9	43	27.3	Soikkeli (1967)
Stilt Sandpiper	*Microphalama himantopus*	S	B	M	F	0.0	11	28.0	Jehl (1973)
Semi-palmated Sandpiper	*Calidris pusilla*	S	B	M	F	20.0	140	30.0	Gratto-Trevor (1991)
Semi-palmated Sandpiper	*Calidris pusilla*	S	B	M	F	32.0	37	38.0	C. S. Moitoret (personal communication)
Western Sandpiper	*Calidris mauri*	S	B	M	—	38.5	13	50.0	Holmes (1971)
Arctic Skua	*Stercorarius parasiticus*	S	B	M	F	17.1	205	11.4	O'Donald (1983)
South Polar Skua	*Catharacta maccormicki*	S	B	M	F	1.5	267	4.6	Wood (1971)
Kittiwake	*Rissa tridactyla*	S	B	M	F	26.1	820	16.5	Coulson and Thomas (1983)
Kittiwake	*Rissa tridactyla*	S	B	M	F	19.3	171	7.5	Hatch *et al.* (1993)
Red-billed Gull	*Larus novaehollandiae*	S	B	M	F	10.5	3903	14.4	Mills *et al.* (Chapter 16)
Ring-billed Gull	*Larus delawarensis*	—	—	—	—	28.0	25	—	Southern and Southern (1982)
Glaucous-winged Gull	*Larus glaucescens*	S	Y	S	F	30.0	10	13.2	Vermeer (1963)
Caspian Tern	*Hydroprogne caspia*	S	B	M	N	53.8	13	25.0	Cuthbert (1985)
Black Guillemot	*Uria grylle*	S	B	R	F	4.5	22	13.0	Peterson (1981)
Common Guillemot	*Uria aalge*	S	B	M	F	11.7	626	5.7	M. P. Harris (personal communication)
Cassin's Auklet	*Ptychoramphus aleuticus*	S	Y	R	F	7.3	220	25.0	Sydeman *et al.* (Chapter 11)
Least Auklet	*Aethia pusilla*	S	B	M	F	36.4	55	19.0	Jones and Montgomerie (1991)
Puffin	*Fratercula arctica*	S	B	M	F	16.0	25	13.0	F. Davidson (personal communication)
Puffin	*Fratercula arctica*	S	B	M	F	5.6	90	5.0	Ashcroft (1976)
Psittaciformes									
Galah	*Cacatua roseicapillus*	S	Y	R	F	6.2	195	20.0	Rowley (1990)
Major Mitchell's Cockatoo	*Cacatua leadbeateri*	S	Y	S	F	1.3	75	12.0	Rowley and Chapman (1991)
Apodiformes									
Little Swift	*Apus affinis*	M	Y	R	F	4.1	73	16.9	Hotta (personal communication)
Coraciiformes									
European Bee-eater	*Merops apiaster*	S	—	M	F/N	19.0	42	50.0	B. J. Lequette and C. M. Lessells (personal communication)
White-fronted Bee-eater	*Meroos bulluckoides*	S	Y	R	N	12.9	62	27.5	Emlen (1990)
Green Woodhoopoe	*Phoeniculus purpureus*	S	B	R	F	1.1	178	11.6	M. du Plessis; J. D. Ligon and S. H. Ligon (personal communication)
Green Woodhoopoe	*Phoeniculus purpureus*	S	B	R	F	0.0	138	15.3	M. du Plessis; J. D. Ligon and S. H. Ligon (personal communication)

Appendix 19.1 Continued

Species	Latin name	Broods	Pair bond	Residency	Site fidelity	Divorce rate %	Pair years	Mortality rate %	Source
Piciformes									
Red-cockaded Woodpecker	*Picoides borealis*	S	Y	R	F	5.0	680	28.0	Waltẽrs (1990)
Passeriformes									
Skylark	*Alauda arvensis*	M	B	M	F	52.8	36	33.5	Delius (1965)
House Martin	*Delichon urbica*	M	B	M	—	100.0	21	57.1	Bryant (1979); Rowley (1983)
Barn Swallow	*Hirundo rustica*	S	B	M	F	73.1	26	58.4	Shields (1984)
Barn Swallow	*Hirundo rustica*	M	B	M	F	8.3	72	69.9	A. P. Møller (personal communication)
Natal Robin	*Cossypha natalensis*	S	Y	R	F	30.8	13	16.1	Farkas (1969)
Black Wheatear	*Oenanthe leucura*	M	Y	R	F	15.4	13	50.5	J. Moreno (personal communication)
European Blackbird	*Turdus merula*	M	B	R	F	11.1	18	33.0	Snow (1958)
European Blackbird	*Turdus merula*	M	B	R	F	31.7	183	30.0	Desrochers and Magrath (1993*b*, Chapter 9)
Styan's Grasshopper-Warbler	*Locustella pleski*	S	B	M	F	79.7	59	43.0	Nagata (1986)
Seychelles Warbler	*Acrocephalus sechellensis*	S	Y	R	F	0.6	525	13.7	J. Komdeur (personal communication)
Superb Fairy-wren	*Malurus cyaneus*	M	Y	R	F	0.0	27	33.0	Rowley (1965)
Splendid Fairy-wren	*Malurus splendens*	M	Y	R	F	2.0	346	35.5	Russell and Rowley (Chapter 8)
Flame Robin	*Petroica phoenicea*	S	B	M	F	28.6	14	25.2	Robinson (1990)
Scarlet Robin	*Petroica multicolor*	S	Y	R	F	15.4	13	22.5	Robinson (1990)
Rufous Whistler	*Pachycephala rufiventris*	S	B	M	F	37.0	27	18.3	L. Bridges (personal communication)
Great Tit	*Parus major*	M	B	R	F	22.5	71	51.0	Kluÿver (1951)
Great Tit	*Parus major*	S	B	R	F	33.5	221	50.5	Perrins and McCleery (1985)
Great Tit	*Parus major*	M	B	R	F	34.7	452	67.8	A. v. N oordwijk (personal communication)
Great Tit	*Parus major*	M	B	R	F	21.3	89	67.8	A. v. Noordwijk (personal communication)
Great Tit	*Parus major*	M	B	R	F	15.9	410	63.5	A. v. Noordwijk (personal communication)
Great Tit	*Parus major*	M	B	R	F	25.0	104	75.9	A. v. Noordwijk (personal communication)
Great Tit	*Parus major*	M	B	R	F	0.0	25	42.0	Dhondt *et al.* (Chapter 13)
Great Tit	*Parus major*	M	B	R	F	12.7	63	45.0	Dhondt *et al.* (Chapter 13)
Great Tit	*Parus major*	M	B	S	F	12.5	24	52.0	Dhondt *et al.* (Chapter 13)
Great Tit	*Parus major*	M	B	S	F	19.4	36	56.0	Dhondt *et al.* (Chapter 13)
Great Tit	*Parus major*	M	B	S	F	46.2	13	48.0	Dhondt *et al.* (Chapter 13)
Great Tit	*Parus major*	M	B	S	F	20.8	24	53.0	Dhondt *et al.* (Chapter 13)

Great Tit	*Parus major*	M	B	R	F	19.6	46	56.0	Dhondt *et al.* (Chapter 13)
Great Tit	*Parus major*	M	B	S	F	49.3	75	57.0	Dhondt *et al.* (Chapter 13)
Blue Tit	*Parus caeruleus*	S	B	R	F	30.8	91	65.7	A. v. Noordwijk (personal communication)
Blue Tit	*Parus caeruleus*	S	B	R	F	56.3	16	82.8	A. v. Noordwijk(personal communication)
Blue Tit	*Parus caeruleus*	S	B	S	F	84.6	26	66.6	Dhondt and Adriaensen (1994)
Blue Tit	*Parus caeruleus*	S	B	R	F	47.6	42	37.2	Dhondt and Adriaensen (1994)
Blue Tit	*Parus caeruleus*	S	B	R	F	25.0	80	52.8	Dhondt and Adriaensen (1994)
Willow Tit	*Parus montanus*	S	Y	R	F	12.0	226	36.2	M. Orell (personal communication)
European Nuthatch	*Sitta europea*	S	Y	R	F	0.0	16	50.0	E. Matthysen (personal communication)
Scarlet-tufted Malachite Sunbird	*Nectarinia johnstoni*	M	B	R	F	0.0	6	37.5	M. R. Evans (personal communication)
Hooded Warbler	*Wilsonia citrina*	M	B	M	F	42.9	7	46.0	Evans Ogden and Stutchbury (in press)
Large Cactus Finch	*Geospiza conirostris*	M	B	R	F	21.7	92	20.5	Grant and Grant (1989)
White-crowned Sparrow	*Zonotrichia leucophrys*	S	B	M	F	34.1	82	50.0	M. L. Morton(personal communication)
Lapland Longspur	*Calcarius lapponicus*	S	B	M	F	67.0	6	65.0	C. S. Moitoret (personal communication)
Indigo Bunting	*Passerina cyanea*	M	B	M	F	50.7	221	53.0	Payne and Payne (Chapter 17)
Little Raven	*Corvus mellori*	S	B	M	F	7.5	107	15.3	Rowley (1973)
Australia Raven	*Corvus coronoides*	S	Y	R	F	0.0	221	23.0	Rowley (1973)
Jackdaw	*Corvus monedula*	S	Y	S	F	6.0	218	20.0	Röell (1978)
Pinyon Jay	*Gymnorhinus cyanocephalus*	S	Y/B	S	N	1.1	279	24.0	Marzluff and Balda (Chapter 7)
Florida Scrub Jay	*Aphelocoma coerulescens*	S	Y	R	F	2.6	960	25.0	Woolfenden and Fitzpatrick (Chapter 7)

[1]Refer to Williams (Chapter 15).
[2]Refer to Black *et al.* (Chapter 5).
[3]Refer to Williams and McKinney (Chapter 4).

Author index

Page numbers in *italic* refer to figures and tables.

Subject index